0410510-5

BLAZYNSKI, T Z
EXPLOSIVE WELDING, FORMING AND
000410510

621.779

B64

EXPLOSIVE WELDING, FORMING AND COMPACTION

EXPLOSIVE WELDING, FORMING AND COMPACTION

Edited by

T. Z. BLAZYNSKI

*Department of Mechanical Engineering,
The University of Leeds, UK*

APPLIED SCIENCE PUBLISHERS
LONDON and NEW YORK

APPLIED SCIENCE PUBLISHERS LTD
Ripple Road, Barking, Essex, England

Sole Distributor in the USA and Canada
ELSEVIER SCIENCE PUBLISHING CO., INC.
52 Vanderbilt Avenue, New York, NY 10017, USA

British Library Cataloguing in Publication Data

Explosive welding, forming and compaction.
1. Explosive welding
I. Blazynski, T. Z.
671.5'2   TS227

ISBN 0-85334-166-4

WITH 16 TABLES AND 156 ILLUSTRATIONS

© APPLIED SCIENCE PUBLISHERS LTD 1983

The selection and presentation of material and the opinions expressed in this publication are the sole responsibility of the authors concerned.

All rights reserved. No part of this publication may be reproduced, stored in a retrieval system, or transmitted in any form or by any means, electronic, mechanical, photocopying, recording, or otherwise, without the prior written permission of the copyright owner, Applied Science Publishers Ltd, Ripple Road, Barking, Essex, England

Photoset in Malta by Interprint Limited
Printed in Northern Ireland at The Universities Press (Belfast) Ltd

PREFACE

The last two decades have seen a steady and impressive development, and eventual industrial acceptance, of the high energy-rate manufacturing techniques based on the utilisation of energy available in an explosive charge. Not only has it become economically viable to fabricate complex shapes and integrally bonded composites—which otherwise might not have been obtainable easily, if at all—but also a source of reasonably cheap energy and uniquely simple techniques, that often dispense with heavy equipment, have been made available to the engineer and applied scientist.

The consolidation of theoretical knowledge and practical experience which we have witnessed in this area of activity in the last few years, combined with the growing industrial interest in the explosive forming, welding and compacting processes, makes it possible and also opportune to present, at this stage, an in-depth review of the state of the art.

This book is a compendium of monographic contributions, each one of which represents a particular theoretical or industrial facet of the explosive operations. The contributions come from a number of practising engineers and scientists who seek to establish the present state of knowledge in the areas of the formation and propagation of shock and stress waves in metals, their metallurgical effects, and the methods of experimental assessment of these phenomena. On the manufacturing side, the contributors are concerned to explain the mechanisms of explosive forming, welding and powder compaction, and to indicate both the operational problems that may exist and the practical opportunities and advantages that these processes offer.

The purpose of the book as a whole is to provide the existing 'practitioner' with, perhaps, more detailed and yet, at the same time, wider knowledge of his field, and to indicate to the 'newcomer' what it is that he can expect to encounter in his work. It is with a view to the particular needs of the latter that each chapter contains an extensive list of references which in themselves form valuable sources of further information.

Finally, I would like to express my thanks to the authors of individual chapters for their contributions and especially to Mr John Pearson who as one of the original pioneers of the explosive processes has brought so much of his personal experience to the writing of the introductory chapter.

T. Z. BLAZYNSKI

CONTENTS

Preface v

List of Contributors xiii

1. Introduction to High-energy-rate Metalworking
   JOHN PEARSON 1

   1.1. Background 1
   1.2. High-energy-rate Processes 2
      1.2.1. Operation Concepts 2
      1.2.2. Basic Types of Operations 3
      1.2.3. Nature of Load Application 4
   1.3. Development of the Field 5
      1.3.1. Early Studies 5
      1.3.2. Explosive Forming 6
      1.3.3. Explosive Hardening 8
      1.3.4. Explosive Compaction 9
      1.3.5. Explosive Welding 10
   1.4. Continued development of the field 12
      1.4.1. Cooperative Effort 12
      1.4.2. Thoughts for the Future 12
   References 14

2. Propagation of Stress Waves in Metals
   M. A. MEYERS and L. E. MURR 17

   2.1. Dynamic Propagation of Deformation 17
   2.2. Elastic Waves 21
      2.2.1. Introduction 21
      2.2.2. Elastic Waves in Isotropic Materials . . . 22
      2.2.3. Elastic Waves in Anisotropic Media . . . 27
   2.3. Plastic Waves 30
      2.3.1. Preliminary Considerations 30
      2.3.2. The von Kármán and Duwez Plastic Wave Theory . 32
      2.3.3. Plastic Shear Waves 35

		2.3.4.	Additional Considerations on Plastic Waves.	36
		2.3.5.	Adiabatic Shear Bands.	37
	2.4.	Shock Waves		39
		2.4.1.	Hydrodynamic Treatment	39
		2.4.2.	More Advanced Treatments and Computer Codes	47
		2.4.3.	Attenuation of Shock Waves.	50
		2.4.4.	Elastic Precursor Waves	58
	2.5.	Defect Generation		60
		2.5.1.	Dislocation Generation.	62
		2.5.2	Point Defects.	69
		2.5.3.	Deformation Twinning.	70
		2.5.4.	Displacive/Diffusionless Transformations	72
		2.5.5.	Other Effects.	76
	Acknowledgements			77
	References			77

3. **Metallurgical Effects of Shock and Pressure Waves in Metals**
   L. E. MURR and M. A. MEYERS. 83

   3.1. Principal Features of High-strain-rate and Shock deformation in Metals 83
   3.2. Permanent Changes: Residual Microstructure–Mechanical Property Relationships 92
       3.2.1. Grain Size Effects. 93
       3.2.2. Shock-induced Microstructures 94
       3.2.3. Shock Deformation Versus Conventional Deformation (Cold Reduction) 104
       3.2.4. Effects of Shock Pulse Duration in Shock Loading . 105
       3.2.5. Effects of Point Defects, Precipitates and Other Second-Phase Particles 106
   3.3. Response of Metals to Thermomechanical Shock Treatment . 108
       3.3.1. Shock–mechanical Treatment, Stress-cycling and Repeated Shock Loading 109
       3.3.2. Microstructural Stability and Thermal Stabilization of Substructure 114
   3.4. Summary and Conclusions 117
   Acknowledgements 118
   References 118

4. **High-rate straining and Mechanical Properties of Materials**
   J. HARDING. 123

   4.1. Introduction 123
   4.2. Testing Techniques at High Rates of Strain . . . 124
       4.2.1. Testing Techniques at Intermediate Rates of Strain . 126
       4.2.2. Testing Techniques at Impact Rates of Strain . 130
       4.2.3. Attempts to Reach Higher Strain Rates . . 137

4.3. Mechanical Properties of Materials at High Rates of Strain . 139
    4.3.1. Theoretical Considerations 142
    4.3.2. Strain-rate Dependence of fcc Materials . . . 143
    4.3.3. Strain-rate Dependence of hcp and Orthorhombic Materials 143
    4.3.4. Strain-rate Dependence of bcc Materials . . . 146
    4.3.5. Mechanical Response at Very High Strain Rates . . 148
    4.3.6. Effect of Changing Strain Rate (Strain Rate History) . 149
4.4. Mechanical Equations of State at High Rates of Strain . . 152
4.5. Summary 154
    References 154

5. Basic Consideration for Commercial Processes
D. B. Cleland 159

5.1. Explosive cladding 159
    5.1.1. Introduction 159
    5.1.2. Cladding Sites and Facilities 162
    5.1.3. Range of Products 164
    5.1.4. Bonding Parameters 165
5.2. Design of Clad Assemblies 166
    5.2.1. General 166
    5.2.2. Shell Plates 167
    5.2.3. Tube Plates 168
    5.2.4. Metal Requirements 169
    5.2.5. Extension Bars 170
    5.2.6. Temperature of Metals 170
5.3. Assembly of Clads 171
    5.3.1. Metal Preparation 171
    5.3.2. Assembly 171
    5.3.3. Protection of Cladding Plate Surface 174
5.4. Explosives 175
    5.4.1. Main Charge 175
    5.4.2. Initiation 176
5.5. Double Sided Clads 177
5.6. Multilayer Clads 177
5.7. Post Cladding Operations 178
    5.7.1. Preliminary Examinations 178
    5.7.2. Stress Relief 178
    5.7.3. Levelling 178
    5.7.4. Cutting and Trimming 180
    5.7.5. Ultrasonic Testing 180
5.8. Destructive Testing 181
5.9. Tubular Components 184
    5.9.1. Nozzles 184
    5.9.2. Other Tubular Components 186
5.10. Explosive Hardening 187

6. Mechanics of Explosive Welding
   H. EL-SOBKY 189

   6.1. Introduction 189
   6.2. The Mechanism of Explosive Welding 190
   6.3. Parameters of the Explosive Welding Process . . . 195
       6.3.1. The Collision Parameters V_p, V_c, β . . . 197
       6.3.2. Limiting Conditions for Welding 200
   6.4. Interfacial Waves 206
       6.4.1. Introduction 206
       6.4.2. Mechanisms of Wave Formation 206
   6.5. Analysis of Flow in the Collision Region 210
       6.5.1. Introduction 210
       6.5.2. The Interfacial Pressure Profile 211
       6.5.3. The Flow Pattern in the Collision Region . . 213
   References 216

7. Explosive Welding in Planar Geometries
   M. D. CHADWICK and P. W. JACKSON 219

   7.1. Introduction 219
   7.2. Material Combinations and Flyer Thicknesses . . . 220
   7.3. Basic Welding Geometries 224
       7.3.1. Parallel Geometries 224
       7.3.2. Inclined Geometries 227
       7.3.3. Parallel/Inclined Geometries 230
       7.3.4. Double Inclined Geometries 231
       7.3.5. Geometries Producing Welding Conditions Transiently 233
   7.4. Selection of Bonding Parameters 234
       7.4.1. General Considerations 234
       7.4.2. Impact Velocity 236
       7.4.3. Explosive Loading 240
       7.4.4. Collision Angle/Collision Point Velocity . . 242
       7.4.5. Stand-Off Distance 246
       7.4.6. Anvil 248
       7.4.7. Surface Finish 249
   7.5. Direct Measurement of Bonding Parameters . . . 251
       7.5.1. Introduction 251
       7.5.2. The Dautriche Method 251
       7.5.3. Wire and Pin Contactor Methods . . . 252
       7.5.4. High Speed Photography 253
       7.5.5. Flash Radiography 254
       7.5.6. Velocity Probe 254
       7.5.7. Slanting Wire Methods 255
   7.6. Miscellaneous Welding Geometries for Sheets and Plates . 257
       7.6.1. Lap Welding of Narrow Plates 258

7.6.2.	Seam/Line Welding of Sheets	258
7.6.3.	Scarf Welding	262
7.6.4.	Butt Welding	262
7.6.5.	Spot Welding	264
7.6.6.	Patch Welding	265
7.6.7.	Channel Welding	265
7.7.	Welding of Foils	265
7.7.1.	Theoretical Considerations	265
7.7.2.	Welding of Single Foils and Simple Multi-Foil Laminates	267
7.7.3.	Wire-Reinforced Composites	268
7.8.	Applications	273
7.8.1.	Introduction	273
7.8.2.	Clad Plate	274
7.8.3.	Dissimilar Metal Joints	276
7.8.4.	Transition Joints	277
7.8.5.	Honeycomb	279
7.9.	Conclusions	280
Acknowledgements		281
References		281

8. Welding of Tubular, Rod and Special Assemblies
T. Z. BLAZYNSKI 289

8.1.	Introduction	289
8.2.	Explosive and Implosive Welding Systems and Bonding Parameters	290
8.2.1.	Welding Systems	290
8.2.2.	Welding Mechanisms	292
8.3.	Welding of Duplex and Triplex Cylinders	296
8.3.1.	Development of Welding Techniques	296
8.3.2.	Characteristics of the Welded Systems	300
8.3.3.	Residual Stresses	304
8.3.4.	Conventional Processing	311
8.4.	Tube-to-tubeplate Welding	313
8.4.1.	Introduction	313
8.4.2.	Geometry and Parameters	313
8.4.3.	System Characteristics	318
8.4.4.	Properties of Joints	320
8.5.	Explosive Plugging of Tubes in Tubeplates	321
8.5.1.	Applications	321
8.5.2.	Plugging Systems	322
8.5.3.	Plugs, Materials and Testing	324
8.6.	Multilayer Foil Reinforced Cylinders	326
8.6.1.	Introduction	326
8.6.2.	Implosive Welding System	327
8.6.3.	Pressures and Stresses	328

	8.6.4. Structural Properties		329
8.7.	Interface Wire Mesh Reinforcement		331
	8.7.1. Welding Systems		331
	8.7.2. Metallurgical Characteristics		331
	8.7.3. Mechanical Properties		335
8.8.	Transition Joints		338
	8.8.1. Applications and Systems		338
	8.8.2. The Machined Joint		339
	8.8.3. Welding of Tubular Joints		340
8.9.	Solid and Hollow Axisymmetric Components		341
	8.9.1. Introduction		341
	8.9.2. Welding Systems		341
	8.9.3. Hydrostatic Extrusion		343
References			343

9. Explosive Forming
J. W. Schroeder 347

9.1.	Introduction	347
9.2.	Formability of Engineering Alloys	348
9.3.	Mechanical Properties of Explosively formed Components	349
9.4.	Air and Underwater Forming Systems	350
9.5.	Die and Dieless Forming	351
9.6.	Analysis of Final Shapes in Free-Forming	355
9.7.	Parameters and Analysis of Die Design	357
9.8.	Forming of Domes and of Elements of Spherical Vessels	360
9.9.	Forming and Punching of Tubular Components	362
9.10.	Miscellaneous Forming Operations	365
9.11.	Conclusion	367
References		367

10. Powder Compaction
R. Prümmer 369

10.1	Introduction	369
10.2	Dynamic Compressibility of Powders	371
10.3.	Type of Shock Wave and Density Distribution	373
10.4.	Temperature and Strain Rate Effects	378
10.5.	Phase Transitions in Shock Loading Mixtures	381
10.6.	General Mechanical Properties of Compacted Powders	385
10.7.	X-ray and Other Methods of Evaluating Residual Stress Distribution	388
10.8.	Basic Problems in Fabricating Semi-finished Parts	389
10.9.	Static and Dynamic Compaction: A Comparison of Material Properties	390
References		392
Index		397

LIST OF CONTRIBUTORS

T. Z. BLAZYNSKI
  Department of Mechanical Engineering, University of Leeds, Leeds LS2 9JT, UK.

M. D. CHADWICK
  International Research and Development Co. Ltd, Fossway, Newcastle upon Tyne NE6 2YD, UK.

D. B. CLELAND
  Metal Cladding Department, Nobel's Explosives Co. Ltd, Nobel House, Stevenston, Ayrshire KA20 3NL, UK.

H. EL-SOBKY
  Department of Mechanical Engineering, University of Manchester Institute of Science and Technology, Sackville Street, Manchester M60 1QD, UK.

J. HARDING
  Department of Engineering Science, University of Oxford, Parks Road, Oxford OX1 3PJ, UK.

P. W. JACKSON
  International Research and Development Co. Ltd, Fossway, Newcastle upon Tyne NE6 2YD, UK.

M. A. MEYERS
  Department of Metallurgical and Materials Engineering, New Mexico Institute of Mining and Technology, Socorro, New Mexico 87801, USA.

L. E. Murr

   Oregon Graduate Center, 19600 NW Walker Road, Beaverton, Oregon 97006, USA.

John Pearson

   Michelson Laboratories, Naval Weapons Center, China Lake, California 93555, USA.

R. Prümmer

   Fraunhofer-Institut für Werkstoffmechanik, Rosastrasse 9, D-7800 Freiburg, Federal Republic of Germany.

J. W. Schroeder

   Foster Wheeler Development Corporation, John Blizard Research Center, 12 Peach Tree Hill, Livingstone, New Jersey 07039, USA.

Chapter 1

INTRODUCTION TO HIGH-ENERGY-RATE METALWORKING

JOHN PEARSON

Michelson Laboratories,
Naval Weapons Center,
China Lake, California, USA

1.1. BACKGROUND

A small underwater explosion forms a large, dish-shaped end closure for a tank car; a thin layer of explosive detonated over the surface of a railroad frog improves the mechanical properties of the high-manganese steel part; the simultaneous detonation of several strips of sheet explosive welds together components of dissimilar metals in the fabrication of a heat exchanger; and in a powder metallurgy process the shock pressures from an explosion produce a high-density preform. All of these examples are but a small sampling of the many and diverse operations now being performed in the industrial field of high-energy-rate metalworking where the constructive use of explosives is being applied in order to broaden the scope of manufacturing techniques, lower the costs, and shorten the lead time required in the production and fabrication of reliable metal parts.

While the ability to perform constructive metalworking operations through the use of explosives has been recognized for about 100 years, it is only within the last 25 years that the use of explosives for metalworking has been developed into a production tool, and even more recently that it has become widely adopted throughout the industrial world. The advanced use of explosives from the World War II era; the extensive and thorough study of underwater explosions and air blast; a better understanding of material behavior under high-pressure and high-strain-rate conditions; and new techniques for studying events in the milli- and microsecond ranges, particularly with high speed photography—all of these advances, and others, came together in the late 1950s to provide the theoretical and experimental base for the start of this new technology.

The need for such a technology was apparent by the mid-1950s when the accelerated efforts of the aerospace industry with its unusual demands were being felt throughout the metals processing field. New manufacturing methods were required for a wide variety of parts, ranging from the low-volume, close-tolerance production of large, complex pieces, to small, precision items of difficult-to-work metals. To meet these requirements companies began to consider techniques which utilized high-energy sources such as explosives, and by the end of that decade the development of the explosive metalworking field was well underway.

During the late 1950s and through most of the 1960s it was the needs of the aerospace industry which dominated this new and emerging field. However, it was quickly recognized that these same techniques could be applied in other industries. These new concepts were then modified and extended to meet the needs of the automotive, shipbuilding, chemical processing, nuclear, mining, construction and other industries. This development into new industrial fields is still continuing, and today the technology of high-energy-rate metalworking with explosives has something to offer, in varying degrees of sophistication, to almost every industry which involves the production and fabrication of metal parts.

1.2. HIGH-ENERGY-RATE PROCESSES

1.2.1. Operation Concepts

An explosive charge represents a small, economical, easy-to-handle package capable of performing a great deal of useful work when designed and applied correctly. The major problem faced by the engineer in the use of such a tool is that of matching the charge design and the manner in which it is to be used to the specific needs of the job, so that the explosion is applied in a controlled manner. While the detonation of 1 g of explosive releases only about 1200 calories, the process is completed in just a few microseconds. Therefore, the engineer should recognize that it is not the total energy released by the explosive that it is so important in its usage, but rather the power of the explosive; and the engineer must allow for this factor in the design of the operation. In a number of applications the use of an explosive may represent the only feasible manner in which a metalworking operation may be performed, thus opening up entirely new areas of industrial effort. More generally, it will serve as an adjunct to existing methods and thus offer the designer a wider range of operations by which to perform the task.

The overall field of working metals with explosives includes changing the shape of a metal part; cutting, shearing and punching; joining, welding, and cladding of metals: the high-density compaction of powders; and improving the engineering and metallurgical properties of metal parts. The workpiece material in such operations might be in plate, sheet, bulk or powder form, with both detonating and deflagrating explosives providing the energy source. Details of major types of operations, the loading processes involved, and the effects on the workpiece materials, are all given in the following chapters.

1.2.2. Basic Types of Operations

Explosive metalworking operations can be divided into two basic categories, depending on the location of the explosive relative to the workpiece. In the first category the explosive is detonated some distance from the workpiece, and the energy is transmitted through an intervening medium, usually water or air. These are classed as 'stand-off operations' and include the forming, sizing, deep-drawing and flanging of metal parts. In the second category the explosive is detonated while in intimate contact with the workpiece. These are classed as 'contact operations' and include explosive hardening, welding and cladding. Some operations such as explosive powder compaction and the contouring of heavy plate historically have been listed under both categories depending on the design of the system. Recent advances in both areas would now place them predominantly in the contact category.

Large differences in pressures, velocities, and metalworking times can occur between stand-off and contact operations. The working pressures can run from several thousand psi at the low end of the pressure range for stand-off charges to several million psi with contact charges, with corresponding times for load application extending from the millisecond to the microsecond range. When applied to the area of deformation processing where metal movement is involved, the range of metal displacement velocities can typically run from about several hundred ft/sec to several thousand ft/sec, with corresponding strain rates being related to the metal displacement velocity and the configuration of the metal part. Thus, in going from a typical stand-off operation to a typical contact operation there is, quite roughly, an increase in the working pressure of about $1000\times$; a decrease in the time for which the working pressure is applied of about $1000\times$; and an increase in the metal displacement velocity of about $10\times$ to $100\times$.[1]

1.2.3. Nature of Load Application

No single, simple description can cover the many types of variations which can occur in the dynamics of the operation for either stand-off or contact methods. For long stand-off operations with open dies the metal working is performed by the interaction of a pressure pulse against the workpiece, whereas if the stand-off distance is sufficiently short the major amount of metal work may relate to the action of the detonation products cloud if in air, or the direct application of bubble energy if in water. For underwater stand-off operations, cavitation fields and cavitation tunnels may be used for energy focusing,[1] while for operations conducted in air, energy focusing can be accomplished by means of shock reflectors.[2] For closed-die systems, the time and degree of pressure confinement become important factors.

For contact operations, the direct coupling of energy from the detonating explosive into the workpiece produces high-intensity, transient stress waves in the metal.[3] The action of these stress waves can change the mechanical and metallurgical properties of the material and produce permanent phase transitions in some materials while in others they leave behind the permanent signature of transient events such as the 130-kilobar transition in iron.[4-6] The controlled reflection, refraction and interaction of such waves can be used for controlled cutting, parts separation, and the high-velocity movement of metal surfaces. The size, shape and possible confinement of the charge, properties of the explosive, initiation geometry, and the use of internal wave guides and external buffers all relate to means of controlling the application of energy to the workpiece.[1] The design use of these many parameters to specific operations, and the effects that intense loading conditions may have on materials, are treated in later chapters.

1.3. DEVELOPMENT OF THE FIELD

1.3.1. Early Studies

The use of explosives for the working of metals, and the appreciation for the unusual behavioral patterns which can occur under intense conditions of loading, date back to the late 1800s and the years preceding World War I. In 1876, Adamson, an English engineer, used detonating charges of gun cotton to study the relative behavior of iron and steel plates when subjected to concussive forces.[7] His experimental arrangements were similar to those currently used in open-die, cupping operations in air. A few years later in 1888, Charles Munroe, an American chemist, detonated gun cotton charges in contact with iron plates to produce finely detailed engravings in the metal, using stencils of leaves and pieces of wire gauze.[8] Thus, while Adamson's experimental method was the forerunner of current standoff operations, Munroe's technique was an early version of today's contact operations.

The behavior of metals under high-strain-rate loading was studied by John and Bertram Hopkinson, father and son. John Hopkinson, an English engineer, was one of the first people to recognize that large differences can exist in the engineering properties of metals under static and dynamic loads, and in 1872 he published his studies on the impact behavior of iron wires.[9] This work was continued by Bertram Hopkinson in the early 1900s, and just prior to World War I he studied the effects of explosively induced stress waves in metals and presented the explanation for back-face spalling, a fracture phenomenon frequently encountered in many of today's contact operations.[10]

During the period between World Wars I and II, detailed studies were conducted on the behavior of materials under high pressures,[11] and an interest developed in the microstructural response of metals to contact explosions, especially the formation of shock twins in iron.[12] For experimental studies with contact explosives, 40% and 60% straight nitrodynamite replaced the gun cotton charges used by earlier investigators. Then by the late 1930s military explosives were being used as field expedients to shape steel plates in the construction of gun emplacements.

The period from the late 1930s to the mid-1950s was notable for an increased understanding of the modern-day role of explosives,[13,14] for in-depth studies of underwater explosions[15] and air blast,[16] for studies in the behavior and effects of stress waves,[17,18] and for the development of new techniques and equipment for studying short-time and high-pressure events. A variety of methods were developed to evaluate the effects on materials of intense loading conditions. These included the use of high-speed tensile impact machines,[19] the firing of high-velocity projectiles,[20] the use of modified Hopkinson pressure bars,[21] explosively activated loading devices, and the direct application of explosives.[4] As a result of such studies a better understanding of plastic strain propagation evolved,[20,22] and led to the concept of a critical particle velocity for failure by fracture.[4,19] The role of explosively induced stress waves in the fracturing of metals was studied extensively,[4] and techniques were devised for either using or eliminating these stress wave effects. This growth to the present day in the understanding of how materials respond to extreme conditions of loading is well documented.[4,5,6,23]

It was not until the mid-1950s that investigations in these many areas of effort were brought together in the process of developing what eventually became the industrial field of explosive metalworking. The early development of the field in this late post-World War II era was based largely on a mixture of theory, experimentation and serendipity. The theory was limited, the experimentation was extensive, and the investigators had the ability to recognize what were then new and unusual patterns of behavior, and to find ways to capitalize on them through the development of completely new concepts in the controlled working of metals.

1.3.2. Explosive Forming

Major interest in the industrial use of explosive metalworking did not develop until the late 1950s when metal parts needed for the space program were frequently beyond the capabilities of conventional metalworking equipment. These interests were originally concentrated in the explosive forming of metal sheet and plate. Once started, the field progressed rapidly and by the mid-1960s the large aerospace companies and their support contractors were producing a massive assortment of parts. These items ranged in weight from a few ounces to several tons, with dimensions from a fraction of an inch to over 20 ft (6m), and with thicknesses from a few thousandths of an inch to about 6 in. (15.24 cm).[24,25] Die assemblies for some of the larger parts weighed 60 tons or more.[26] Items were produced in almost the complete range of present day metals. The unusual and more complicated shape was the rule rather than the exception, for it was in this area that explosive operations demonstrated their commercial capabilities.

High explosives and low explosives were utilized with both closed and open dies in a variety of designs and sizes. Closed dies and open split dies were commonly used for the forming of complex cylindrical and conical shapes, respectively. However, the use of large open-die systems with high explosives was of particular values in the aerospace field as a means for fabricating large structures without having to procure large, precision machine tools, or massive, complicated die sets. It was rapidly demonstrated that it was possible to shorten lead times for tooling and equipment, and that there was real economic advantage for making unusually large metal bodies and parts of unusual configuration, especially for short run items. By the late 1960s large structural components had been successfully formed for use in vehicles which would operate in the extreme environments of the ocean depths[27] and outer space.[25] Today the explosive forming of large parts is of particular value in the construction of ships, tank cars, and storage and reaction vessels.

The open-die forming of large parts was performed both in air and underwater. Originally, the use of underwater forming predominated in the United States, while air-forming was preferred in some European countries. Each type of operation had distinct advantages. Air-forming offered a lower initial capital outlay since the forming tank and die retrieval equipment were not required. Also, the handling time per operation was considerably shorter. With underwater forming, smaller charges were required, the noise factor was essentially eliminated, and many investigators claimed that better control could be exercised over the impulse factor in the load. Forming tanks became fairly common[24-26] with some of the larger ones featuring tracked runways to lower the die assembly into the water, bubble screens to reduce shock loading of the tank wall, and means for heating the water where temperature control for the workpiece was required. For the production of some items, companies found that advantages of both types of forming were available through the use of expendable water containers placed over the workpiece and die assembly. For the forming of moderately-sized parts, emphasis today frequently is placed on tooling designed to use preforms and explosively size the part to tolerance without the use of a water tank.

In programs conducted for the aerospace industry it was found that for the production of a small number of parts the unit cost usually favored the explosive forming operation due to the lower initial cost of tooling, capital equipment, etc. However, for larger numbers of parts, the conventional forming provided the lower unit cost, usually due to a more rapid production rate and lower unit labor costs which eventually offset the initial advantage of lower tooling costs when using explosives.[24] This is still the case for many operations. However, one of the more recent advances in this field has been the development of automated, closed-die, multi-stage operations which minimize the labor cost, and which can be used economically for large production runs. These automated operations are now used for the production of a variety of parts, including such large items as axle housings for heavy-duty automotive equipment.[28] One other approach that has found industrial favor is the use of explosive forming with soft (inexpensive) tooling to produce prototype parts for test and evaluation. Then, after the design has been finalized, conventional forming with hard tooling is used for the production runs. Today, explosive forming operations run the gamut from highly sophisticated, rapid production, factory-conducted automated operations to essentially one-of-a-kind contouring of large, heavy-plate elements which are formed outdoors using hand-emplaced layers of sand and explosive.

While the early forming efforts were largely experimental, by the early 1970s considerable theory had been developed regarding the energy requirements for new operations,[25,29] and the use of subscale models to provide full-scale data for large parts was being employed successfully.[25] Today, the explosive forming field offers extensive experience, reasonably well-developed theory, and to some extent the use of computer programming for operation design.

1.3.3. Explosive Hardening

The first type of contact operation to receive industrial acceptance was explosive hardening which is used to improve the mechanical properties of high-manganese steel parts. Based on studies conducted during the late 1940s,[30] this process was introduced commercially in the mid-1950s,[31] and was quickly accepted as an industrial tool. The major industrial use of explosive hardening is with parts of austenitic manganese steel which are subjected to severe impact and abrasion, such as railroad frogs, rock-crusher jaws, grinding mills, power shovel buckets, and coal mining machines. Major users of these improved parts have been the railroad, mining, quarrying, and construction industries. For some applications, such as the hardening of railroad frogs, parts are now being mass produced in specially constructed blast chambers.[32]

In explosive hardening, a thin sheet of explosive is detonated while in contact with the surface to be hardened, usually at grazing incidence. The propagation of a high-intensity shock pulse through the metal increases the BHN of the metal both at the surface and in depth.[3] In doing so, it increases the yield and tensile strength so that the piece offers more resistance to impact-type wear and has less tendency to deform. The use of contact charges can produce a greater depth of hardening than conventional methods and does so with much smaller deformation than is required for conventional cold-work hardening.[26] Explosive hardening is a process which for many applications cannot be performed by conventional methods. In addition, usually it is a relatively quick and simple operation, and one which can be performed in remote locations without heat treating furnaces or heavy-duty deformation processing equipment. It can be used for sizes and configurations of parts which are inappropriate for conventional methods and, if desired, the process can be applied to only specific areas of the workpiece.

The initial development of explosive hardening and other contact operations was speeded by the availability of plastic bulk explosives such as Composition C-3 and Composition C-4, which had been developed for military demolition, and which could be hand-shaped to the desired configurations and thicknesses. Later, flexible sheet explosives were developed commercially and currently are used for all types of industrial contact operations.

1.3.4. Explosive Compaction

Interest in the use of explosives to compact powders started in the late 1950s and was initiated by the requirements for high-density parts from powdered materials for use in aerospace and atomic energy applications. The techniques developed were used later to produce items such as heat shields, bearings, electronic components, transmission filters, and small precision parts for high-speed equipment.

In compaction operations explosive loads are applied either by direct contact or through a loading system, with both deflagrating and detonating explosives used.[24] One of the earlier methods consisted of placing the powder in a thin cloth or plastic container and subjecting it to the air blast or underwater shock from an explosion. Later designs employed closed systems based on the use of gun powder cartridges. Explosively actuated presses were designed which used detonating explosives to drive one or more pistons into the powder compaction chamber.[33] Other methods involved semi-implosion concepts and the direct pressure from contact charges using appropriately configured buffers.[24,26] All of these concepts in appropriate design versions currently are used for experimental studies[34,35] and the production of parts.

Explosive compaction is an operation which offers unusual advantages and has opened doors to new areas of effort. Powder compacts can be fabricated which approach the theoretical density of bulk materials.

Parts can be produced from materials not suitable for conventional pressing techniques. Composites of traditionally incompatible metals, and of ceramic materials and low-ductility metals, can be compacted. Improved, and even unique, material properties can be obtained by explosive compaction. However, present production methods need to be modified to handle larger parts. Also, explosive compaction needs to be incorporated more fully into future applications which utilize conventional compaction. For example, conventional powder metallurgy technology now is being applied with increasing frequency by both airframe and aircraft engine manufacturers for the production of parts from 'superalloy' powders. By incorporating explosive techniques into these manufacturing processes, shorter production times and improved properties might result.

1.3.5. Explosive Welding

Explosive welding is a solid phase welding process in which the detonation of high explosives is used to bring the weld surfaces together in a high velocity oblique collision, which produces severe, but localized plastic flow at the interacting surfaces. While the process is usually described as a cold technique, meaning that no external heat is used to promote the bonding, high localized temperatures are normally generated at the weld interface due to the dynamics of the process. The combination of pressure, heat and flow can produce an interface bond having a strength equal to or greater than that of the parent metal.[24] While the process can be used to weld together two or more pieces of the same metal, its major industrial potential lies in the fact that it can be used to weld together combinations of different metals, many of which are impossible by any other means.

This process was first noted in ordnance tests where warhead fragments would weld to steel target plates, or the simultaneous detonation of stacked ordnance items such as bombs would produce welded sections between adjacent bomb casings. In early studies of explosive forming it was painfully discovered that an over-charged system could result in portions of the preform or blank being welded to the die cavity, resulting in costly rework of the die. Also, in early powder compaction studies with explosive presses,[33] it was found that misalignment of press components could produce accidental welds. It was from incidents such as these, and the need to understand how to prevent welding from occurring in other explosive metalworking operations, that the study of explosive welding started in the late 1950s.

Originally the process was viewed mainly as a replacement for conventional welding in remote locations, and as a means of fabricating difficult to-get-at configurations of standard welding materials. However, industrial interest developed rapidly when it was demonstrated that strong welds could be obtained when working with metal combinations which possessed widely different melting points, greatly different thermal expansions, and large hardness differences[25]—property variations which do not lend themselves to conventional welding techniques. By 1970 the literature contained information on the successful bonding of more than 260 combinations of similar and dissimilar metals,[25] and the list has grown during the last decade. While the initial studies treated mainly the welding and cladding of flat surfaces, commercial developments quickly included the lining of tubes, the cladding of curved surfaces, the butt-welding of pipes, the manufacture of transition joints, and numerous other applications.

In a controlled welding operation layers or strips of explosive are detonated in contact with either the weld plates or intermediate buffers using a system geometry such that the surfaces to be bonded are brought together at high velocities in a progressive, oblique collision. The system may involve two or more weld surfaces, with multilayer combinations of metals frequently involved. The weld normally progresses from the apex of collision along the collision interface by a process not too dissimilar from the collapse process of a shaped-charge liner. As with the wartime development of the anti-armor, shaped-charge warhead, explosive welding was initially applied with considerable success before a good theoretical understanding of the process was available. The dynamic behavior of the process was first studied by means of submicrosecond flash radiography[36] and then by various theoretical[37] and experimental models.[38] Extensive studies have been conducted during the past decade and the theory of the process is now reasonably well developed.[39,40] Design parameters for an explosive welding operation include the physical and mechanical properties of the metals to be bonded, the type and amount of explosive, the geometry of initiation, possible use of buffer layers, and the overall initial geometry of the weld system.

Explosive welding has become an exciting field of study, with applications in the nuclear, aerospace, chemical processing, electronics, shipbuilding, power transmission, and other industries. Its applications are usually novel and unique, and represent operations which in many instances were not available prior to the development of the process. The future limits in its use are bounded only by the imagination and ingenuity of the designer.

1.4. CONTINUED DEVELOPMENT OF THE FIELD

1.4.1. Cooperative Effort

One aspect that is frequently overlooked in discussing the development of the explosive metalworking field has been the interest and participation of military laboratories, and the fact that much of the original experimental data in the use of explosives came from government supported research facilities in many countries. Initially at least, this is where the ordnance expertise, the firing facilities, and the unique and expensive equipment for studying these events were to be found. At a later date, industrial and university laboratories took over the major effort in developing processes and theory, but the cooperation among military, industrial, and university laboratories has continued. This continuing cooperation represents one of the major strengths for the future development of the field.

In the working of metals with explosives it is interesting to note the parallel interests which exist between military and industrial applications, but which lead to different product use. It might be said that industrial applications are used to put things together, while military applications are used to take things apart. But, it is the same basic understanding and knowledge that is used in both product areas; hence, the need and desirability for continued cooperation and interaction.

1.4.2. Thoughts for the Future

The future of explosive metalworking will continue to depend, as it has in the past, on the imagination and ingenuity of the user. For standoff operations, which are still largely looked upon as alternative methods to conventional processes, the designer should look for such advantages as reduced lead time, lower overall costs, and possibly a better designed product, with the unique capabilities of the method stressed in terms of part design and configuration. Contact operations, on the other hand, are being viewed more and more not just as alternative methods, but as unique operations capable of doing a job, or producing a part, which is not possible by any other means.

The development of the field to date has been highlighted by unusual areas of effort, and it is expected that this will continue and will be one of the identified strengths of the field. Examples of such past effort might include the use of explosive forming to produce stainless denture bases and metal implants for orthopedic and dental surgery; the artists' use of explosive free-forming to produce large metal sculptures; the use of explosive welding to fabricate honeycomb structures, and to produce transition pieces whereby light-metal superstructures can be fabricated to the steel body of a ship; and the use of explosive compaction and explosive welding to produce unusual laminates and sandwich structures for the nuclear industry.

The near-term development of the field covers all types of operations. With continued improvements in multi-stage, automated operations, explosive forming should find greatly increased use for large production runs in the automotive and other mass-production industries. The unique capabilities of explosive welding and compaction which allow these operations to interwork combinations of metals and non-metals provide a potential for increased growth in nearly every industry. The design of all types of operations will benefit from the increased use of computer programming, and the theoretical treatment of contact operations, where heavy plastic flow occurs, will profit from additional studies using hydrodynamic modeling.

Three of the problems which faced the field during its development were: (1) a hesitancy to use explosives, (2) the multi-disciplinary nature of the required effort, and (3) the fact that explosive metalworking and conventional metalworking were usually performed by different organizations. The hesitancy to use explosives, while still present to a lesser degree, has been largely overcome through 20 years of expanding usage of the field. Problems stemming from the multi-disciplinary nature of explosive metalworking and the associated shortage of skilled personnel have been largely solved through the cooperative effort of military, industrial, and university laboratories; the training of people by specialized university programs; and the results of on-the-job training. The different types of metalworking are still performed largely by separate organizations, a fact which has slowed the growth of the field. However, full-spectrum metalworking in one organization is a concept which is growing in many countries,[32] and one which will be implemented more and more in the years ahead.

In addition to using explosive metalworking to support current industrial requirements, the engineer should plan now on how to use it to meet the needs of future areas of technological effort. To give just one example, entirely new areas of effort may soon be required in terms of space application. In a sense this was the driving force for the opening of the field when new techniques were needed to build the parts for large missiles and space vehicles. Those operations, however, were conducted in a known and friendly environment. A logical extension of this earlier effort is the use of explosives in space for the construction of satellites, space vehicles, manned communication centers, space laboratories, etc., and eventually for whatever else man may do in the exploration and utilization of space. The use of explosives in a space environment may present both unusual risks and high rewards, and present the designer with the challenge of operational patterns quite different from what he has come to recognize and understand. Now is the time to research this and other unusual but potential areas of application, so that the technology base exists when the need arises.

REFERENCES

1. PEARSON, J. In *Proc. Behavior and Utilization of Explosives in Engineering Design*, 1972, ASME, Albuquerque, New Mexico, 69–84.
2. BRUNO, E. J., Ed. *High-Velocity Forming of Metals*, Rev. Edn., 1968, ASTME, Dearborn, Michigan.
3. JONES, O. E. In *Proc. Behavior and Utilization of Explosives in Engineering Design*, 1972, ASME, Albuquerque, New Mexico, 125–148.
4. RINEHART, J. S., and PEARSON, J. *Behavior of Metals Under Impulsive Loads*, 1954, Am. Soc. Metals, Cleveland, Ohio.
5. SHEWMAN, P. G., and ZACKAY, V. F., Ed. *Response of Metals to High Velocity Deformation*, 1961, Interscience, New York.
6. MEYERS, M. A., and MURR, L. E., Ed. *Shock Waves and High-Strain-Rate Phenomena in Metals*, 1981, Plenum, New York.
7. ADAMSON, D. *J. Iron Steel Inst.* (1878), 383.
8. MUNROE, C. E. *Scribners Magazine*, **3** (1888), 563–567.
9. HOPKINSON, J. *Proc. Manchester Lit. and Phil Soc.*, **11** (1872), 40–45 and 119–121.
10. HOPKINSON, B. *Scientific Papers*, 1921, Cambridge University Press, Cambridge.
11. BRIDGMAN, P. W. *The Physics of High Pressure*, 1949, G. Bell and Sons, Ltd., London.
12. FOLEY, F. B., and HOWELL, S. P. *Trans. AIME*, **68** (1923), 891–915.
13. TAYLOR, J. *Detonation in Condensed Explosives*, 1952, Clarendon Press, Oxford.
14. COOK, M. A. *The Science of High Explosives*, 1958, Reinhold, New York.
15. COLE, R. H. *Underwater Explosions*, 1948, Princeton University Press, Princeton.
16. STONER, R. G., and BLEAKNEY, W. *J. Appl. Phys.*, **19** (1948), 670–678.
17. KOLSKY, H. *Stress Waves in Solids*, 1953, The Clarendon Press, Oxford.
18. DAVIS, R. M. In *Surveys in Mechanics*, 1956, The University Press, Cambridge, 64–138.
19. CLARK, D. S., and WOOD, D. S. *Trans. ASM*, **42** (1950), 45–74.
20. TAYLOR, G. I. *J. Inst. Civil Engrs.*, **26** (1946), 486–519.

21. DAVIS, R. M. *Trans. Royal Society (London)*, **240A** (1946–48), 375–457.
22. VON KÁRMÁN, T., and DUWEZ, P. *J. Appl. Phys.*, **21** (1950), 987–994.
23. HARDING, J., Ed. *Mechanical Properties at High Rates of Strain 1979*, 1980, Institute of Physics, London.
24. RINEHART, J. S., and PEARSON, J. *Explosive Working of Metals*, 1963, Pergamon, London.
25. EZRA, A. A. *Principles and Practices of Explosive Metalworking*, 1973, Industrial Newspapers, London.
26. NOLAND, M. C., GADBERRY, H. M., LOSER, J. B. and SNEEGAS, E. G. *High Velocity Metalworking*, 1967, NASA, Washington, D.C.
27. JUE, L. S., and GIANNOCCOLO, S. *Marine Technology*, **3** (1966), 99–105.
28. STEINICKE, H. Presentation at the Int. Conf. on the Metallurgical Effects of High Strain-Rate Deformation and Fabrication, June 1980, TMS-AIME, Albuquerque, New Mexico.
29. JOHNSON, W. *Impact Strength of Materials*, 1972, Edward Arnold, London.
30. MACLEOD, N. A. US Patent No. 2,703,297.
31. HARPER, W. A. In *High Energy Rate Working of Metals*, Vol. 1, 1964, Central Inst. for Industrial Research, Oslo, 247–254.
32. DERIBAS, A. A. In *Shock Waves and High-Strain-Rate Phenomena in Metals*, 1981, Plenum, New York, 915–939.
33. LA ROCCA, E. W., and PEARSON, J. *Rev. Sci. Inst.*, **29** (1958), 848–851.
34. PRUEMMER, R. A. In *Proc. Fourth Int. Conf. Center for High Energy Forming*, 1973, University of Denver, Denver, Colorado, 9.2.1–9.2.27.
35. STAVER, A. M. In *Proc. Fifth Int. Conf. on High Energy Rate Fabrication*, 1975, University of Denver, Denver, Colorado, 2.1.1–2.1.31.
36. PEARSON, J. In *Advanced High Energy Rate Forming*, 1961, ASTME, Detroit, Michigan, SP60-159.
37. COWAN, G. R., and HOLTZMAN, A. H. *J. Appl. Phys.*, **34** (1963), 928–939.
38. EL-SOBKY, H., and BLAZYNSKI, T. Z. In *Proc. Fifth Int. Conf. on High Energy Rate Fabrication*, 1975, University of Denver, Denver, Colorado, 4.5.1–4.5.21.
39. CROSSLAND, B. In *Mechanical Properties at High Rates of Strain 1979*, 1980, Institute of Physics, London, 394–409.
40. CARPENTER, S. In *Shock Waves and High-Strain-Rate Phenomena in Metals*, 1981, Plenum, New York, 941–959.

Chapter 2

PROPAGATION OF STRESS WAVES IN METALS

M. A. MEYERS

Department of Metallurgical and Materials Engineering,
New Mexico Institute of Mining and Technology,
Socorro, New Mexico, USA
and
L. E. MURR

Oregon Graduate Center,
Beaverton, Oregon, USA

2.1. DYNAMIC PROPAGATION OF DEFORMATION

The application of an external force to a body is, by definition a dynamic process. However, when the rate of change of the applied forces is low, one can consider the process of deformation as a sequence of steps in which the body can be considered in static equilibrium. Figure 2.1 shows how the distance between the atoms changes upon the application of an external force F. For each of the stages of deformation shown in Figs. 2.1(b) and 2.1(c), the body can be considered under static equilibrium and one can apply the methods of mechanics of materials to determine the internally-resisting stresses (by the method of sections). Hence, a section made at AA or BB will yield identical stresses.

However, the internal stresses are not instantaneously transmitted from the force-application region to the different regions of the body. The stresses (and strains) are transferred from atom to atom at a certain specific velocity. Figure 2.2(a) shows the application of a force at a rate dP/dt such that the stresses (and attendant strains) vary from section to section. Section BB has not 'seen' the application of the force at time t_1, while, at section AA, the separation between the atoms varies from point to point along the bar. One could establish a preliminary criterion for

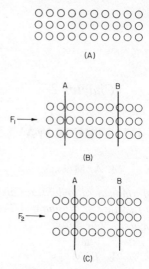

FIG. 2.1 Effect of application of force F on structure of solid (elastic deformation) under quasi-static conditions.

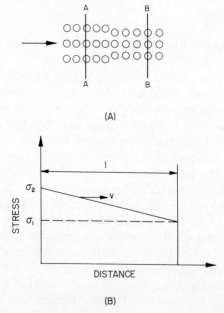

FIG. 2.2 (a) Wave propagation (elastic) in solid when rate of application of force is high. (b) Stress versus distance showing propagation of a general disturbance with velocity v.

'dynamic' deformation by stating that it requires a variation of stress (from one end to the other) of 10 per cent. If the velocity of the stress pulse is v and the length of the bar is l, one has, from Fig. 2.2(b):

$$\frac{d\sigma}{dt} = \frac{\sigma_2 - \sigma_1}{t} \tag{2.1}$$

$$v = \frac{l}{t} \tag{2.2}$$

Equating t in eqns. (2.1) and (2.2):

$$\frac{\sigma_2 - \sigma_1}{\left(\dfrac{d\sigma}{dt}\right)} = \frac{l}{v} \tag{2.3}$$

The criterion of 10 per cent variation in σ can be expressed as:

$$\sigma_2 - \sigma_1 \leqslant 0.1 \sigma_{max} \tag{2.4}$$

Substituting this into eqn. (2.3), one has:

$$\frac{d\sigma}{dt} \leqslant \frac{0.1 v \sigma_{max}}{l} \tag{2.5}$$

This is, of course, a somewhat arbitrary criterion, but it establishes the value of the rate of load application at which the 'dynamic' aspects or wave-propagation effects become important. In the wave-propagation regime, deformation is localized at the wavefront and release part and the behavior of the material is not only quantitatively but also qualitatively different.

At an atomic level, one may envisage the wave as a succession of impacts between adjacent atoms. Each atom, upon being accelerated to a certain velocity, transmits its (or part of) momentum to its neighbor(s).

The mass, separation between, and forces of attraction and repulsion of atoms determine the way in which the stress pulse is carried from one point to the other. Of importance also is the stress state established by the pulse, which determines the relative direction of motion of atoms and stress pulse, and the extent of motion of atoms. As a result of these differences, one can classify the stress pulses into three categories: (a) elastic, (b) plastic and (c) shock waves. Elastic waves produce only elastic deformation in the material. On an atomic scale, all atoms return to the original position in relation to their neighbors. There are two classes of elastic waves: longitudinal (or irrotational, dilatational, P) and shear (or equivoluminal, transverse, distortional, S) waves. They travel at velocities that are determined by their elastic constants. One may add a third class of elastic wave to the two above: surface, Love, or Raleigh waves, which travel at the surface. These are the commonly known waves propagating in water. Elastic waves are treated in Section 2.2. When the amplitude of the elastic wave exceeds a critical value for the yield stress of the material, at that specific strain rate (and we know that the yield stress of most metals is strain-rate dependent) the atoms undergo permanent changes in position with respect to their neighbors: macroscopically, this entails a change in the dimensions of the body. These are called plastic (or elastoplastic) waves. Depending upon their nature, one may also have plastic longitudinal or shear waves. These waves are treated in Section 2.3. If the geometry of the body is such that it allows a strain state called 'uniaxial strain', the propagation velocity of the plastic wave increases with increasing pressure because there cannot be any lateral flow of material (perpendicular to the direction of propagation of the waves). The wave takes the configuration shown in Fig. 2.3(b): the sharp front is

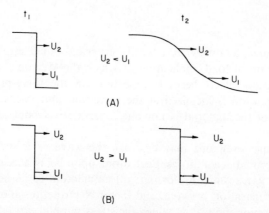

FIG. 2.3 Propagation of disturbances. (a) When $U_2 < U_1$ disturbance front slope decreases with propagation distance. (b) When $U_2 > U_1$ 'shock' front forms and remains stable.

the characteristic that defines a shock wave. The shock waves are treated in Section 2.4. The most simplified treatment (hydrodynamic theory) is introduced first; then, some more contemporary considerations and techniques such as attenuation of shock waves, method of characteristics, computer codes (incorporating the von Neumann-Richtmyer artificial viscosity) are discussed (Section 2.5). The fundamental aspects of the metallurgical effects of shock waves are discussed in Section 2.6. The specific residual microstructural and mechanical effects of a number of metals are treated in Chapter 3.

The treatment of this chapter will be kept at a level consistent with the 'engineering' approach of this book. There have been significant advances in the theoretical treatment of disturbances in solids in the past twenty years; these treatments, incorporated into computer programs, have dramatically increased the predictive capability. In-depth studies of disturbances in solids are numerous and the reader is referred to references 1 to 21, and 25 to 31.

2.2. ELASTIC WAVES

2.2.1. Introduction

Three types of elastic waves can propagate in solids: longitudinal (or dilatational) waves, distortional (or equivoluminal) waves and surface (or Raleigh) waves. A brief concept of these waves is given below.

(a) *Longitudinal or dilatational waves.* In longitudinal waves, the particle and wave velocity have the same direction. If the wave is compressive, they have the same sense; if it is tensile, they have the opposite senses.

(b) *Distortional or equivoluminal waves.* In this case the displacement of the 'particles' and wave are perpendicular. There should be no change in density and all longitudinal strains ε_{11}, ε_{22}, ε_{33} are zero.

(c) *Surface waves.* The most obvious example of this type of wave are the waves in the sea. They only happen at interfaces. This type of wave is restricted to the region adjacent to the interface, and 'particle' velocity decreases very rapidly (exponentially) as one moves away from it. The particles describe elliptical trajectories. The Raleigh wave is the slowest of the three waves; the fastest is the longitudinal wave as will be seen from the derivations that follow. When the elastic limit (or the critical shear stress for plastic flow on the plane with highest Schmid factor) is reached, the elastic wave is followed by either a plastic or a plastic shock wave, depending on the state of stress (uniaxial stress and uniaxial strain, respectively.).

2.2.2. Elastic Waves in Isotropic Materials

Figure 2.4 shows a unit cube which is not in equilibrium. Consequently, the stresses acting on opposite faces are not identical. Newton's second law can be expressed, in relation to the three axes, as:

$$\sum F x_1 = m a x_1,$$
$$\sum F x_2 = m a x_2, \qquad (2.6)$$
$$\sum F x_3 = m a x_3,$$

All stresses acting in the direction $0x_1$ are represented in the cube. It is considered that at the center of the cube (with dimensions $\delta x_1, \delta x_2, \delta x_3$) the stresses have the values of $\sigma_{11}, \sigma_{22}, \sigma_{33}$ (normal) and $\sigma_{12}, \sigma_{13}, \sigma_{23}$ (shear). In the derivations that follow the symbols defined by Nye[144] will be used.

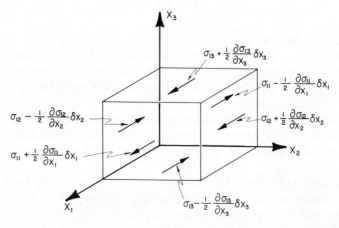

FIG. 2.4 Unit cube when stresses vary from one face to the other; only stresses in $0x_1$ direction shown.

The term $\frac{1}{2}(\partial \sigma_{11}/\partial x_1)\delta x_1$ expresses the change in σ_{11}, with respect to $0x_1$, as one moves from the center of the cube to one of the faces perpendicular to $0x_1$. The other terms have similar meanings. So, neglecting the effects of gravitational forces and body moments, one has, in the direction $0x_1$:

$$\left(\sigma_{11} + \frac{1}{2}\frac{\partial \sigma_{11}}{\partial x_1}\delta x_1 - \sigma_{11} + \frac{1}{2}\frac{\partial \sigma_{11}}{\partial x_1}\delta x_1 \right)\delta x_2 \delta x_3$$

$$+\left(\sigma_{12}+\frac{1}{2}\frac{\partial\sigma_{12}}{\partial x_2}\delta x_2-\sigma_{12}+\frac{1}{2}\frac{\partial\sigma_{12}}{\partial x_2}\delta x_2\right)\delta x_1 \delta x_3$$

$$+\left(\sigma_{13}+\frac{1}{2}\frac{\partial\sigma_{13}}{\partial x_3}\delta x_3-\sigma_{13}+\frac{1}{2}\frac{\partial\sigma_{13}}{\partial x_3}\delta x_3\right)\delta x_1 \delta x_2$$

$$=\rho\delta x_1 \delta x_2 \delta x_3 \frac{\partial^2 u_1}{\partial t^2}$$

u_i is the displacement in the direction x_i. The derivations that follow are based on Kolsky's[1] and Rinehart's[2] presentations. A similar account is given by Wasley.[10] So:

$$\frac{\partial\sigma_{11}}{\partial x_1}\delta x_1 \delta x_2 \delta x_3+\frac{\partial\sigma_{12}}{\partial x_2}\delta x_1 \delta x_2 \delta x_3+\frac{\partial\sigma_{13}}{\partial x_3}\delta x_1 \delta x_2 \delta x_3$$

$$=\rho\delta x_1 \delta x_2 \delta x_3 \frac{\partial^2 x_1}{\partial t^2}.$$

For the other eqns. (2.6):

$$\frac{\partial\sigma_{11}}{\partial x_1}+\frac{\partial\sigma_{12}}{\partial x_2}+\frac{\partial\sigma_{13}}{\partial x_3}=\rho\frac{\partial^2 u_1}{\partial t^2};$$

$$\frac{\partial\sigma_{21}}{\partial x_1}+\frac{\partial\sigma_{22}}{\partial x_2}+\frac{\partial\sigma_{23}}{\partial x_3}=\rho\frac{\partial^2 u_2}{\partial t^2};$$

$$\frac{\partial\sigma_{31}}{\partial x_1}+\frac{\partial\sigma_{32}}{\partial x_2}+\frac{\partial\sigma_{33}}{\partial x_3}=\rho\frac{\partial^2 u_3}{\partial t^2};$$

In dummy suffix notation

$$\frac{\partial\sigma_{ij}}{\partial x_j}=\rho\frac{\partial^2 u_i}{\partial t^2} \qquad (2.7)$$

The solution of this system of differential equations will yield, once the stresses are replaced by strains, the equations of the wave. For an isotropic material, the calculations are simplified. There are only two independent elastic constants instead of 36. One has, for an isotropic material:

$$C_{12}=C_{13}=C_{23}=\lambda$$
$$C_{44}=C_{55}=C_{66}=\mu$$
$$C_{11}=C_{22}=C_{33}=\lambda+2\mu \quad (2.8)$$
$$C_{14}=C_{15}=C_{16}=C_{24}=C_{25}=C_{26}=C_{34}=C_{35}=C_{36}$$
$$=C_{45}=C_{46}=C_{56}=0$$

Notice that the stiffnesses are represented by their matrix notation and *not* by their tensor notation. So:

$$\sigma_{11}=\lambda\Delta+2\mu\varepsilon_{11}, \quad \sigma_{23}=2\mu\varepsilon_{23}, \quad \sigma_{22}=\lambda\Delta+2\mu\varepsilon_{22},$$
$$\sigma_{13}=2\mu\varepsilon_{13}, \quad \sigma_{12}=2\mu\varepsilon_{12}, \quad \sigma_{33}=\lambda\Delta+2\mu\varepsilon_{33} \quad (2.9)$$

λ and μ are Lamé's constants and Δ is the dilatation: $\Delta=\varepsilon_{11}+\varepsilon_{22}+\varepsilon_{33}$ Replacing the values of σ_{ij} in eqns. (2.6) by their values in eqn. (2.9).

$$\frac{\partial}{\partial x_1}(\lambda\Delta+2\mu\varepsilon_{11})+\frac{\partial}{\partial x_2}(2\mu\varepsilon_{12})+\frac{\partial}{\partial x_3}(2\mu\varepsilon_{13})=\rho\frac{\partial^2 u_1}{\partial t^2}$$

$$2\mu\frac{\partial\varepsilon_{11}}{\partial x_1}+\mu\frac{\partial\varepsilon_{12}}{\partial x_2}+\mu\frac{\partial\varepsilon_{13}}{\partial x_3}=\rho\frac{\partial^2 u_1}{\partial t^2}$$

But, from the definition of strains, one has:

$$\varepsilon_{11}=\frac{\partial u_1}{\partial x_1}; \quad \varepsilon_{12}=\frac{1}{2}\left(\frac{\partial u_1}{\partial x_2}+\frac{\partial u_2}{\partial x_1}\right) \quad \text{and} \quad \varepsilon_{13}=\frac{1}{2}\left(\frac{\partial u_1}{\partial x_3}+\frac{\partial u_3}{\partial x_1}\right)$$

So

$$2\mu\frac{\partial^2 u_1}{\partial x_1^2}+\mu\frac{\partial^2 u_1}{\partial x_2^2}+\mu\frac{\partial^2 u_2}{\partial x_1 \partial x_2}+\mu\frac{\partial^2 u_1}{\partial x_3^2}+\mu\frac{\partial^2 u_3}{\partial x_1 \partial x_3}+\frac{\partial\lambda\Delta}{\partial x_1}=\rho\frac{\partial^2 u_1}{\partial t^2}$$

We will define the operator ∇^2 as:

$$\nabla^2=\frac{\partial^2}{\partial x_1^2}+\frac{\partial^2}{\partial x_2^2}+\frac{\partial^2}{\partial x_3^2} \quad (2.10)$$

So

$$\lambda\frac{\partial\Delta}{\partial x_1}+\Delta\frac{\partial\lambda}{\partial x_1}+\mu\frac{\partial^2 u_1}{\partial x_1^2}+\mu\nabla^2 u_1+\mu\frac{\partial^2 u_2}{\partial x_1 \partial x_2}+\mu\frac{\partial^2 u_3}{\partial x_1 \partial x_3}=\frac{\partial^2 u_1}{\partial t^2}$$

and

$$(\lambda+\mu)\frac{\partial\Delta}{\partial x_1}+\mu\nabla^2 u_1=\rho\frac{\partial^2 u_1}{\partial t^2} \quad (2.11)$$

Repeating the procedure for the other two eqns. (2.6):

$$(\lambda+\mu)\frac{\partial \Delta}{\partial x_2}+\mu\nabla^2 u_2 = \rho\frac{\partial^2 u_2}{\partial t^2} \quad (2.12)$$

$$(\lambda+\mu)\frac{\partial \Delta}{\partial x_3}+\mu\nabla^2 u_3 = \rho\frac{\partial^2 u_3}{\partial t^2} \quad (2.13)$$

Still, the displacements have to be replaced by strains and the three equations have to be grouped into one. For this, we take the derivative with respect to x_1 of eqn. (2.11), with respect to x_2 of eqn. (2.12), and with respect to x_3 of eqn. (2.13).

$$(\lambda+\mu)\frac{\partial^2 \Delta}{\partial x_1^2}+\mu\frac{\partial}{\partial x_1}\nabla^2 u_1 = \rho\frac{\partial^3 u_1}{\partial t^2 \partial x_1}$$

$$(\lambda+\mu)\frac{\partial^2 \Delta}{\partial x_1^2}+\mu\nabla^2\frac{\partial u_1}{\partial x_1} = \rho\frac{\partial^2 \varepsilon_{11}}{\partial t^2} \quad (2.14)$$

$$(\lambda+\mu)\frac{\partial^2 \Delta}{\partial x_1^2}+\mu\nabla^2\varepsilon_{11} = \rho\frac{\partial^2 \varepsilon_{11}}{\partial t^2}$$

Repeating the procedure for eqns. (2.12) and (2.13):

$$(\lambda+\mu)\frac{\partial^2 \Delta}{\partial x_2^2}+\mu\nabla^2\varepsilon_{22} = \rho\frac{\partial^2 \varepsilon_{22}}{\partial t^2} \quad (2.15)$$

$$(\lambda+\mu)\frac{\partial^2 \Delta}{\partial x_3^2}+\mu\nabla^2\varepsilon_{33} = \rho\frac{\partial^2 \varepsilon_{33}}{\partial t^2} \quad (2.16)$$

Adding eqns. (2.14), (2.15) and (2.16):

$$(\lambda+\mu)\nabla^2\Delta+\mu\nabla^2(\varepsilon_{11}+\varepsilon_{22}+\varepsilon_{33}) = \rho\frac{\partial}{\partial t^2}(\varepsilon_{11}+\varepsilon_{22}+\varepsilon_{33})$$

and

$$(\lambda+2\mu)\nabla^2\Delta = \rho\frac{\partial^2 \Delta}{\partial t^2}$$

or

$$\frac{\partial^2 \Delta}{\partial t^2} = \left(\frac{\lambda+2\mu}{\rho}\right)\nabla^2\Delta \quad (2.17)$$

A dimensional analysis shows that the units of the coefficient $(\lambda+2\mu)/\rho$ are (distance/time)^2. A closer comparison with equations of waves given in basic textbooks shows that:

$$V_{\text{long}} = \left(\frac{\lambda + 2\mu}{\rho}\right)^{1/2} \tag{2.18}$$

If one assumes that $v = 0.3$, one obtains

$$V_{\text{long}} = \left(\frac{1.35E}{\rho}\right)^{1/2}$$

Indeed, Kreyzig[22] has obtained the following equations for one-dimensional (vibrating string) and two-dimensional waves:

$$\frac{\partial^2 u_1}{\partial t^2} = C^2 \frac{\partial^2 u_1}{\partial x_1^2} \quad \text{(for 1-D wave)} \tag{2.19}$$

$$\frac{\partial^2 u}{\partial t^2} = C^2 \nabla^2 u$$

The elimination of Δ between eqns. (2.11) and (2.12) yields the shear wave. For this, we differentiate eqn. (2.11) with respect to x_2 and eqn. (2.12) with respect to x_1:

From eqn. (2.11)

$$(\lambda + \mu)\frac{\partial^2 \Delta}{\partial x_1 \partial x_2} + \mu \frac{\partial}{\partial x_2}\nabla^2 u_1 = \rho \frac{\partial^3 u_1}{\partial x_2 \partial t^2} \tag{2.20}$$

From eqn. (2.12)

$$(\lambda + \mu)\frac{\partial^2 \Delta}{\partial x_2 \partial x_1} + \mu \frac{\partial}{\partial x_1}\nabla^2 u_2 = \rho \frac{\partial^3 u_2}{\partial x_1 \partial t^2} \tag{2.21}$$

$$\rho \frac{\partial^2}{\partial t^2}\left(\frac{\partial u_1}{\partial x_2} - \frac{\partial u_2}{\partial x_1}\right) = \mu \nabla^2 \left(\frac{\partial u_1}{\partial x_2} - \frac{\partial u_2}{\partial x_1}\right) \tag{2.22}$$

But, by definition, the rigid body rotations are given by:

$$\omega_{ij} = \frac{1}{2}\left(\frac{\partial u_i}{\partial x_j} - \frac{\partial u_j}{\partial x_i}\right) \tag{2.23}$$

And hence,

$$\frac{\partial^2 \omega_{12}}{\partial t^2} = \frac{\mu}{\rho}\nabla^2 \frac{\partial \omega_{12}}{\partial x_i} \tag{2.24}$$

Thus, the rotation ω_{12} propagates at the velocity $(\mu/\rho)^{1/2}$. The same procedure can be repeated to obtain rotations ω_{13} and ω_{23}.

$$V_{\text{shear}} = (\mu/\rho)^{1/2} \tag{2.25}$$

For the treatment of the surface waves the reader is referred to Kolsky.[1] For steel ($v = 0.29$) a calculation reported by Kolsky[1] shows the velocity is:

$$V_{\text{Raleigh}} = 0.9258\, V_{\text{shear}}$$

Typical velocities of longitudinal and shear waves are reported in Table 2.1.

TABLE 2.1
VELOCITIES OF ELASTIC WAVES

Material	$V_{\text{long}}(m/sec)$	$V_{\text{shear}}(m/sec)$
Air	340	None
Aluminum	6 100	3 100
Steel	5 800	3 100
Lead	2 200	700

2.2.3. Elastic Waves in Anisotropic Media

In anisotropic media the situation complicates itself. Because deformation mechanisms operate on a microscopic scale, and because metals are, on a microscopic scale, anisotropic, wave anisotropy should be understood. The derivation of wave velocities for crystals exhibiting cubic symmetry is presented by Ghatak and Kothari[3] and is summarized here.

When eqn. (2.7) is substituted into eqns. (2.4), the stiffnesses of cubic symmetry have to be used. In a cubic crystal, there are three independent elastic constants: C_{11}, C_{12} and C_{44}.
So

$$\begin{aligned}
\sigma_1 &= (C_{11} - C_{12})\varepsilon_1 + C_{12}(\varepsilon_1 + \varepsilon_1 + \varepsilon_3) \\
\sigma_2 &= (C_{11} - C_{12})\varepsilon_2 + C_{12}(\varepsilon_1 + \varepsilon_2 + \varepsilon_3) \\
\sigma_3 &= (C_{11} - C_{12})\varepsilon_3 + C_{12}(\varepsilon_1 + \varepsilon_3 + \varepsilon_3) \\
\sigma_4 &= C_{44}\varepsilon_4 \\
\sigma_5 &= C_{44}\varepsilon_5 \\
\sigma_6 &= C_{44}\varepsilon_6
\end{aligned} \tag{2.26}$$

We should remember that

$$\varepsilon_m = 2\varepsilon_{ij} = \left(\frac{\partial u_i}{\partial x_j} + \frac{\partial u_j}{\partial x_i}\right) \quad (2.27)$$

So, we have

$$(C_{11} - C_{12})\frac{\partial^2 u_1}{\partial x_1^2} + C_{12}\left(\frac{\partial^2 u_1}{\partial x_1^2} + \frac{\partial^2 u_2}{\partial x_1 \partial x_2} + \frac{\partial^2 u_3}{\partial x_1 + \partial x_3}\right)$$
$$+ C_{44}\left(\frac{\partial^2 u_1}{\partial x_2^2} + \frac{\partial^2 u_1}{\partial x_3^2} + \frac{\partial^2 u^2}{\partial x_1 \partial x_2} + \frac{\partial^2 u_3}{\partial x_1 \partial x_3}\right) = \rho\frac{\partial^2 u_1}{\partial t^2} \quad (2.28)$$

and two other, similar equations.

We now have to group these three equations into one, and at the same time eliminate the displacements u_i. By making the following substitution:

$$u_i = A_i \exp(-ij(\omega t - \mathbf{q}\cdot\mathbf{v})) \quad (2.29)$$

we are assuming that we have a plane wave with angular frequency ω; the wave vector is $\mathbf{q}$·ij should not be confused with i. ij is the imaginary number. We end up with a set of three homogeneous equations, whose solutions are given by the secular equation or determinant below.

$$\begin{bmatrix}(C_{11}-C_{44})q_1^2+C_{44}q^2-\rho\omega^2 & (C_{12}+C_{44})q_1q_2 & (C_{12}+C_{44})q_1q_3 \\ (C_{12}+C_{44})q_2q_1 & (C_{11}-C_{44})q_2^2+C_{44}q^2-\rho\omega^2 & (C_{12}+C_{44})q_2q_3 \\ (C_{12}+C_{44})q_3q_1 & (C_{12}+C_{44})q_2q_3 & (C_{11}-C_{44})q_3^2+C_{44}q^2-\rho\omega^2\end{bmatrix} \quad (2.30)$$

We have $q^2 = q_1^2 + q_2^2 + q_3^2$

We can determine the solution of this equation for certain orientations of the wave, e.g.

(a) [100] $q_1 = q$, $q_2 = 0 = q_3$

$$(C_{11}q^2 - \rho\omega^2)(C_{44}q^2 - \rho\omega^2)^2 = 0, \quad C_{11}q^2 - \rho\omega^2 = 0,$$
and
$$C_{44}q^2 - \rho\omega^2 = 0 \quad (2.31)$$

The following are solutions:

$$U_1 = \frac{\omega}{q} = (C_{11}/\rho)^{1/2}, \quad \text{and} \quad U_2 = U_3 = (C_{44}/\rho)^{1/2} \quad (2.32)$$

The ratio ω/q is the velocity of the wave. (See eqns. (2.34) in ref. 3) U_1 is the longitudinal wave; U_2 and U_3 are the shear waves.

(b) [110] For this orientation of **q**, we have $q_1 = q_2 = q/\sqrt{3}$; $q_3 = 0$. The secular equation becomes:

$$(C_{44}q^2 - \rho\omega^2)\left[(\tfrac{1}{2}C_{11}q^2 + C_{44}q^2 - \rho\omega^2)(\tfrac{1}{2}C_{11}q^2 + C_{44}q^2 - \rho\omega^2) \right.$$
$$\left. - (C_{11} + C_{44})(C_{12} + C_{44})\frac{q^4}{4}\right] = 0 \quad (2.33)$$

This equation accepts the following solutions:

$$C_{44}q^2 - \rho\omega^2 = 0, \qquad U_3 = (C_{44}/\rho)^{1/2} \quad (2.34)$$

$$(\tfrac{1}{2}C_{11}q^2 + C_{44}q^2 - \rho\omega^2)^2 - (C_{11} + C_{44})(C_{12} + C_{44})\frac{q^4}{4} = 0$$

$$\tfrac{1}{4}C_{11}^2 q^4 + C_{44}^2 q^4 - \rho^2\omega^4 + C_{11}C_{44}q^4 - C_{11}\rho\omega^2 q^2$$
$$- 2C_{44}\rho\omega^2 q^2 - (C_{11} + C_{44})(C_{12} + C_{44})\frac{q^4}{4} = 0 \quad (2.35)$$

Dividing by q^4 the two solutions are:

$$U_1 = \left(\frac{C_{11} + C_{12} + 2C_{44}}{2\rho}\right)^{1/2}, \quad \text{and} \quad U_2 = \left(\frac{C_{11} - C_{12}}{2\rho}\right)^{1/2} \quad (2.36)$$

Where U_1 is the longitudinal wave, U_2 is the shear wave with vibration along the x_1, x_2 axes, and U_3 the shear wave with vibration along the x_3 axis.

(c) [111]
We find that

$$U_1 = \left(\frac{2C_{11} + C_{12} + 2C_{44}}{3\rho}\right)^{1/2},$$

and,

$$U_2 = U_3 = \left(\frac{C_{11} - C_{12} + C_{44}}{3\rho}\right)^{1/2} \quad (2.37)$$

It is interesting to observe that, if a material is isotropic, the velocities of the waves reduce themselves to the ones calculated for isotropic materials. An isotropic material has the anisotropy ratio equal to 1

$$A = 1 = \frac{2C_{44}}{C_{11} - C_{12}}$$

therefore

$$C_{44} = \frac{C_{11} - C_{12}}{2}$$

Substituting this into eqns. (2.32), (2.34), (2.36) and (2.37), one obtains:

$$U_1[100] = U_1[110] = U_1[111] = (C_{11}/\rho)^{1/2} \quad (2.38)$$

$$U_2[100] = U_3[100] = U_2[110] = U_3[110]$$
$$= U_2[111] = U_3[111] = (C_{44}/\rho)^{1/2} \quad (2.39)$$

Lamé's constants have the following values:

$$C_{44} = \mu, \quad C_{12} = \lambda \quad (2.40)$$

So $C_{11} = 2C_{44} + C_{12} = 2\mu + \lambda$. Consequently, eqns. (2.38) and (2.39) reduce themselves to:

$$U_1 = \left(\frac{2\mu + \lambda}{\rho}\right)^{1/2}, \quad U_2 = \left(\frac{\mu}{\rho}\right)^{1/2} \quad (2.41)$$

These are, in fact, eqns. (2.18) and (2.25).

An idea of the dependence of elastic wave velocity upon orientation for an FCC metal can be obtained from the calculations performed by Meyers and Carvalho.[23]

For Ni, at ambient temperature (ref. 3): ($\rho = 8.9$)

$$C_{11} = 2.508 \times 10^{11} \text{ Pa}, \quad C_{12} = 1.500 \times 10^{11} \text{ Pa}, \quad C_{44} = 1.235 \times 10^{11} \text{ Pa}$$

Hence

$$U_{[100]} = 5.31 \times 10^3 \text{ m/sec}, \quad U_{[110]} = 6.03 \times 10^3 \text{ m/sec},$$
$$U_{[111]} = 5.80 \times 10^3 \text{ m/sec}$$

2.3. PLASTIC WAVES

2.3.1. Preliminary Considerations

Every plastic deformation propagates within a solid as a disturbance in the same way as elastic deformation. However, when the rate of application of the load is low with respect to the velocity of propagation of the disturbance, one can consider the plastic strain as uniformly distributed over the whole extent of the solid body (at the macroscopic level, because microscopically plastic deformation is inherently inhomogeneous). The velocity is given by:

$$v \sim \left(\frac{d\sigma}{dt}\right)^{1/2} \qquad (2.42)$$

The work hardening rate is lower than Young's modulus. For the uniaxial stress configuration, $d\sigma/d\varepsilon$ decreases with ε. This, in turn will produce a plastic wave whose velocity decreases with strain. Hence, the wave front will have a lower and lower slope as it propagates through the metal (Fig. 2.3(a)). If no lateral flow of material is allowed (flow perpendicular to the direction of propagation of the wave), the stress–strain curve takes a concave appearance (in contrast to a convex shape for uniaxial, biaxial, and similar stress states). The Hugoniot pressure versus volume curve can be converted into a stress versus strain curve (Fig. 2.8(b)) where the increase in slope with strain can be seen. This configuration defines the shock wave, which will be treated separately in Section 4. It is seen that between the state of uniaxial stress established in a thin wire and the state of uniaxial strain set up in an infinite (laterally) plate, one has a whole spectrum of stress regimes, leading from a gently sloping stress front to a discontinuous shock front. In the former case the hydrostatic component of the stress is zero; in the latter, it is very high (as high as allowed by the uniaxial strain).

Although one can consider the plastic waves to be of a pure shock nature in carefully controlled shock-loading experiments, explosive forming, welding and compaction present a mixture of these waves. In explosive forming, one has plastic waves propagating *along* the sheet; shock waves could only exist while passing through it. Nevertheless, this would require simultaneous arrival of the shock front over the whole surface of the sheet, which is not the case. In shock-wave compaction, one can distinguish two regions: (a) the core of the particles, which do not undergo any macroscopic deformation and are consequently subjected to a shock wave, and (b) the particle periphery, which is formed in such a way as to fill the gaps between particles. Figure 2.5 shows how a shock front compacts initially spherical particles. There has to be a continual (plastic wave)⇆(shock wave) conversion as the wave travels through the system. The regions undergoing residual plastic deformation are responsible for a dispersion of the wave. Additional complications are the friction between particles. The increased attenuation rate of the plastic wave (resulting in increased heating) contributes significantly to the interparticle bonding. Melting often occurs at the interfaces.

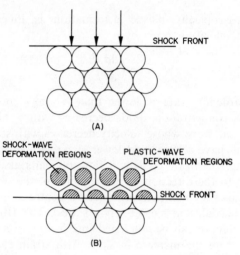

FIG. 2.5 Propagation of disturbance in metal powder producing consolidation. Only the internal portion of particle undergoes a shock deformation. External portions suffer heavy residual deformation and are traversed by plastic waves.

2.3.2. The von Kármán and Duwez Plastic Wave Theory

Von Kármán and Duwez,[5] G. I. Taylor[32] and Rakhmatulin[33] independently developed the theory of plastic waves. The von Kármán and Duwez treatment is presented here. Two frames of reference used when dealing with disturbance-propagation problems are: (a) Lagrangian, when one considers a particle in the material and observes the change of position of this particle with time; and (b) Eulerian, when one considers a certain region in space and observes the flow of material in and out of it. Hence, a property F, which varies with time, spatial position (X) and particle position (x), can either be expressed as:

$$F = f(x, t), \quad \text{or} \quad F = f(X, t) \qquad (2.43)$$

These two approaches will be discussed again in Section 2.5, in connection with numerical methods of solution.

Von Kármán and Duwez considered the simplest possible plastic wave propagation problem: a semi-infinite thin wire being impacted at a certain velocity generating a downward motion at a velocity V_1. The initial position of the extremity of the wire is taken as the origin, and we observe the displacement of a particle situated at the position x. At time

t, it will be displaced by u. Hence

$$dF = dm \frac{\partial^2 u}{\partial t^2} = \rho_0 A_0 \, dx \frac{\partial^2 u}{\partial t^2}$$

therefore

$$\frac{d\sigma}{dx} = \rho_0 \frac{\partial^2 u}{\partial t^2} \tag{2.44}$$

where ρ_0 and A_0 are the initial density and area, and σ is the stress. Since we have a state of plastic deformation, and assuming that one has a univalent relationship between stress and strain in *loading* (not in unloading because of the irreversibility of the process), one can write,

$$\rho_0 \frac{\partial^2 u}{\partial t^2} = \frac{d\sigma}{d\varepsilon} \frac{\partial \varepsilon}{\partial x}$$

and with $\varepsilon = \partial u/\partial x$, we have

$$\frac{\partial^2 u}{\partial t^2} = \frac{1}{\rho_0} \frac{d\sigma}{d\varepsilon} \frac{\partial^2 u}{\partial x^2} \tag{2.45}$$

One can see the similarity with eqn. (2.19). The velocity of the plastic wave can be seen to be given by

$$\left(\frac{1}{\rho_0} \frac{d\sigma}{d\varepsilon} \right)^{1/2}$$

and the comparable elastic wave velocity can also be determined.[34,35]

The application of the boundary conditions allows the determination of the wave profile. The boundary conditions are

$$u = V_1 t \text{ at } x_1 = 0, \quad \text{and} \quad u = 0 \text{ at } x = \infty \quad \text{(any } t > 0)$$

There are two solutions to eqn. 2.45. The first (found by inspection) is

$$u = V_1 t + \varepsilon_1 x$$

A second solution is found when

$$\frac{\left(\dfrac{d\sigma}{d\varepsilon}\right)}{\rho_0} = \left(\frac{x}{t}\right)^2$$

$d\sigma/d\varepsilon$ being a function of ε, the strain ε has to be a function of the velocity x/t, which we will call β

$$u = \int_{\infty}^{x} \frac{\partial u}{\partial x} dx = \int_{\infty}^{x} f(\beta) dx = t \int_{-\infty}^{\beta} f(\beta) d\beta \qquad (2.46)$$

Differentiating twice with respect to t, we get:

$$\frac{\partial^2 u}{\partial t^2} = \frac{\beta^2}{t} f'(\beta) \qquad (2.47)$$

where $f'(\beta)$ is the derivative of $f(\beta)$ with respect to β. This leads to the following solution:

(a) $x=0$ to $x=Ct$ (C is the velocity of propagation of the plastic wavefront); the strain is constant at ε_1.

(b) $Ct < x < C_0 t$ (C_0 is the velocity of elastic longitudinal waves in bars). In this interval, one has:

$$\frac{x}{t} = \left(\frac{d\sigma/d\varepsilon}{\rho_0}\right)^{1/2} \qquad (2.48)$$

(c) $x > C_0 t$, $\varepsilon = 0$.

Figure 2.6 shows graphically how the strain varies as a function of x/t.

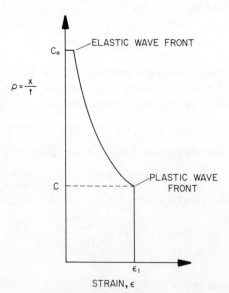

FIG. 2.6 Strain versus x/t for disturbance propagating in wire (adapted from von Kármán and Duwez[5])

From eqn. (2.46) for $u(0,t)=V_1 t$, we have

$$V_1 = -\int_0^{\varepsilon_1}\left(\frac{d\sigma/d\varepsilon}{\rho_0}\right)^{1/2} d\varepsilon \tag{2.49}$$

For a known impact velocity and a known σ versus ε relationship, one can determine ε. Assuming a power law:

$$\sigma = k\varepsilon^n, \quad V_1 = -\frac{k\varepsilon_1^n}{\rho_0} \tag{2.50}$$

and

$$\varepsilon_1 = \left(\frac{\rho_0 V_1}{k}\right)^{1/n} \tag{2.51}$$

The maximum impact velocity can be easily found by setting it equal to the velocity that will produce necking on the specimen. Applying Considère's criterion $\varepsilon = n$,

$$V_1 = -(kn^n/\rho_0) \tag{2.52}$$

2.3.3. Plastic Shear Waves

One can have plastic shear waves, in an analogous way to elastic shear waves. The particles undergo plastic displacement perpendicular to the direction of propagation of the wave. It is very difficult to generate exclusively plastic shear waves, and a wave in a rod would be a plastic shear wave; however, the amplitude of motion would vary with the distance from the longitudinal axis of the rod. Hence, it is easier to generate a plastic shear wave simultaneously with a longitudinal stress wave. For the mathematical treatment of plastic shear waves see references 32, 36 and 37. However, it is Clifton's group[25-27,39] that has investigated them—both experimentally and theoretically—in greater depth. Using a parallel inclined impact of a projectile with a target, it is seen that the impact generates a pressure wave moving perpendicular to the interface and a shear wave, producing lateral displacement, also moving perpendicular to it. Measurement of the lateral displacement determines the rate of arrival of this shear disturbance at the back face of the target. Figure 2.7 shows the transverse displacement as a function of time after impact for 6061-T6 aluminum. It can be seen that the slope of the line is constant over the interval. Hence, the shear wave differs from a shock wave in the gradually increasing front. For the situation analysed,[25,26] the velocities of the pressure front and shear front are 5.24

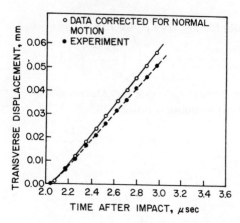

FIG. 2.7 Experimental results obtained by Abou-Sayed et al.[27]

and 3.10 mm/μs, respectively. Hence, the plastic shear wave has a velocity slightly higher than half the plastic pressure (shock) velocity. This is about the same ratio as the one obtained between shear and longitudinal elastic waves. Plastic shear waves are certainly of great importance in dynamic deformation.

2.3.4. Additional Considerations on Plastic Waves

The theory of von Kármán and Duwez[5] is only the first step towards an understanding of plastic waves in solids. A considerable theoretical effort has been devoted, over the past twenty years, to develop models for the propagation of plastic waves. At the same time, it has been realized that the hydrodynamic treatment of shock waves is over-simplified and fails to predict a number of phenomena; even at high pressures strength effects are of importance. A number of mathematical formulations have been proposed incorporating the material properties by a constitutive equation. These relationships are, for the most part, non-linear. Lee[38] developed a theory which he called elasto-plastic with finite deformation. Hermann and Nunziato[9] divide the response into several categories, depending on the wave-propagation characteristics, and analyse them in terms of linear viscoelastic infinitesimal elastic-perfectly plastic, non-linear viscoelastic, and thermoelastic, behaviors. Clifton[39] and Hermann,[7] describe the analytical approach, and a recent report by the National Materials Advisory Board[41] makes a critical analysis of the state of the art and of the areas that need development.

An experimental technique that has been used often to obtain constitutive relations at high strain rates is the split Hopkinson bar. It is, in its present form, a modified version of the bar originally used by Hopkinson,[42] Davies,[43] and Kolsky,[44] who were responsible for its development (see Chapter 4).

2.3.5. Adiabatic Shear Bands

Adiabatic shear bands are a common feature of dynamic deformation. They are not generated by shock waves but have been reported in a range of high-strain-rate deformation processes, such as ballistic impact, high-velocity forming and forging, explosive fragmentation and machining. They consist of narrow bands where intense shearing took place and the effect can be qualitatively explained as follows. Plastic deformation generates heat. As the temperature of the deformed region increases, its flow stress decreases. This leads to a concentration of the deformation along that region, with further heat generation. This self-accelerating process might lead to eventual melting. Morphologically, adiabatic shear bands can be of two types: 'deformation' bands and 'transformation' bands. The latter ones are usually observed in steel, and etch as white streaks in metal. It seems that their structure is different from that of the surrounding material, and attempts (so far, unsuccessful) have been made to identify it.[41] It is composed of very small grain sizes (0.1–1 μm). There is evidence for four possibilities: ferrite (BCC), martensite (BCT), austenite (FCC), and an unnamed BCC phase with a lattice parameter smaller than the normally encountered ferrite. It is certain that adiabatic shear bands are present in high-energy-rate fabrication processes such as explosive forming, welding, and even compaction. Virtually nothing is known about the micromechanical deformation processes leading to thermoplastic band formation. The mechanical aspects and requirements will be very briefly reviewed here.

Zener and Hollomon[83] were the first to report these bands. They suggested that they would form when the decrease in flow stress, due to temperature rise, offset the increase in flow stress due to work hardening. Recht, in 1964,[84] expressed this criterion in a more quantitative way, starting from a simplified mechanical equation of state:

$$\sigma = f(\varepsilon, T),$$

$$d\sigma = \left(\frac{\partial \sigma}{\partial \varepsilon}\right)_T d\varepsilon + \left(\frac{\partial \sigma}{\partial T}\right)_\varepsilon dT$$

where σ is the shear stress, and ε and T are the strain and temperature, respectively. Recht assumed that catastrophic (unstable) flow would occur when $d\sigma/d\varepsilon$ would become equal to zero; or:

$$\left(\frac{\partial \sigma}{\partial \varepsilon}\right)_T = -\left(\frac{\partial \sigma}{\partial T}\right)_\varepsilon \frac{dT}{d\varepsilon} \tag{2.53}$$

He derived an expression for the increase in temperature with strain for a narrow band of thickness t, in a block of thickness L, of the form:

$$\frac{dT}{d\varepsilon} = \frac{\sigma_y L}{2W}\left[\frac{\dot{\varepsilon}}{\pi k \rho c(\varepsilon - \varepsilon_y)}\right] \tag{2.54}$$

where σ_y and ε_y are the stress and strain of yield (initial), k is the thermal conductivity, W is the work equivalent of heat, c is the specific heat, and ρ is the density. Substituting eqn. (2.54) into eqn. (2.53) one arrives at the critical strain rate:

$$\dot{\varepsilon}_c = 4\pi k \rho c(\varepsilon - \varepsilon_y) \left[\frac{\left(\frac{\partial \sigma}{\partial \varepsilon}\right)_T}{\left(\frac{\partial \sigma}{\partial T}\right)_\varepsilon}\right] \frac{W^2}{\sigma_y^2 L^2} \tag{2.55}$$

More sophisticated criteria have been proposed but Recht's analysis seems to predict the correct trends. Additional work on adiabatic shear bands is reported in references 85 to 88 and reference 21.

An interesting concept that was recently introduced[87] is wave trapping. One can see from eqn. (2.48) that the wave velocity varies with the square root of the work-hardening rate. Thus, when $d\sigma/d\varepsilon \to 0$, the plastic wave should cease to propagate, and deformation will become localized.

The authors would like to speculate on a mechanism for the formation of 'transformation' shear bands. These regions, also called 'white streaks', would be due to localized melting caused by high levels of strain and/or fracture. The solid metal surrounding the molten layer is an almost ideal heat sink and, once the deformation is completed, would provide very high cooling rates. Recent studies on rapidly-solidified metals[40] show that one has the following sequence of morphologies; as the cooling rate is increased: microdendritic structure, microcrystalline structure and amorphous structure. The microcrystalline structure requires cooling rates of 10^4 K s^{-1} or higher. Concomitant with melting, one would have dissolution of the carbides; during subsequent solidification, there would be no time for segregation and a supersaturated solution of carbon in iron would result. This, combined with the extremely small grain size (of the order of fractions of a micrometer) would be responsible for the high hardness.

2.4. SHOCK WAVES

Shock waves are characterized by a steep front, and require a state of uniaxial strain which allows the build-up of the hydrostatic component of stress to high levels. When this hydrostatic component reaches levels that exceed the dynamic flow stress by several factors, one can, to a first approximation, assume that the solid has no resistance to shear ($G=0$). The treatment developed by Hugoniot and Rankine for fluids is commonly applied to the treatment of shock waves.

2.4.1. Hydrodynamic Treatment

The calculation of shock-wave parameters is based, in its simplest form, on the Rankine–Hugoniot[45,46] equations. Essentially, it is assumed that the shear modulus of the metal is zero and that it responds to the wave as a liquid. Hence, the theory is restricted to higher pressures. At pressures close to the dynamic yield strength of metals, more complex computations have to be used. However, it will suffice here to derive the equations for hydrodynamic behavior. The *fundamental* requirement for the establishment of a shock wave is that the velocity of the pulse increases with increasing pressure. This is shown in Fig. 2.3(b). The velocity of the front will be that of the particles subjected to the highest pressure.

Ahead of the front, the pressure is P_0 and density ρ_0; behind they are P and ρ respectively. The velocity of the front is U_s; the particles (or atoms) are stationary ahead of the front. At the front and behind it, they are moving at a velocity u_p. This displacement of the particles is responsible for the pressure build-up. If one considers the center of reference as the shock front and moving with it, and sets up the equation for the conservation of mass, one has: material moving towards front: $A\rho_0 U_s dt$, material moving away from front: $A\rho(U_s-u_p)dt$. Hence

$$\rho_0 U_s = \rho(U_s - u_p) \tag{2.56}$$

The conservation of momentum can be expressed likewise.

Unit impulse = (momentum in–momentum out) = $(\rho_0 U_s dt A)u_p$

Where $\rho_0 U_s dt A$ is the mass, while u_p is the particle velocity change. The

impulse is
$$(P_0 - P)\,dt\,A$$
Hence
$$(P - P_0) = \rho_0 u_p U_s \qquad (2.57)$$

The quantity $\rho_0 u_p U_s$ is usually called the shock impedance. The conservation of energy is obtained by setting up an equation in which the work done is equal to the difference in the total energy of the two sides.

$$A\,dt(Pu_p - P_0 \times 0) = 1/2 m u_p^2 A\,dt + mA\,dt(E - E_0)$$
$$Pu_p = 1/2 \rho_0 U_s u_p^2 + \rho_0 U_s (E - E_0) \qquad (2.58)$$

where E and E_0 are the internal energies.

However, since there are five variables in eqns. (2.56) to (2.58) an additional equation is needed. This fourth equation is experimentally determined and is given as a relationship between shock and particle velocities.

$$U_s = C_0 + S_1 u_p + S_2 u_p^2 \qquad (2.59)$$

S_1 and S_2 are empirical parameters and C_0 is the sound velocity in the material at zero pressure. For most metals $S_2 = 0$ and eqn. (2.59) is reduced to

$$U_s = C_0 + S_1 u_p \qquad (2.60)$$

With the knowledge of the values of C_0 and S_1 for a given material and applying eqns. (2.56), (2.57), (2.58) and (2.60), it is possible to calculate the required quantities.

Figure 2.8(a) shows that U_s versus u_p curve for nickel; as expected, the relationship is linear. By applying eqns. (2.56) to (2.58) one can plot the pressure versus the volume (V/V_0) (Fig. 2.8(b)), and versus the particle velocity (u_p), as shown in Fig. 2.8(c). Specific information on the design of shock recovery systems is given by deCarli and Meyers.[29] Three good sources of information are refs. 47 to 51.

The Mie–Grüneisen–Debye theory is well known for its application in a somewhat more refined treatment of shock waves. The Mie–Grüneisen equation of state is:

$$P = P_k(V) + NkT\Gamma(V)D(\theta/T)/V + (2/3)\alpha V^{-1/3} T^2 \qquad (2.61)$$

where $D(\theta/T)$ is the Debye function, $\Gamma(V)$ is the Grüneisen parameter and other symbols have their usual value.

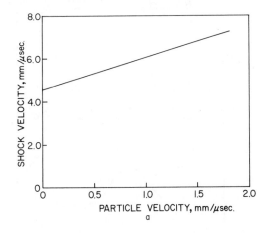

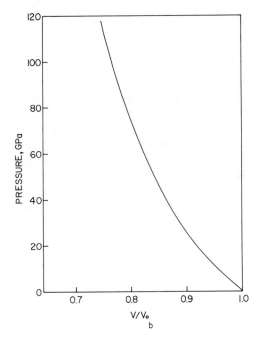

FIG. 2.8 (a) Particle velocity versus shock velocity for nickel (adapted from Meyers[53]). (b) Hugoniot curve for nickel (adapted from Meyers[53]). (c) (overleaf) Pressure versus particle velocity for nickel (adapted from Meyers[53]).

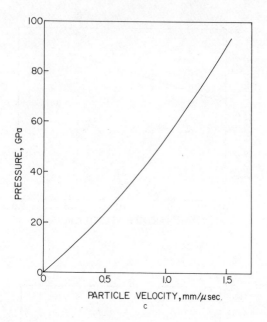

FIG. 2.8 contd.

In order to obtain the pressure generated by an impact of a flyer plate on a target plate at a known velocity, one plots the pressure, P, versus particle velocity, u_p, for both materials and applies the impedance matching technique, as described in reference 29. The equations required to obtain the density (ρ), the compressibility ratio (V/V_0), the shock velocity (U_s), and particle velocity (u_p) can be obtained by manipulating the Rankine–Hugoniot relations (eqns. (2.56), (2.57), (2.58), (2.60)). The coefficients in the empirical linear relationship between u_p and U_s are taken from Table C-1 (ref. 21). The sound velocity C as a function of pressure was obtained from eqn. (33) in ref. 52.

$$C^2 = -V^2 \frac{dP_H}{dV}\left[1 - \frac{\gamma}{V}\frac{V_0-V}{2}\right] + V^2\frac{\gamma}{V}\left[\frac{P_H-P_0}{2}\right]$$
$$+ V^2(P-P_H)\left[\frac{\gamma}{V} + \frac{d\ln(V/\gamma)}{dV}\right] \quad (2.62)$$

The two derivatives can be evaluated and are:

$$\frac{dP_H}{dV} = \frac{-\rho_0^2 C_0^2}{[1-S(1-\rho_0 V)]^2} - \frac{2S\rho_0^2 C_0^2(1-\rho_0 V)}{[1-S(1-\rho_0 V)]^3} \quad (2.63)$$

$$\frac{d \ln (V/\gamma)}{dV} = \left(\frac{\gamma}{V}\right) d\left(\frac{V}{\gamma}\right)/dV = \frac{1}{V} \qquad (2.64)$$

where γ is the Grüneisen constant and P_H is the pressure at the Hugoniot. Table 2.2 presents the shock wave parameters for a number of metals at the pressures of 10, 20, 30, 40 GPa. The values of γ were taken from ref. 52.

When the solid being shocked undergoes a pressure-induced phase transformation, the P versus V/V_0 plot exhibits a discontinuity in the slope. Figure 2.9(a) shows this effect for iron. At 13 GPA iron undergoes an $\alpha(BCC) \to \varepsilon(HCP)$ transformation. Upon unloading, the reverse transformation does not occur at the same pressure, producing the hysteresis behavior shown. Duvall and Graham[54] have presented a comprehensive treatment of shock-induced phase transformations. The velocity of the shock wave is proportional to the square root of the slope of the Raleigh line. The Raleigh line is defined as the line passing through the points $(V/V_0, P)$ and $(1, 0)$. Two Raleigh lines R_1 and R_2, corresponding to pressures P_1 and P_2, are shown in Fig. 2.9. One can see that the slope of R_2 is lower than R_1, although $P_2 > P_1$. Hence, the wave will decompose itself into two waves, as shown in Fig. 2.9(b).

The rarefaction—or release—part of the shock wave is the region beyond the peak pressure, where the pressure returns to zero. The attenuation of a wave, on the other hand, is the decay of the pressure pulse as it travels through the material. Figure 2.10 shows schematically how a shock wave is changed as it progresses into the material. At t_1, the wave has a definite peak pressure, pulse duration and rarefaction rate (mean slope of the back of the wave). The inherent irreversibility of the process is such that the energy carried by the shock pulse continuously decreases. This is reflected by a change of shape of the pulse. If one assumes a simple hydrodynamic response of the material, the change of shape of the pulse can be simply seen as the effect of the differences between the velocities of the shock and rarefaction part of the wave. It can be seen in Fig. 2.10 that the rarefaction portion of the wave has a velocity $u_p + C$, where u_p is the particle velocity and C the sound velocity at the pressure. As the wave progresses, the rarefaction part of the wave overtakes the front, because $u_p + C > U_s$. This will reduce the pulse duration to zero. After it is zero, the peak pressure starts to decrease. As this peak pressure decreases, so does the velocity of the shock front: $U_{s4} < U_{s3} < U_{s2} = U_{s1}$. This can be easily seen by analysing the data in Table 2.2. By appropriate computational procedures one can calculate the change in pulse shape based on the velocities of the shock and

TABLE 2.2
SHOCK-WAVE PARAMETERS FOR SOME REPRESENTATIVE METALS (ADAPTED FROM REF. 21)

Pressure GPa	ρ g/cm^3	V/V_0	U_s km/s	u_p km/s	C km/s
2024 Al					
0	2.785	1.0	5.328	0.0	5.328
10	3.081	0.904	6.114	0.587	6.220
20	3.306	0.842	6.751	1.064	6.849
30	3.490	0.798	7.302	1.475	7.350
40	3.647	0.764	7.794	1.843	7.774
Cu					
0	8.930	1.0	3.940	0.0	3.94
10	9.499	0.940	4.325	0.259	4.425
20	9.959	0.897	4.656	0.481	4.808
30	10.349	0.863	4.950	0.679	5.131
40	10.668	0.835	5.218	0.858	5.415
Fe					
0	7.85	1.0	3.574	0.0	3.574
10	8.479	0.926	4.155	0.306	4.411
20	8.914	0.881	4.610	0.550	5.054
30	9.258	0.848	4.993	0.759	5.602
40	9.543	0.823	5.329	0.945	6.092
Ni					
0	8.874	1.0	4.581	0.0	4.581
10	9.308	0.953	4.916	0.229	5.005
20	9.679	0.917	5.213	0.432	5.357
30	9.998	0.888	5.483	0.617	5.661
40	10.285	0.863	5.732	0.786	5.933
304SS					
0	7.896	1.0	4.569	0.0	4.569
10	8.326	0.948	4.950	0.256	5.051
20	8.684	0.909	5.283	0.479	5.439
30	8.992	0.878	5.583	0.681	5.770
40	9.264	0.852	5.858	0.865	6.061
Ti					
0	4.528	1.0	5.220	0.0	5.220
10	4.881	0.928	5.527	0.4	5.420
20	5.211	0.869	5.804	0.761	5.578
30	5.525	0.820	6.059	1.094	5.708
40	4.826	0.777	6.296	1.403	5.815
W					
0	19.224	1.0	4.029	0.0	4.029
10	19.813	0.970	4.183	0.124	4.207
20	20.355	0.944	4.326	0.240	4.365
30	20.849	0.922	4.462	0.350	4.508
40	21.331	0.901	4.590	0.453	4.638

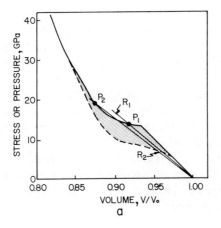
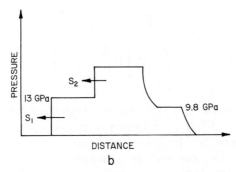

FIG. 2.9 (a) Pressure versus volume (V/V_0) for iron, showing the effects of transformation. (b) Shock pulse configuration for iron above 13 GPa.

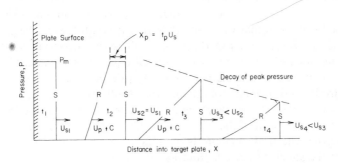

FIG. 2.10 Progress of a shock pulse through metal; rarefaction front steadily overtakes shock front.

rarefaction portion of the wave. This is done below, in a very simplified manner. In order to calculate the rarefaction rate of the pressure pulse as it enters the material, it is best to use Fig. 19 of ref. 29. The rarefaction rate is the slope of the wave tail. Although the curve is concave upwards, we can, as a first approximation, assume it to be a straight line and calculate an average dP/dt. The average rarefaction rate can be obtained by dividing the peak pressure by the difference between the time taken for the head and the tail of the rarefaction wave to pass through a certain point. The head and tail of the rarefactions are shown and travel with velocities $u_p + C$ and C_0, respectively. If one wants to determine the rarefaction rate at the collision interface, one has to find the difference $t_3 - t_2$. One should notice that when the flyer plate is under compression, its thickness is ρ_0/ρ. Hence

$$t_3 - t_2 = t^F \left(\frac{\rho_0}{\rho}\right)\left(\frac{1}{c}\right) - t^F\left(\frac{1}{c_0}\right) \tag{2.65}$$

$$\dot{P} = \frac{dP}{dt} = P_m \left[t^F\left(\frac{\rho_0}{\rho c} - \frac{1}{c_0}\right)\right]^{-1} \tag{2.66}$$

The rate of rarefaction is very sensitive to pressure. So, if either the impact velocity or the flyer-plate thickness is changed, for the same target-projectile system, different rarefaction rates will result.

The attenuation rate (or decay rate) measures the rate at which the pressure pulse dissipates itself as it travels through the material. The energy carried by the pressure pulse is dissipated as heat, defects generated, and other irreversible processes. Figure 2.10 shows schematically how the energy carried by the wave decreases as it travels from the front to the back face of the target. Up to a certain point the pressure remains constant; it can be seen that at t_3 the pressure has already decreased from its initial value and that at t_4 it is still lower. The greater the initial duration of a pulse, the greater will be the energy carried by it, and consequently, its ability to travel throughout the material. The simplest approach to calculating the decay rate of a pulse is to assume the hydrodynamic response of the material. As illustrated in Fig. 2.10 the relative velocities of the shock and release waves will determine the attenuation. The head of the release wave travels at a velocity $u_p + C$; the distance that the peak pressure is maintained is given by the difference between the shock velocity U_s and $u_p + C$. Hence:

$$S = \frac{t_p U_s^2}{u_p + C - U_s} \tag{2.67}$$

Beyond this point numerical techniques have had to be used to compute the pressure decay. This can be done in an approximate way by drawing the pulse shape at fixed intervals, assuming the shock-wave velocity constant in each of them.

2.4.2. More Advanced Treatments and Computer Codes

The hydrodynamic theory refers only to pressures, since it assumes that $G=0$ and that the material does not develop shear stresses. However, the state of uniaxial strain generates shear stresses, and these cannot be ignored in a more detailed account.

One can define the pressure as the hydrostatic component of the strain:

$$dP = \frac{E}{3(1-2v)} d\varepsilon_1$$

Assuming that E and v are not dependent upon pressure, one has:

$$P = \frac{E}{3(1-2v)} d\varepsilon_1$$

The maximum shear stresses can be obtained from the deviatoric stresses, which are given with respect to their principal axes:

$$\tau_{max} = \frac{\sigma_1 - \sigma_3}{2} = \frac{3(1-2v)}{2(1+v)} P \qquad (2.68)$$

The derivation is given in greater detail by Meyers and Murr.[55]

The recognition of the existence of material strength has led to a number of proposals. Another great problem in the mathematical treatment of shock waves is the discontinuity in particle velocity, density, temperature and pressure across the shock front. The differential equations describing these processes are non-linear and trial-and-error computations are required at each step (in time). For this reason, von Neumann and Richtmyer[56] proposed in 1949 a method for treatment of shock waves which circumvented these discontinuous boundary conditions, and, as a result, lent itself much better to mathematical computations. In essence they introduced an *artificial viscosity* term. This artificial viscosity term had the purpose of smoothing the sharp shock front and rendering it tractable in differential equations and finite difference techniques. The shock front was made somewhat larger than the grid in the finite element network. The physical explanation for the introduction of a viscosity form[56] is reasonable: dissipative mechanisms take place at the shock front and they can be represented by a mathematical viscosity term. Hsu et al.[68] studied in detail the specific dissipative mechanisms. These are discussed in the next section. The artificial term used by von Neumann and Richtmyer[56] was:

$$q = -\frac{(c\Delta x)^2}{V} \frac{\partial u_p}{\partial x} \left| \frac{\partial u_p}{\partial} \right| \qquad (2.69)$$

where x designates the position in Lagrangian coordinates, and u_p is the particle velocity.

The differential equations describing the progress of a shock wave can be second-order partial ones. If there are two independent variables (for example, the position, x, and the time, t), one has a hyperbolic differential equation if two characteristic curves pass through each point of the space. A detailed description of the method of characteristics is given by Karpp and Chou.[19] These characteristic curves are obtained from characteristic directions. The application of finite difference techniques to these characteristic curves produces what is known as 'the method of characteristics'. In its graphical representation, one sees two fans of characteristic curves mutually normal. The shock-wave parameters can be determined at these points. Figure 2.11 shows an example.[57] The detonation of an explosive (at grazing incidence) in contact with a metal block generates in the latter a pressure pulse, whose front will be curved because the shock-wave velocity depends upon pressure, which decreases as the wave attenuates itself. Figure 2.11(a) shows the situation schematically. The characteristic curves are shown in Fig. 2.11(b). One can determine the state properties of the expanding gas at all intersections of the curves of two families; the pressures within the slab can also be found. This method of analysis lends itself to a variety of problems involving dynamic propagation of disturbances. Specific examples are: the impact of a plate against a target (NIP code), the motion of compressible flat plates and cylinders drawn by detonation waves (ELA code), plane supersonic gas flow and impact of a cylinder on a plate.

An alternative approach is the finite difference method incorporating an artificial viscosity term. Hence, one would have the following example given by Walsh,[18] for the wave equation:

$$\frac{\partial^2 u}{\partial t^2} = c^2 \frac{\partial^2 u}{\partial x^2} = c \frac{\partial}{\partial x}\left(c \frac{\partial u}{\partial x}\right)$$

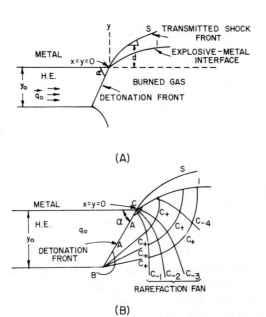

FIG. 2.11 (a) Detonation of explosive in contact with metal block at grazing incidence. (b) Characteristic fans that are used in the calculation of the state variables at the intersection of the lines, both inside the metal and in the detonation-gases regions (adapted from Drummond.)[57]

Making

$$\sigma = c\frac{\partial u}{\partial x}, \text{ and } \omega = \frac{\partial u}{\partial x}$$

One arrives at the equivalent equations:

$$\frac{\partial \omega}{dt} = c\frac{\partial v}{\partial x};$$
$$\frac{\partial v}{\partial t} = c\frac{\partial \omega}{\partial x} \quad (2.70)$$

If one expresses these equations in forms of finite differences, one has

$$\frac{\omega_{j+1}^{(n+1)} - \omega_{j+1}^{(n-1)}}{2\Delta t} = c(x_{j+1})\frac{v_{j+2}^n - v_j^n}{2\Delta x};$$
$$\frac{v_j^{(n+2)} - v_j^n}{2\Delta t} = c(x_j)\frac{\omega_{j+1}^{(n+1)} - \omega_{j-1}^{(n+1)}}{2\Delta x}$$

The application of the finite difference method to shocks produces a great deal of 'noise'. The introduction of an artificial viscosity form greatly improves the solution. The coefficient of viscosity smoothes out the pulse; for q^2 equal to 4, the representation is satisfactory. Figure 2.12 shows another example of the propagation of a shock wave as it travels through a metal. This sequence was generated by Wilkins.[16] If residual plastic deformation takes place, one cannot use the hydrodynamic assumption. Wilkins describes the problem in detail. In order to incorporate plastic flow he applies the von Mises yield criterion. Since deformation is three-dimensional, one has a yield surface. For large deformations one can neglect the elastic strains which rarely exceed 0.5 pct. Figure 2.13 shows in sequential form the detonation of composition B explosive in contact with a metal block. Both the expansion of the detonation gases and the propagation of the shock wave into the copper plate can be seen. However, if the hydrodynamic assumption is made (no material strength), the cratering effect is much larger (and not realistic). Hence a yield surface has to be incorporated in this type of problem.

A variety of codes have been developed both with Lagrangian and Eulerian coordinates. Examples of some Lagrangian finite difference codes are TOODY,[58] HEMP,[16] WONDY,[59] while OIL[60] uses Eulerian coordinates. The NAG (nucleation and growth code) was developed by Curran, et al.[61] and is intended for the study of spalling. Seaman et al.[62] developed a code (PEST) to be used in porous materials.

Although most of these codes have been used exclusively in the development of weapons, warheads, and for gaining a better understanding of nuclear explosions, they could be very helpful in aiding the engineer in high-energy-rate forming applications. The tendency in military research has been to replace a great number of experiments by a few carefully controlled and instrumented tests. Explosive forming, welding and compaction could be simulated in computers, and the problems of wave reflections and spalling could be predicted and avoided. Indeed Hoenig et al.[63] have applied the HEMP code to explosive compaction of powders.

2.4.3. Attenuation of Shock Waves

If one looks at a slightly more realistic representation of a shock pulse in a metal, one can distinguish features not represented in Fig. 2.3 and not treated in the hydrodynamic theory. Figure 2.14(a) illustrates such a situation. The material strength is incorporated into the model. The shock pulse is preceded by an elastic precursor wave with amplitude

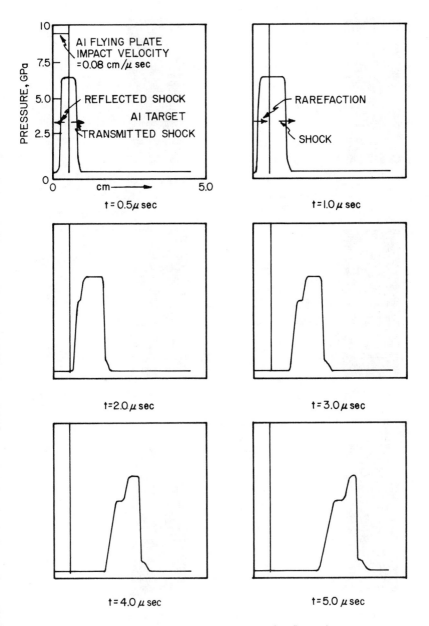

FIG. 2.12 Stress profile produced by the impact of a flyer plate on target, as simulated, using Wilkins[16] code (adapted from ref. 16).

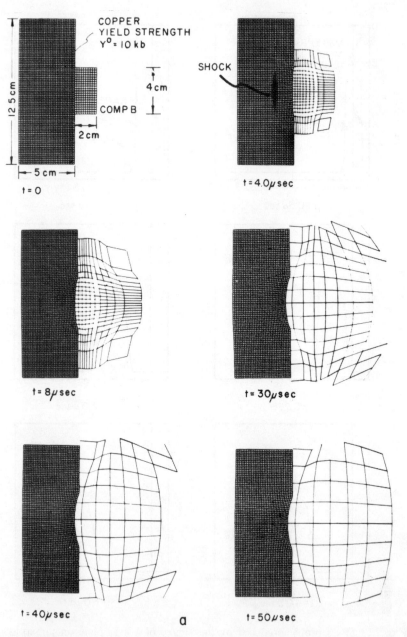

FIG. 2.13 (a) Simulation of detonation of a high explosive (Comp B) in contact with copper block. (b) Same as (a), assuming hydrodynamic behavior (from ref. 16).

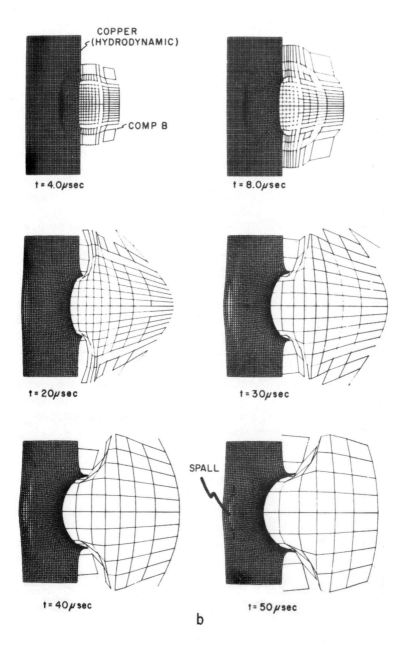

FIG. 2.13 contd.

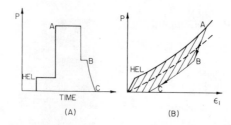

FIG. 2.14 (a) Shock-wave profile including elastic precursor and incorporating strength effects. (b) Pressure versus strain (or volume V/V_0) for hypothetical solid including strength effects; cross-hatched region represents energy dissipated in cycle.

equal to the 'Hugoniot Elastic Limit' (HEL). The rarefaction portion of the wave exhibits a plateau. These two features are better understood if one refers to the pressure versus V/V_0 plot of Fig. 2.14(b). The initial response of the material (at low pressures) is elastic; therefore, it deviates from the hydrodynamic V/V_0 curve. When the Hugoniot elastic limit is reached, the dynamic compressibility curve follows the Hugoniot curve. Taking the slopes of the elastic and plastic curves (Raleigh lines) one can see that the elastic precursor travels faster than the travelling shock waves. It is only for very high pressures that the Raleigh line of the shock pulse has a larger slope than that of the HEL region. Elastic precursors are discussed in Section 2.4.4. Upon unloading, one has first the relieving of the elastic stresses (A to B); then, the curve follows the Hugoniot P versus V/V_0 response up to pressure zero. At point B, the rarefaction part of the wave shows a plateau. The very important feature of Fig. 2.14(b) is the hysteresis behavior. The cross-hatched area indicates the amount of irreversible work done in the process. This irreversibility has a direct bearing on the attenuation of the shock pulse.

Curran[64] and later Erkman et al.[65-67] investigated the attenuation of planar shock waves and compared it with the hydrodynamic treatment. They found much higher observed attenuation. This is to be expected, if one looks at the energy dissipated in Fig. 2.14(b). They found a much better agreement incorporating the artificial viscosity into their treatment of shock waves. Rempel et al.[66] observed that the pressure starts to decay almost immediately after it propagates, contrary to the hydrodynamic calculations (Section 2.4.2). The problem with the artificial viscosity is that it does not have a clearly defined physical meaning in terms of micromechanical frictional processes. In view of this, Hsu et al.[68] recently investigated the attenuation of shock waves in nickel and proposed a model incorporating defect generation and motion. Figure 2.15 shows the calculated (hydrodynamic theory) and observed peak pressures at varying distances from the peak interface. Two different initial pressures were used: 10 and 25 GPa. The peak pressures observed at 10 cm from the interface are clearly lower than the observed ones. They found no effect of metallurgical variables (substructure and grain size) on the attenuation rate of the pulse. An attempt was made to physically explain the dissipative mechanisms responsible for the attenuation of the pulse. This model is called the 'accumulation model' because it is based on the conservation of energy law:

$$\text{INPUT} - \text{OUTPUT} = \text{ACCUMULATION}$$

The following calculation provides a further understanding. Assume that a shock wave with a peak pressure of 10 GPa and 2 μs pulse duration travels 1 cm through a nickel plate of 1 cm^2 cross-section area. According to the Rankine–Hugoniot theory, one can calculate internal energy per unit volume, E.

In this case, the energy of the incoming pulse with 2 μs pulse duration is 233 J/cm^2. From a metallurgical or microstructural viewpoint, point defects, line defects, twinning, precipitates, martensitic transformations, and heating are dissipative processes causing the attenuation of a shock wave. Since no twinning, precipitation, or phase transformation occurs in the pressure range of 10 and 25 GPa, point and line defect generation, hydrodynamic residual rise in temperature, and temperature rise due to dislocation motion are the significant mechanisms responsible for the attenuation of shock waves in nickel. Dieter[69] estimated the total strain energy of a dislocation to be 1.36×10^{-18} J per atomic plane. The lattice parameter of nickel is 3.52 A. Multiplying by the dislocation density of 3.1×10^{10} cm/cm^3, the total strain energy of dislocations is 1.2 J/cm^3. The vacancy concentration at 10 GPa is about 3.85×10^5/cm^2.[70] The formation energy being 1.6×10^{-19} J per vacancy, one obtains 6.2×10^{-14} J/cm^3 for vacancies. This value is so small that it can be neglected. According to the hydrodynamic Rankine–Hugoniot theory, the residual temperature rise is 2 K for nickel at 10 GPa. The heat capacity of nickel is 39 J/cm^3. The thermal energy change is 7.8 J/cm^3. The total accumulation energy above is about 9 J/cm^3. Subtracting this value from the input, one obtains an output of 224 J/cm^2 after the shockwave travels 1 cm within the nickel plate. Figure 2.16 shows a comparison of the 'accumulation model' with the hydrodynamic theory and experimental results.

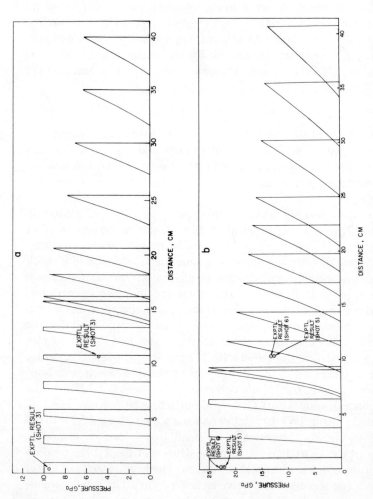

FIG. 2.15 (a) Comparison of attenuation rate predicted by hydrodynamic theory with observations made in model, with manganin piezo-resistive gauges at initial pressure of 10 GPa. (b) Comparison of attenuation rate predicted by hydrodynamic theory with observations made on nickel with manganin piezo-resistive gauges at initial pressure of 25 GPa (from ref. 68).

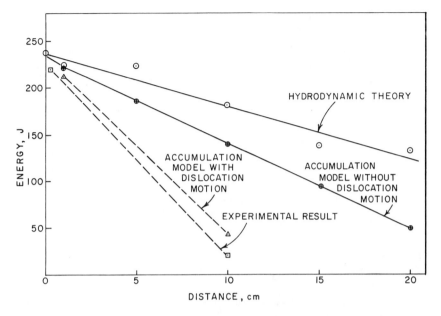

FIG. 2.16 Pulse energy in its downward trajectory (from ref. 68).

One can see that there is still some discrepancy between the 'accumulation' model and the experimental results. This difference is about 120 J at a distance of 10 cm from the top, or an average of 12 J/cm. Hence, dislocation motion was introduced into the model to take into account the additional energy dissipation. This can be done assuming that the work required to move a dislocation is totally converted into heat:

$$W = \tau b l \rho \tag{2.71}$$

where τ is the applied shear stress, b is the Burgers vector (3.5×10^{-10} m), l is the distance moved by each dislocation, and ρ the dislocation density. Setting this work equal to 12 J and assuming, to a first approximation, that each dislocation moves 0.7 µm, one can compute τ; ρ is about 3.1×10^{10} cm^{-2} (Table IV of ref. 68). One finds that τ is approximately 0.16 GPa. From the stress versus velocity plot presented by Meyers (Fig. 4. of ref. 71) for nickel, one can estimate the velocity at which the dislocations would have to move to dissipate the required amount of energy (12 J/cm); it is around 500 m/s. This is clearly a subsonic and reasonable value. Hence, the 'accumulation model' with dislocation motion is in a reasonable agreement with the experimentally determined attenuation of the shock pulse.

2.4.4. Elastic Precursor Waves

As seen in the preceding section, shock waves are preceded by an elastic precursor when their Raleigh slope is less than that of the precursor. At a high enough stress, called 'overdrive stress', the velocity of the shock wave becomes larger than that of the precursor. Figure 2.17 shows the

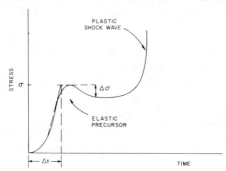

FIG. 2.17 Schematic representation of elastic precursor (from ref. 81).

shape of a general precursor pulse. One has a rise time Δt, a peak stress σ, and a stress drop $\Delta \sigma$. Although the elastic precursor is unimportant in relation to the plastic pulse that succeeds it, it can provide important information on the nature of dynamic deformation. It is for this reason that it has received considerable attention in the past. Davison and Graham[12] provide a comprehensive review of the subject. Some of the work is described below.

Taylor and Rice[72] first observed the decay of the elastic precursor amplitude with depth of penetration into the target; Taylor[73] later explained it successfully in terms of the Johnson–Gilman[74] expression for dislocation velocity. Another feature observed is yield point formation[73,75,76] and, consequently, stress relaxation behind the elastic precursor; Barker et al.[75] attributed it to dislocation effects predicted by the Johnson–Gilman model. Kelly and Gillis[77] showed that thermal activation models (e.g., ref. 7) for dislocation dynamics could explain the observed decay behavior as well as the Johnson–Gilman model. Johnson[79] extended the Taylor[73] interpretation of precursor decay to polycrystalline metals. Rohde[80] studied the precursor decay in iron shock-loaded at temperatures ranging from 76 to 573 K and found that the data did not satisfy entirely any of the following models: the Johnson–Gilman model,[74] the activation energy model,[78] or the linear damping model. Meyers[81] studied the effect of polycrystallinity on the

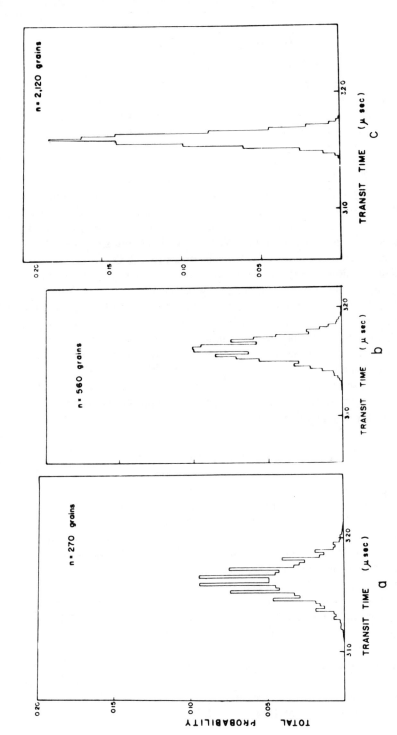

FIG. 2.18 Distribution of transit times of elastic precursor due to velocity anisotropy of the wave, after penetration of 19.05 mm into iron specimen with grain sizes of (a) 70.3, (b) 34, and (c) 9 μm (from ref. 81).

configuration of the elastic precursor wave. Referring to Section 2.2.3, one can see that the velocity of elastic waves is strongly dependent upon crystallographic orientation. Thus, as the precursor travels through different grains of a polycrystalline aggregate, the front undergoes changes as it is reflected at the boundaries and as different parts travel at different velocities. Consequently, the rise time (Δt in Fig. 2.17) is affected; Fig. 2.18 shows how Meyers' model predicts different distributions of the precursor front for different grain sizes, after a distance of 19.05 mm as has been traversed. The greater the grain size, the greater the spread in the front and, consequently, the greater the rise time Δt. This effect is particularly applicable to low shock pressures (<5 GPa) but decreases in significance with peak pressures >5 GPa.

Of great significance from the point of view of fundamental understanding of the behavior of metals, was the discovery by Jones and Holland.[82] The height of the precursor pulse (σ in Fig. 2.17) was found to be independent of grain size. In conventional deformation, on the other hand, the flow stress of iron is significantly dependent upon grain size, the Hall–Petch relationship being the well-known equation relating these two parameters. This finding indicates that the mechanisms of dislocation generation and plastic flow at the precursor front *are different* from those in conventional deformation, and that grain boundaries do not play any significant role.

2.5. DEFECT GENERATION

The reader is referred to ref. 55 for a more in-depth treatment of this section. Metallurgical microdefects are classified into four groups: point, line, interfacial or planar, and volume defects. They are responsible for the great variety of mechanical properties and can be used to strengthen the material. A substantial amount of knowledge and understanding on defect generation and their effects on the mechanical properties of shock-loaded metals has been gained over the past thirty years. The deformation regimes between conventional and shock wave strain rates have not been so exhaustively and systematically investigated. Effects of plastic waves and thermoplastic shear instabilities are not well understood. The detailed mechanical and metallurgical effects are treated in Chapter 3; the fundamental mechanisms are discussed here. Except for this paragraph, the discussion will be restricted to shock waves. An increase in strain rate produces, in general, an increase in flow stress; this is shown in Fig. 3.1 of Chapter 3. Edington[142,143] systematically investigated the effect of strain rate on the dislocation substructure. For both niobium[143] and copper,[142] he observed the classical flow-stress increase with strain rate increase. He studied the interval 10^{-4}–10^{3} s^{-1} for niobium and 10^{-4}–10^{4} s^{-1} for copper, using a Hopkinson bar for the higher strain rates. For niobium, he found a more uniform dislocation distribution at 1.5×10^{3} s^{-1} than at 1.2×10^{-4} s^{-1}. On the other hand, for copper no significant difference in the distribution and density of dislocations deformed at 6.5×10^{3} s^{-1} and 4×10^{-4} s^{-1} was found. In order to understand the effect of strain rate on the final substructure, one has to recognize the importance of the stages leading to this final distribution; dislocation generation, interaction, movement. The dissipative processes involved in dislocation motion are responsible for the requirement of increasingly higher stresses to move the dislocations at increasingly high velocities. The nature of this dependence is very important in establishing the flow stress and the dislocation multiplication and reaction processes. These, in turn, determine the final distribution. In the intermediate velocity range, the velocity of dislocations has been found to be linearly related to the stress:

$$\tau = Bv$$

where B is the damping constant; at ambient temperature phonon 'viscosity' seems to be the principal damping mechanism.

The importance of separating deviatoric from hydrostatic stresses in the treatment of shock waves cannot be over emphasized. Different phenomena are controlled by different stresses. Hence, one has:

Dislocations: generation and motion controlled by deviatoric stresses, stacking-fault energy affected by hydrostatic stresses.

Dispersed particles: they are a source of dislocations due to the different compressibilities; hence, this is an effect of hydrostatic stresses.

Individual grains: in materials that do not exhibit cubic symmetry, individual grains have anisotropic compressibilities and hydrostatic stresses will establish compatibility stresses at their interfaces.

Displacive/diffusionless phase transformation: a number of phase transformations are induced in materials by the hydrostatic component of stress. Martensitic transformation can also be induced by shear stresses or strains.

Twinning: activated by shear stresses. The hydrostatic stresses might have an indirect effect.

Point defects: their generation is due to shear stresses; their diffusion rate is affected by hydrostatic stresses.

Recovery and melting point: affected by hydrostatic stress.

Shock and residual temperatures: affected by both hydrostatic and deviatoric stresses, but for different reasons.

2.5.1. Dislocation Generation

The dislocation substructures generated by shock loading depend on a number of shock-wave and material parameters. Among the shock-wave parameters, the pressure is the most important one. As the pressure is increased, so does the dislocation density.[21] As the dislocation density increases for high stacking-fault energy FCC metals, the cell size decreases. Murr and Kuhlmann-Wilsdorf[90] found that the dislocation density varies as the square root of pressure ($\rho \propto p^{1/2}$). This dependence breaks down at pressures close to 100 GPa, due to shock-induced heating.

The effect of pulse duration has been and remains, to some extent, the object of controversy. Its effect for various alloys is reviewed in reference 91 and discussed in detail in reference 92. Appleton and Waddington[93] were the first to suggest its importance.

The effect of pulse duration is principally to allow more time for dislocation reorganization. The cell walls become better defined as the pulse duration increases, because there is more time for dislocation reorganization. This is shown in Fig. 2.19 which also shows the important microstructural changes, particularly dislocation density, which result when the pressure increases. Experimental observations at very low pulse durations do not seem to be in line with the above rationalization. Marsh and Mikkola[94] have observed that dislocation density increases with increasing pulse duration in the sub-microsecond range. It is possible that these effects are due to subtle pressure variations.

There are some systematic differences and similarities between shock-induced and conventionally-induced dislocations; some of these will be briefly reviewed here. In FCC metals, the stacking-fault energy determines the substructure to a large extent. In any case, however, the dislocations seem to be more uniformly distributed in shock than in conventional deformation. In high-stacking-fault energy alloys, the cell walls tend to be less well-developed after shock loading than after conventional deformation, especially by creep or fatigue, which allow time for dislocations to equilibrate into more stable configurations.

In addition, if the shock pulse duration is low, the substructures are more irregular because there is insufficient time for the dislocations

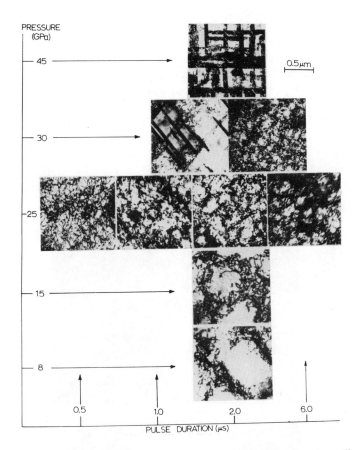

FIG. 2.19 Composite of bright-field electron micrographs showing effect of pressure and pulse duration on the shock-wave response of nickel. All surface orientations are (001), and at 30 GPa, 2 μs both twin and cells exist in (001) orientation. The preponderance of twins also increases in (001) orientations at 30 GPa above 2 μs pulse duration.

generated by the peak pressure (in the shock front) to equilibrate[90] (Fig. 2.19). There is usually a preponderance of dislocation loops associated with the residual shock microstructures, and this is especially unique to shock loading at high stacking-fault free energy metals and alloys. At shock pressures above about 10 GPa, most FCC metals having stacking-fault free energies above about $50\,\text{mJ/m}^2$ tend to exhibit dislocation cell structures as shown in Fig. 2.19. Between 50 and about $40\,\text{mJ/m}^2$ a transitional range gives rise to tangles of dislocations, poorly formed cells, and sometimes more planar arrays of dislocations (associated with the {111} slip planes).

For lower stacking-fault free energy metals and alloys, there is a tendency toward planar dislocation arrays below about 40 mJ/m^2; with stacking faults and twin faults becoming prominent for stacking-fault free energies below about 25 mJ/m^2. This is illustrated in Fig. 2.20 for nickel and 304 stainless steel shock loaded to the same pressure, at the same pulse duration.

The effect of shock loading on BCC metals is similar to that for nickel in Fig. 2.20(a) but lacks well defined dislocation cells. Shock loaded iron

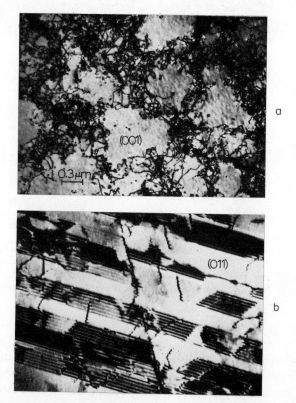

FIG. 2.20 Comparison of substructures in high stacking-fault free energy nickel (128 mJ/m^2) in (a) and in low stacking-fault free energy type 304 stainless steel 21 mJ/m^2) in (b) (15 GPa, $t = 2$ μs).

is characterized, at pressures below 13 GPa, by arrays of straight and parallel screw dislocations, in the properly oriented grains. In molybdenum, the substructure is one of homogeneously distributed dislocations.

Shock loaded HCP metals have not been extensively studied by transmission electron microscopy. Koul and Breedis[96] found, at 7 GPa, dislocation arrays that they described as being intermediary between those of FCC metals such as Cu and Ni, and BCC metals such as iron and Fe_3Al. The substructure also exhibited twins and phase transformations, at higher pressures. Murr and Galbraith[97] studied shock-loaded beryllium which exhibited dislocation substructures similar to those of BCC metals. No twins were observed up to a pressure of 0.9 GPa. A strong dependence of dislocation density was noted for grain boundary structure. This feature is consistent with the establishment of compatibility stresses at the interfaces due to the anisotropy of compressibility, as noted above, giving rise to a preponderance of dislocations from grain-boundary sources as well as alterations in the boundary structure.

The limitations of Smith[98] and Hornbogen's[99] proposals led Meyers[71,100] to propose a model whose essential features are:

(a) Dislocations are homogeneously nucleated at (or close to) the shock front by the deviatoric stresses set up by the state of uniaxial strain; the generation of these dislocations relieves the deviatoric stresses.
(b) These dislocations move short distances at subsonic speeds.
(c) New dislocation interfaces are generated as the shock wave propagates through the material.

This model presents, with respect to its predecessors, the following advantages:

(a) No supersonic dislocations are needed.
(b) It is possible to estimate the residual density of dislocations.

Figure 2.21 shows the progress of a shock wave throughout the material in a highly simplified manner. As the shock wave penetrates into the material, high deviatoric stresses effectively distort the initially cubic lattice into a monoclinic lattice. When these stresses reach a certain threshold level, homogeneous dislocation nucleation can take place. Hirth and Lothe[101] estimate the stress required for homogeneous dislocation nucleation. The nucleation mechanism at the shock front is

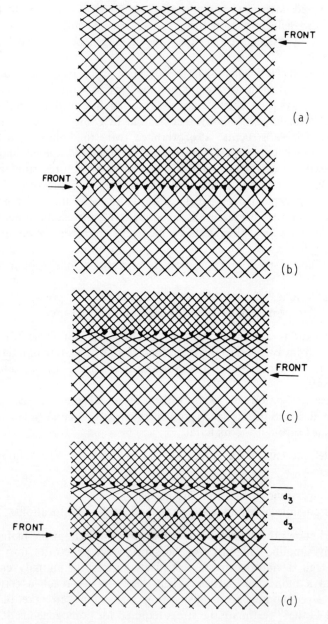

FIG. 2.21 Progress of shock front according to Meyers model.

unique, and different from homogeneous nucleation in conventional deformation. In shock loading, the dislocation interface separates two lattices with different parameters. However, it will be assumed that the stress required is the same, (as a first approximation). From Hirth and Lothe[101] one has:

$$\tau_h/G = 0.054 \tag{2.72}$$

where τ_h is the shear stress required and G is the shear modulus, which is pressure-dependent. When the maximum shear stress becomes equal to τ_h (and is acting in the correct orientation), homogeneous dislocation nucleation takes place. Substituting eqn. (2.72) into eqn. (2.68) one obtains:

$$P = 0.027 \, K$$

This value can be obtained from Fig. 2.8(b) by trial and error. It corresponds to a pressure of approximately 6 GPa. Figure 2.21(b) shows the wave as the front coincides with the first dislocation interface. The density of dislocations at the interface depends on the difference in specific volume between the two lattices and can be calculated therefrom. In Fig. 2.21(c) the front has moved ahead of the interface and the deviatoric stresses build up again; other layers are formed in Fig. 2.21(d). It should be noticed that since the macroscopic strain is ideally zero after the passage of the wave, the sum of the Burgers vectors of all dislocations has to be zero. This is accomplished, in the simplified model presented here, by assuming that adjacent dislocation layers are made of dislocations with opposite Burgers vectors. Figure 2.22 shows two adjacent layers under the effect of shear stresses still existing in the lattice after the dislocations were nucleated; a group of dislocations move away from it. It is possible to estimate the velocity at which these dislocations move if one knows τ_{res}. As these dislocations move, they locally accommodate and decrease τ_{res}. The total amount of internal friction and heat generation due to dislocation motion can be calculated by knowing the difference between the measured and the thermodynamically-calculated residual temperature. A simplified calculation is presented by Hsu et al,[68] Greater details of the model as well as comparisons of calculations with measured dislocation densities are presented elsewhere.[71,100]

Recent experimental results[68,102] lead the authors to believe that the rarefaction part of the wave plays only a minor role in dislocation generation. The main reason for this is that the rarefaction part of the wave enters into a material that is already highly dislocated. It was found

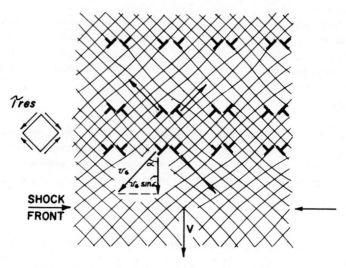

FIG. 2.22 Movement of dislocation generated at the shock front (from ref. 55).

that when nickel is shock-loaded repeatedly, the increase in dislocation density is much less pronounced for the succeeding events.[68,102] The shock wave passing through a highly dislocated material is not such an effective dislocation generator. This is consistent with Meyers' predictions[100] ... 'that if a pre-strained material is shock-loaded, part of the deviatoric stresses at the shock front could be accommodated by existing dislocations; in this case the number of dislocations that would be generated at the front would be reduced. The same argument can be extended to the rarefaction portion of the wave; it can accommodate the deviatoric stresses by the movement of the existing dislocations. Additionally, the time interval in which attenuation takes place is much higher than in which the shock front rises; 200 ns versus 1 ns'.

By using a different approach, Mogilevsky[103,104] independently reached some conclusions similar to those presented here. Computer calculations using a Born–Meyer potential for the atoms allowed Mogilevsky to follow the position of atoms with time. Although a perfect lattice of copper remained elastic up to pressures of 30 GPa, the introduction of point defects allowed the deviatoric stresses to be relaxed by stacking-fault (and, possibly, dislocation) generation at pressures as low as 5 GPa.

An alternative model for dislocations at the shock front was recently proposed by Weertman.[105] For a strong shock, pulse, he concluded that the front is composed of supersonic and trailing subsonic dislocations. For weaker shock pulses no supersonic interface would be required; arrays of subsonic dislocations, trailing the front (similar to Meyers' dislocations) would attenuate the deviatoric stresses.

2.5.2. Point Defects

Shock loading is also responsible for a high density of point defects. The dynamic strain induced is represented by:

$$\varepsilon_s = \frac{4}{3} \ln\left(\frac{V}{V_0}\right)$$

This corresponds to the sum of the strains imparted by the shock front and the rarefaction part of the wave. In the first systematic comparison between point defects generated by shock loading and cold rolling, Kressel and Brown[70] reported vacancy and interstitial concentrations three to four times higher after shock loading than after cold rolling. However, direct quantitative evidence of vacancies and vacancy-type defects was obtained for the first time by Murr et al.[107] Figure 2.23 shows some of the vacancy loops in shock-loaded molybdenum and nickel.[107] Field-ion microscopy showed that these loops in molybdenum accounted for only a small portion of the shock-induced vacancies; the majority exist as single vacancies or small clusters difficult to resolve by conventional electron microscopy.

It is relatively simple to understand why shock loading induces a large concentration of point defects. The primary source of point defects is the non-conservative motion of jogs. These jogs are generated by the intersection of screw or mixed dislocations. Figure 2.22 shows the direction of motion of dislocations under the effect of the residual shear stresses. As they move, τ_{res} decreases, but in the process the dislocations intersect each other, generating jogs. The non-conservative motion of these jogs produces strings of either vacancies or interstitials. These can also occur as dislocation loops when observed in the electron microscope as shown in Fig. 2.23. The subject is treated in detail by Hirth and Lothe.[101] Meyers and Murr[55] describe the model based on non-conservative motion of jogs in detail. For a pressure of 20 GPa in nickel, they find a point-defect concentration of 7×10^{-5}. This compares favorably with results reported by Kressel and Brown,[70] and Graham[108] presents an excellent overview of shock induced point defects.

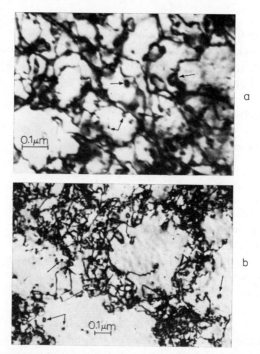

FIG. 2.23 Vacancy-type dislocation loops (arrows) in shock-loaded metals (a) · molybdenum shock loaded at 14 GPa, 2 μs; (b) nickel shock loaded at 20 GPa.

2.5.3. Deformation Twinning

The most important and self-consistent comment that can be made about deformation twins is that twinning is a highly favored deformation mode under shock loading. Metals that do not twin by conventional deformation at ambient temperature can be made to twin by shock loading. In this respect, as in the morphology of dislocation substructures, shock deformation resembles conventional deformation at low temperature: loose cell walls and a greater tendency towards twinning. The ease of twinning depends on several factors:

(a) *Pressure*—Nolder and Thomas[109,110] found that twinning occurred, in nickel, above 35 GPa pressure. This was generally confirmed by Greulich and Murr.[111] DeAngelis and Cohen[112] found the same effect in copper.

(b) *Crystallographic orientation*—it is the deviatoric component of stress that induces twinning. Hence, when the resolved shear stress in the twinning plane and along the twinning direction reaches a critical level, twinning should occur. DeAngelis and Cohen[112] found an orientation dependence for the threshold stress; copper single crystals twinned at 14 GPa when the shock wave traveled along [100] and at 20 GPa when it travelled along [111]. Greulich and Murr[111] found, for nickel, that at and above about 35 GPa, twinning occurred preferentially for [100] grains (Fig. 2.19). As the pressure was increased, the preponderance of twins increased along orientations other than [100].

(c) *Stacking-fault energy*—as the SFE of FCC metals is decreased, the incidence of twinning increases. As a corollary, the threshold stress for twinning should decrease.

(d) *Pulse duration*—the effect of pulse duration, first explored by Appleton and Waddington,[93] was systematically investigated by Champion and Rohde[113] for an austenitic (Hadfield) steel. They found striking differences in twin densities for different pulse durations, at 10 GPa. Numerous twins were observed at $2\,\mu s$, while no twinning was present at $0.065\,\mu s$. They concluded that there must be a threshold time for twinning. Staudhammer and Murr[114] investigated the effect of pulse duration (0.5, 1, 2, 6, 14 μs) on the substructure of AISI 304 stainless steel. They found an increase in twin density up to about 2 μs; beyond that the twin density seemed to be essentially constant. Stone et al.[115] found an increase in twin density as the pulse duration was increased from 0.5 to 1.0 μs, in both the AISI 1008 steel and Armco magnetic ingot iron. The twins generated by the shock pulse should not be confused with the ones formed by the elastic precursor wave, in iron; the latter ones were investigated by Rohde.[116] Although twins are generated by the elastic precursor waves, the volume percent of twins generated by the shock wave is an order of magnitude higher. While the elastic precursor may produce a twin density of 3 vol pct, a shock wave of 30 GPa peak pressure and 1 μs pulse duration has been shown to generate about 50 vol pct of twins.

(e) *Existing substructure*—Rohde et al.[117] found profuse twinning upon shock loading titanium-gettered iron in the annealed condition. However, predeformed samples exhibiting a reasonable density of dislocations did not twin. The same results were obtained by Mahajan[118] for iron. Hence, if one looks at dislocation generation and motion, and twinning as competing mechanisms, one can rationalize this response. The deviatoric stresses generated by a shock wave are accommodated by twinning when no dislocations are available and by motion of the already existing dislocations, if iron is predeformed.

(f) *Grain size*—Wongwiwat and Murr[119] were able to explain conflicting data reported in the literature[107,120] on the incidence of twinning in molybdenum by showing that, at a certain pressure, large-grain sized specimens twinned more readily than small-grain sized ones. However, it should be emphasized that this response is not unique to shock loading; indeed, iron-3 pct silicon[121] and chromium[122] have been shown to exhibit a strong grain-size dependence of the twinning stress (in conventional deformation). Kestenbach and Meyers[106,123] investigated the effect of grain size on the substructure of AISI 304 stainless steel.

Two fundamentally different mechanisms have been proposed to account for twin formation. The first involves a pole mechanism proposed by Cottrell and Bilby[124] for BCC metals, and extended to FCC metals by Venables.[125] The pole mechanism involves dislocation motion which sweeps out the twin spiraling around a dislocation pole. This requires a rather longer time than generally available in shock loading, and this velocity or time limitation led Cohen and Weertman[126] to propose a much simpler model, especially applicable to FCC metals and alloys. This latter model involves the production and systematic glide motion of Shockley partial dislocations on every [111] plane to produce a twin which propagated with the velocity of propagation of the partial dislocation. Sleeswyk[127] has also proposed a model for twin formation in BCC metals which is phenomenologically identical to that of Cohen and Weertman[126] in FCC materials. Sleeswyk's model involves the systematic glide of dislocations on [112] planes in the BCC structure.

2.5.4. Displacive/Diffusionless Transformations

There are numerous instances in which a shock wave induces a phase transformation. A very detailed review is presented by Graham and Duvall;[128] for this reason, only a classification scheme will be presented here. The phase transformations involving diffusion are excluded from this discussion. Stein[129] has shown that precipitation is induced by shock loading.

Cohen *et al.*[130] recently proposed a classification scheme for displacive/diffusionless transformations. The first division is between shuffle and lattice-distortive transformations; the latter group is divided into two sub-groups; dilatation-dominant and deviatoric-dominant transformations. The term martensitic (and quasi-martensitic) is reserved for the lattice distortive transformations in which the deviatoric component of stress is dominant. Hence only the FCC→BCC (or BCT) and FCC→HCP transformation in the Fe-base alloys and the BCC→ close packed transformations in the noble metal alloys can be called martensitic. A whole range of phase transformations, such as the beta–omega in Ti alloys and the tetragonal–cubic transformation in tin, cannot be called martensitic.

The effect of a shock pulse on a displacive/diffusionless transformation has to be analysed from three points of view: (a) pressure; (b) shear stresses and (c) temperature. The changes in these parameters are not independent; there are specific temperature rises and deviatoric stresses associated with a certain pressure level. Nevertheless, they have different effects on the thermodynamics of phase transformation. A phase transformation resulting in a reduction of volume, for example is thermodynamically favored at high pressure, because it will tend to decrease the pressure in that region. On the other hand, a phase transformation in which the product has a lower density is not favored by the pressure. Figure 2.24 shows the pressure–temperature diagram for iron. At a certain critical pressure, the BCC (α) phase transforms to either HCP (ε) or FCC (γ), depending upon the temperature. Both these phases are more closely packed than α(BCC). Figure 2.24 also shows that the pressure increase is coupled to a temperature increase. Hence, the pressure path

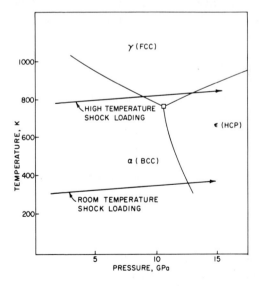

FIG. 2.24 Temperature–pressure diagram for iron.

has a slight slope. Plutonium is another metal that is very interesting from a metallurgical point of view. It undergoes six different phase transformations, some with large differences in density.[131] The room temperature monoclinic phase has a high density (19.86); hence the pressure will not induce phase transformation. On the other hand, if the shock wave passes through one of the lower density phases, it will favor its transformation into the higher-density phases. Plutonium is unique in that melting is accompanied by a volume decrease. Hence, the pressure has the effect of reducing the melting point; in metals where the melting point is associated with an expansion, pressure increases the melting point. Some of the transformations in Pu alloys are, due to the large density changes, considered as dilatation-dominant, and are therefore not considered as martensitic. Patel and Cohen[132] have established a rationale for the effect of stresses on the M_s temperature in martensite transformations. They found that, in Fe-30 pct Ni alloy, the hydrostatic pressure decreased the M_s temperature. In these alloys, there is a dilatation of 5 pct associated with the martensitic phase. Hence, a pressure pulse should not favor the transformation, and this is reflected in the decrease in M_s. On the other hand, an alloy in the martensitic form should revert to austenite, if a pressure pulse were applied, because this would result in a contraction of the lattice.[123] A negative pressure pulse inducing negative hydrostatic pressures would be the converse situation and the $\gamma(FCC) \to \alpha(BCC$ or $BCT)$ transformation would be favored, with an increase in the transition temperature. Meyers and Guimarães[134] were able to produce a tensile pulse and generate martensite in an Fe-31 pct Ni-0.1 pct alloy. Figure 2.25 shows the martensite tube generated by tensile waves; this tensile wave was produced by a compressive shock wave, as it reflected at a free surface. The region of the material traversed by the compressive wave exhibited only a dense array of dislocations organized in cells and occasional twins. This phenomenon was used to calculate a nucleation time for the martensitic transformation.[135]

The effect of shear stresses always associated with the pressure is more difficult to assess. In deviatoric-dominant transformations the externally applied shear stresses can play an important role in the initiation of transformation. The transformation is favored along the crystallographic orientations in which the transformation shear will tend to decrease the externally-applied shear, and will tend to decrease the overall internal energy. Martensite can be considered as a deformation mechanism competing with slip and twinning, and externally applied shear stresses

FIG. 2.25 Martensite generated by tensile hydrostatic stresses produced by a reflected pressure pulse in an iron–nickel alloy (from ref. 134). Bar = 100 μm.

increase M_s. The shear stresses introduced by the shock waves tend to favor the generation of martensite. However, these shear stresses are much lower than the hydrostatic stresses for most shock waves, because dislocation nucleation and twinning will attenuate them. Hence, they will only generate martensite when the temperature at the shock front is slightly higher than M_s. In this case martensite transformation can effectively compete with twinning and slip. Olson and Cohen[136] have shown that, during conventional deformation at temperatures slightly above M_s, yielding is produced by stress-induced martensite. The same phenomenon was confirmed by Guimarães et al.[137] Up to 20 K above M_s, yielding was initiated by martensitic transformation. Hence, one can conclude that the shear stresses associated with shock loading might be important in martensite nucleation if the temperature at the shock front is within 50 K of the M_s under the imposed conditions (at the level of pressure established by the shock pulse). These features are demonstrated by Staudhammer et al.,[138] who also show that martensite forms exclusively in 304 stainless steel at the intersections of twin-faults by a strain-induced process described previously.[136] This process involves the selective movement and interaction of groups of partial dislocations, at temperatures well above the M_s temperature for the stainless steel (which is below 4 K). When Shockley partial dislocations of the type $a/6\langle 112\rangle$ are created on every other $\{111\}$ plane, a packet of HCP (ε) martensite is created. When these bundles intersect with more complex faulting arrays (such as $3a/8\langle 112\rangle$ fault arrays), α' (BCC) martensite is created within the intersection volume. Once nucleated, strain-induced intersection martensite (α) can grow by coalescence of the intersection volumes, or by stress-assisted processes. In other words, once formed, martensite growth can be somewhat catastrophic when a stress assist is imposed. For example, having created some martensite by an initial shock pulse, a subsequent shock may induce considerably more martensite. In addition, shock pressure applied in longer pulses at high pressure can have a similar effect.[89]

2.5.5. Other Effects

There are a number of other shock effects which are mentioned only briefly in concluding this section. These relate mainly to the impedance differences encountered by the shock wave which can manifest themselves in velocity differences, etc. The realization of the anisotropy of elastic and plastic properties of the individual grains in a polycrystalline material led to a 'wavy-mode' model which attempted to account for such differences.[139] However, at higher pressures ($P \gtrsim 10$ GPa) velocity differences due to crystallographic orientations are practically imperceptible, and in general the shock wave is not significantly affected. Wave reflection and refraction at grain boundaries and other internal asperities which arise by difference in structure and composition can also occur by impedance discontinuities. These effects are, however, also of secondary importance at high pressures (>10 GPa). None the less, second-phase particles can play a significant role in the generation of defects, particularly dislocations. This feature is especially prominent at coherent precipitates which lack coherence with the shock wave passage, and the loss of coherence is accommodated by the creation of dislocations which can also multiply. Das and Radcliffe[140] have observed the punching out of dislocations at precipitates and such phenomena can also lead to nucleation of other defects, such as twins as illustrated by Leslie et al.[141]

ACKNOWLEDGEMENTS

The writing of this chapter was conducted while one of us (M.A.M.) was supported by the Research and Development Division of New Mexico Tech. The help of Dr. M. Brook in making this support available is greatly acknowledged. Appreciation is extended to Professor G. Purcell, Chairman of the Department of Metallurgical and Materials Engineering, for provision of facilities.

REFERENCES

1. KOLSKY, H. *Stress Waves on Solids*, Dover, New York, 1963.
2. RINEHART, J. S. *Stress Transients in Solids*, Hyperdynamics, Santa Fe, New Mexico, P.O. Box 392, 1975.
3. GHATAK, A. K., and KOTHARI, L. S. *An Introduction to Lattice Dynamics*, Addison-Wesley, Reading, Mass., 1972, p. 70.
4. LEE, E. H. In *Shock Waves and the Mechanical Properties of Solids*, eds. J. J. Burke and V. Weiss, Syracuse U. Press, Syracuse, 1971, p. 3.
5. VON KÁRMÁN, T., and DUWEZ, P. *J. Appl. Phys.*, **21** (1950), 987.
6. CLIFTON, R. J. Source cited in ref. 4, p. 73.
7. HERRMANN, W. In *Propagation of Shock Waves in Solids*, ed. E. Varley, ASME, AMD-17, New York, 1976, p. 1.
8. CHOU, P. C. In *Dynamic Response of Materials to Intense Impulsive Loading*, eds. P. C. Chou and A. K. Hopkins, Air Force Materials Laboratory, WPAFB, 1972, p. 55.

9. HERRMANN, W., and NUNZIATO, J. W. Source cited in ref. 8, p. 123.
10. WASLEY, R. J. *Stress Wave Propagation in Solids*, M. Dekker, New York, 1973.
11. RICE, M. H., MCQUEEN, R. G., and WALSH, J. M. *Solid State Physics*, **6**, (1958), 1.
12. DAVISON, L., and GRAHAM, R. A. *Phys. Rep.*, **55** (1979), 257.
13. BRADLEY, J. N. *Shock Waves in Chemistry and Physics*, J. Wiley, New York, 1962.
14. KINSLOW, R., ed. *High-Velocity Impact Phenomena*, Academic, New York, 1970.
15. ZELDOVICH, Ya. B., and RAIZER, Yu. P. *Physics of Shock Waves and High-Temperature Hydrodynamic Phenomena*, Academic, New York, 1966.
16. WILKINS, M. In *Methods in Computational Physics*, Vol. 3, eds. B. Alder, S. Fernbach, and M. Rotenberg, Academic, New York, 1964, p. 211.
17. HERRMANN, W., HICKS, D. L., and YOUNG, E. G. Source cited in ref. 4, p. 23.
18. WALSH, R. T. Source cited in ref. 8, p. 363.
19. KARPP, R., and CHOU, P. C. Source cited in ref. 8, p. 283.
20. ROHDE, R. W., BUTCHER, B. M., HOLLAND, J. R., and KARNES, C. H. *Metallurgical Effects at High Strain Rates*, Plenum, New York, 1973.
21. MEYERS, M. A., and MURR, L. E., eds. *Shock Waves and High-Strain-Rate Phenomena in Metals: Concepts and Applications*, Plenum, New York, 1981.
22. KREYZIG, E. *Advanced Engineering Mathematics*, 2nd ed., Wiley, 1971, pp. 490, 512.
23. MEYERS, M. A., and CARVALHO, M. S. *Mat. Sci. Eng.*, **24** (1976), 131.
24. CRITESCU, N. *Dynamic Plasticity*, Interscience Publishers, New York, 1967.
25. ABOU-SAYED, A. S. *Analysis of Combined Pressure-Shear Waves in an Elastic/Visco-Plastic Material*, M.Sc. Thesis, Brown U., 1972.
26. ABOU-SAYED, A. S. *Analytical and Experimental Investigation of Pressure-Shear Waves in Solids*, Ph.D. Thesis, Brown U., 1975.
27. ABOU-SAYED, A. S., CLIFTON, R. J., and HERMANN, L. *Exptl. Mech.*, (1976) 127.
28. GRAHAM, R. A., and ASAY, J. R. *High Temp–High Press.*, **10** (1978), 355.
29. DECARLI, P. S., and MEYERS, M. A. Source cited in ref. 21, p. 341.
30. ORAVA, R. N., and WITTMAN, R. H. *Proc. 5th Intl. Conf. High Energy Rate Fabrication*, U. of Denver, Colorado, 1975, P. 1.1.1.
31. EZRA, A. A. ed., *Principles and Practice of Explosive Metalworking*, Industrial Newspapers Ltd., London, 1973.
32. TAYLOR, G. I. *J. Inst. Civil Engrs.*, **26** (1946), 486.
33. RAKHMATULIN, K. A. *Appl. Math. and Mech.*, **9** (1945), No. 1.
34. POCHAMMER, L. *J. reine angew. Math.*, **81** (1876), 324.
35. CHREE, C. *Trans. Camb. Phil. Soc.*, **14** (1889), 250.
36. TING, T. C. T., and NAN, N. *Trans. ASME, J. Appl. Mech.*, **36**, (1969), 189.
37. BLEICH, H. H., and NELSON, I. *Trans. ASME, J. Appl. Mech.*, **33**, (1966), 149.
38. LEE, E. H. *J. Appl. Mech.* **36** (1969), 1.
39. CLIFTON, R. J. Plastic waves: theory and experiment, in *Mechanics Today*, ed. S. Nemat-Nasser, Vol. 1, Pergamon Press, 1972.

40. HIRTH, J. P. *Met Trans.*, **9A** (1978), 401.
41. National Materials Advisory Board Report NMAB-356, 1980, National Academy of Sciences, 1980.
42. HOPKINSON, B. *Roy. Soc. Phil. Trans.*, **A213** (1914), 437.
43. DAVIES, R. M. *Roy. Soc. Phil. Trans.*, **A240** (1948), 375.
44. KOLSKY, H. *Proc. Roy. Soc. London*, **62B** (1949), 676.
45. RANKINE, W. J. M. *Phil. Trans. Roy. Soc., London*, **160** (1870), 270.
46. HUGONIOT, H. J. J. *L'Ecole Polytechnique*, **58** (1889), 3.
47. RICE, M. H., MCQUEEN, R. J. and WALSH, J. M. Compression of Solids by Strong Shock Waves, in *Solid State Physics*, Vol. 6, p. 1, Academic Press, New York (1958).
48. VAN THIEL, M. *Compendium of Shock Wave Data*, UCRL-50108, Lawrence Radiation Laboratory, Univ. of California, 1966.
49. MCQUEEN, R. J. Source cited in ref. 14, p. 293.
50. MCQUEEN, R. J. and MARSH, S. P. *J.A.P.*, **31**, (1960), 1253.
51. DUVALL, G. E. Source cited in ref. 8, p. 481.
52. MCQUEEN, R. J., MARSH, S. P., TAYLOR, J. W., FRITZ, J. N. Source cited in ref. 14, p. 293.
53. MEYERS, M. A. *Thermomechanical Processing of a Nickel-Base Superalloy by Cold Rolling and Shock-Wave Deformation*, Ph.D. Thesis, U. of Denver, Colorado, 1974.
54. DUVALL, G. E., and GRAHAM, R. A. *Rev. Modern Phys.*, **49** (1977), 523.
55. MEYERS, M. A., and MURR, L. E. Source cited in ref. 21, p. 487.
56. VON NEUMANN, J., and RICHTMYER, R. D. *J. Appl. Phys.*, **21** (1950), 2322.
57. DRUMMOND, W. E. *J. Appl. Phys.*, **29** (1958), 167.
58. BERTHOLF, L. D., and BENZLEY, S. E. *TOODY II, A Computer Program for Two-Dimensional Wave Propagation*, Sandia Laboratories Research Laboratory SC–RR–68–41 (1968).
59. LAWRENCE, R. J. *WONDY III – A Computer Program for One-Dimensional Wave Propagation*, Sandia Laboratories Development Report SC–DR–70–715 (1970).
60. DIENES, J. K., and WALSH, J. M. Source cited in ref. 14, p. 45.
61. CURRAN, D. R., SEAMAN, L., and SHOCKEY, D. A. *Physics Today*, Jan., 1977, 46.
62. SEAMAN, L., TOKHEIM, R., and CURRAN, D. *Computational Representation of Constitutive Relations for Porous Material*, S.R.I. Report DNA 3412F, May 1974.
63. HOENIG, C., HOLT, A., FINGER, M., and KUHL, W. *Proc. 5th Intl. Conf. High Energy Rate Fabrication*, U. of Denver, Colorado, June 1974, p. 6.3.1.
64. CURRAN, D. R. *J.A.P.*, **34**, 2677.
65. ERKMAN, J. O., CHRISTENSEN, A. B., and FOWLES, G. R. *Attenuation of Shock Waves in Solids*, Technical Report No. AFWL–TR–66–72, Stanford Research Institute, Air Force Weapons Laboratory, May, 1966.
66. REMPEL, J. R., SCHMIDT, D. N., ERKMAN, J. O., and ISBELL, W. M. *Shock Attenuation in Solid and Distended Materials*, Standford Research Institute, Technical Report No. WL–TR–65–119, Air Force Weapons Laboratory, February, 1966.
67. ERKMAN, J. O., and CHRISTENSEN, A. B. *J.A.P.*, **38** (1967) 5395.

68. HSU, C. Y., HSU, K. C., MURR, L. E. and MEYERS, M. A. Source cited in ref. 21, p. 433.
69. DIETER, G. E. *Mechanical Metallurgy*, 2nd ed., McGraw-Hill (1976), p. 169.
70. KRESSEL, H., and BROWN, N. J. *J. Appl. Phys.*, **38** (1967), 1618.
71. MEYERS, M. A. In *Strength of Metals and Alloys*, Vol. I., eds. P. Haasen, V. Gerold, and G. Kostorz, Pergamon Press, New York, (1979), p. 549.
72. TAYLOR, J. W., and RICE, M. H. *J. Appl. Phys.* **34** (1963), 364.
73. TAYLOR, J. W. *J. Appl. Phys.* **36** (1965), 3146.
74. JOHNSON, W. G., and GILMAN, J. J. *J. Appl. Phys.*, **30** (1959), 129.
75. BARKER, L. M., BUTCHER, B. M., and KARNES, C. H. *J. Appl. Phys.*, **37** (1966), 1989.
76. HOLLAND, J. R. *Acta Met.*, **15** (1967), 691.
77. KELLY, J. M., and GILLIS, P. P. *J. Appl. Phys.*, **38** (1967), 4044.
78. CONRAD, H., and WIEDERSCH, H. *Acta Met.* **8** (1960), 128.
79. JOHNSON, J. N. *J. Appl. Phys.*, **40** (1969), 2287.
80. ROHDE, R. W. *Acta Met.*, **17** (1969), 353.
81. MEYERS, M. A. *Mat. Sci. Eng.*, **30** (1977), 99.
82. JONES, O. E., and HOLLAND, J. R. *Acta Met.* **16** (1968), 1037.
83. ZENER, C., and HOLLOMON, J. H. *J. Appl. Phys.*, **15** (1944), 22.
84. RECHT, R. F. *J. Appl. Mech.*, **31** (1964), 189.
85. CULVER, R. S. Source cited in ref. 20, p. 519.
86. ROGERS, H. C. *Ann. Rev. Mater. Sci.*, **9** (1979), 283.
87. OLSON, G. B., MESCALL, J. F., and AZRIN, M. Source cited in ref. 21, p. 221.
88. YELLUP, J. M., and WOODWARD, R. L. *Res Mechanica*, **1** (1980), 41.
89. MEYERS, M. A. *Mater. Sci. Eng.*, **51** (1981), 261.
90. MURR, L. E., and KUHLMANN-WILSDORF, D. *Acta Met.*, **26** (1978), 849.
91. MEYERS, M. A. *Met. Trans.*, **8A** (1977), 1641.
92. MURR, L. E. Source cited in ref. 21, p. 753.
93. APPLETON, A. S., and WADDINGTON, J. S. *Acta Met.*, **12** (1963), 681.
94. MARSH, E. T., and MIKKOLA, D. E. *Scripta Met.*, **10** (1976), 851.
95. MEYERS, M. A., KESTENBACH, H-J., and SOARES, C.A.O. *Mat Sci. Eng.*, **45** (1980), 143.
96. KOUL, M. K., and BREEDIS, J. F. In *The Science, Technology, and Application of Titanium*, eds. R. I. Jaffee and N. E. Promisel, Pergamon, Oxford, 1978, p. 817.
97. MURR, L. E., and GALBRAITH, J. *J. Mtls. Sci.*, **10** (1975), 2025.
98. SMITH, C. S. *Trans. AIME*, **212** (1958), 574.
99. HORNBOGEN, E. *Acta Met.*, **10** (1962), 978.
100. MEYERS, M. A. *Scripta Met.*, **12** (1978), 21.
101. HIRTH, J. P. and LOTHE, J. *The Theory of Dislocations*, McGraw-Hill, New York, (1968), p. 689.
102. KAZMI, B., and MURR. L. E. Source cited in ref. 21, p. 733.
103. MOGILEVSKY, M. A. *Proc. Symp. on High Dynamic Pressure*, Paris, August (1978).
104. MOGILEVSKY Source cited in ref. 21, p. 531.
105. WEERTMAN, J. Source cited in ref. 21, p. 469.
106. MEYERS, M. A. *Proc. ICM II*, Boston, Mass., August 16–20, 1976, p. 1804.

107. MURR, L. E. INAL, O. T., and MORALES, A. A. *Appl. Phys. Letters*, **28** (1976), 432.
108. GRAHAM, R. A. Source cited in ref. 21, p. 375.
109. NOLDER, R. L., and THOMAS, G. *Acta Met.*, **11** (1963), 994.
110. NOLDER, R. L., and THOMAS, G. *Acta Met.*, **12** (1964), 227.
111. GREULICH, F., and MURR, L. E. *Mater. Sci. Engr.*, **39** (1978), 81.
112. DEANGELIS, R. J., and DOHEN, J. P. *J. Metals*, **15** (1963), 681.
113. CHAMPION, A. R. and ROHDE, R. W. *J. Appl Phys.*, **41** (1970), 2213.
114. MURR, L. E., and STAUDHAMMER, K. P. *Mater. Sci. Engr.*, **20** (1974), 95.
115. STONE, G. A., ORAVA, R. H., GRAY, G. T., and PELTON, A. R. *An Investigation of the Influence of Shock-Wave Profile on the Mechanical and Thermal Responses of Polycrystalline Iron*, Final Technical Report, U.S. Army Research Office, Grant No. DAA629-76-0181, p. 30, 1978.
116. ROHDE, R. W. *Acta Met.*, **17** (1969), 353.
117. ROHDE, R. W., LESLIE, W. C., and GLENN, R. C. *Met. Trans.*, **3A** (1972), 363.
118. MAHAJAN, S. *Phys. Stat. Sol.*, **33a** (1969), 291.
119. WONGWIWAT, K., and MURR, L. E. *Mater. Sci. Engr.*, **35** (1978), 273.
120. VERBRAAK, C. A. *Science and Technology of W, Ta, Mo, Nb, and Their Alloys*, Pergamon Press, N.Y., 1964, p. 219.
121. MARCINKOWSKI, M. J., and LIPSITT, H. A. *Acta Met.*, **10** (1962), 951.
122. HALL, D. *Acta Met.*, **9** (1961), 191.
123. KESTENBACH, H-J., and MEYERS, M. A. *Met. Trans.*, **7A** (1976), 1943.
124. COTTRELL, A. H., and BILBY, J. *Phil. Mag.*, **42** (1951), 573.
125. VENABLES, J. A. *Phil. Mag.*, **6** (1961), 379.
126. COHEN, J. B., and WEERTMAN, J. *Acta Met.*, **11** (1963), 997.
127. SLEESWYK, A. W. *Acta Met.*, **10** (1962), 803.
128. DUVALL, G. E., and GRAHAM, R. A. *Reviews of Modern Physics*, **49** (1977), 523.
129. STEIN, C. *Scripta Met.*, **9** (1975), 67.
130. COHEN, M., OLSON, G. B., and CLAPP, P. C. *Proceedings ICOMAT 1979*, MIT Press, Cambridge, Mass. 1980, p. 1.
131. HECKER, S. S. Private communication, Los Alamos National Laboratory (1981).
132. PATEL, J. R., and COHEN, M. *Acta Met.*, **1** (1953), 531.
133. ROHDE, R. W., HOLLAND, J. R. and GRAHAM, R. A. *Trans. Met. Soc. AIME*, **242** (1968), 2017.
134. MEYERS, M. A., and GUIMARÃES, J. R. C. *Mater. Sci. Engr.*, **24** (1976), 289.
135. MEYERS, M. A. *Met. Trans.*, **10A** (1979), 1723.
136. OLSON, G. B., and COHEN, M. *J. Less Common Metals*, **28** (1972), 107.
137. GUIMARÃES, J. R. C., GOMES, J. C., and MEYERS, M. A. *Suppl. to Trans. J.I.M.*, 1976 (1741).
138. STAUDHAMMER, K. P., FRANTZ, C. E., HECKER, S. S., and MURR, L. E. Source cited in ref. 21, p. 91.
139. MEYERS, M. A. *Mater. Sci. Engr.*, **30** (1977), 99.
140. DAS, G., and RADCLIFFE, S. V. *Phil. Mag.*, **20** (1969), 589.
141. LESLIE, W. C., STEVENS, D. W., and COHEN, M. In *High-Strength*

Materials, V. F. Zackay (ed.), John Wiley and Sons, New York, 1965, p. 382.
142. EDINGTON, J. W. *Phil. Mag.*, **19** (1969), 1189.
143. EDINGTON, J. W. In *Mechanical Behavior of Materials Under Dynamic Loads*, ed. U.S. Lindholm, Springer, Berlin, 1968, 191.
144. NYE, J. F. *Physical Properties of Crystals*, Oxford U. Press, London, 1959.

Chapter 3

METALLURGICAL EFFECTS OF SHOCK AND PRESSURE WAVES IN METALS

L. E. MURR

*Oregon Graduate Center,
Beaverton, Oregon, USA*

and

M. A. MEYERS

*Department of Metallurgical and Materials Engineering,
New Mexico Institute of Mining and Technology,
Socorro, New Mexico, USA*

3.1. PRINCIPAL FEATURES OF HIGH-STRAIN-RATE AND SHOCK DEFORMATION IN METALS

It is a well-known fact that the mechanical response of metals depends upon the temperature, the velocity of deformation, the previous deformation undergone, and the stress state, among other parameters. As a result, various attempts have been made to incorporate these parameters into a single equation that would have the capability of predicting the response of a specific metal under a wide range of circumstances. By analogy with thermodynamics, this equation is referred to as the 'mechanical equation of state':[1]

$$f(\sigma, \varepsilon, \dot{\varepsilon}, T) = 0 \qquad (3.1)$$

Changes in stress can be expressed as:

$$d\sigma = \left(\frac{\partial \sigma}{\partial \varepsilon}\right)_{\dot{\varepsilon},T} d\varepsilon + \left(\frac{\partial \sigma}{\partial \dot{\varepsilon}}\right)_{\varepsilon,T} d\dot{\varepsilon} + \left(\frac{\partial \sigma}{\partial T}\right)_{\varepsilon,\dot{\varepsilon}} dT \qquad (3.2)$$

However, plastic deformation is, by definition, an irreversible process, and the thermodynamic equation of state, from which eqn. (3.2) was obtained, only strictly applies to reversible processes. Hence, eqn. (3.2) can only be accepted as an approximate formulation, and the term 'mechanical modelling' is perhaps more appropriate than mechanical equation of state. Both Hart[2] and Rohde[3] have proposed models.

The general trend observed is that the flow stress increases with strain rate increase and temperature decrease. The reason for this flow stress increase seems to be connected to a decreasing ability of thermal energy to aid dislocation movement at higher strain rates and lower temperatures. Figure 3.1 shows a stress–strain–strain rate plot for mild steel.[4]

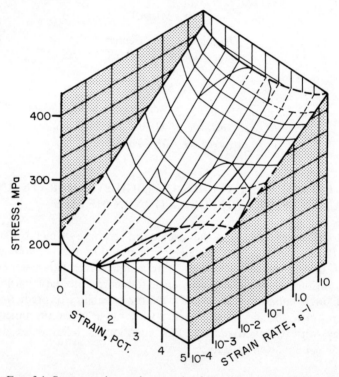

FIG. 3.1 Stress–strain, strain-rate plot for mild steel (from Ref. 4).

The yield stress increases from 215 to 410 MPa, as the strain rate is varied from 10^{-4} to 10^2. However, the strain-rate sensitivity of different materials is different, and the exponent in the equation below has been found to vary significantly

$$\sigma = k(\dot{\varepsilon})^m \tag{3.3}$$

where m is called the strain-rate sensitivity. This dependence of mechanical properties upon strain rate is described in detail by Campbell.[4] An example of a material that does not virtually exhibit any strain-rate sensitivity is given in Fig. 3.2. A 6061-T6 aluminum was tested at strain rates of 9×10^{-3}, 15×10^{-2}, 2, 40, and $210\,\mathrm{s}^{-1}$ and the stress–strain curves are identical.[5] A number of experimental techniques have been designed to determine the mechanical response of metals over this wide strain-rate range.[6,7] For rates of $1\,\mathrm{s}^{-1}$ or lower, the quasi-static universal testing machines are appropriate.[9] In the $1-10^4\,\mathrm{s}^{-1}$ range, special testing machines have been developed (see Chapter 4).

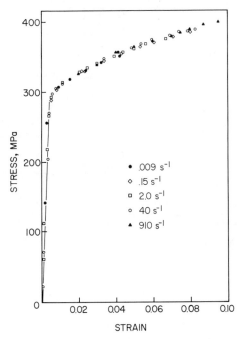

FIG. 3.2 Effect of strain rate on the compressive strength of 6061-T6 aluminum (from Ref. 5).

It is important to know the strain-rate dependence of the mechanical response of metals because a large number of mechanical fabrication techniques use strain rates much above those usually employed in conventional tensile testing. Campbell[4] cites some typical strain rates; in machining it is about $10^5\,\mathrm{s}^{-1}$; in sheet, rod or wire drawing, it varies between 1 and $10^3\,\mathrm{s}^{-1}$ in deep drawing, it may be as high as $10^2\,\mathrm{s}^{-1}$.

However, it is in high-energy rate processes that the strain-rates are typically and systematically high (10^2 to $10^4 \, \text{s}^{-1}$). Figure 3.3 shows the various explosive metal-working processes, which comprise an important part of the high energy rate fabrication processes. One has, in essence, three types of operations: stand-off, contact, and impact. In the stand-off

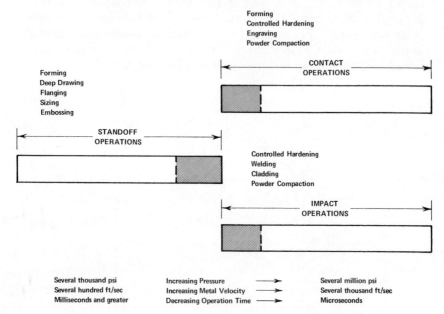

FIG. 3.3 Range of explosive fabrication processes (from Ref. 13).

operations, there is a medium that transfers the energy from the explosive to the workpiece (usually water). The magnitude of the pressure generated in the workpiece is not high enough to generate shock waves (in general). Contact and impact operations, on the other hand, generate high pressures and shock waves in the workpiece. One can see that one can have a whole range of strain rates and deformation regimes in explosive metal working.

A phenomenon of considerable practical importance in explosive forming is the increased ductility exhibited by a number of alloys at high strain rates. This results in an improved formability; the retardation of necking occurs at a certain deformation velocity, as shown in Fig. 3.4. Especially noteworthy is the increased ductility of 17–7 PH stainless steel. Orava[14] interpreted this increased ductility as related to the onset of

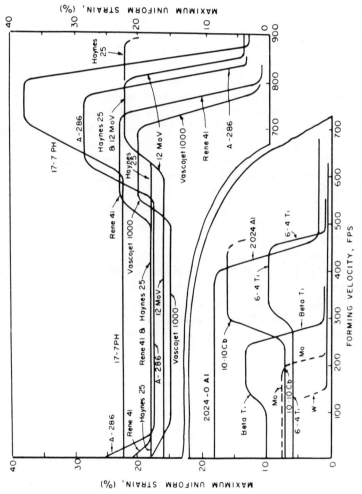

FIG. 3.4 Effect of forming velocity on the maximum uniform strain during the explosive bulging of domes (from Ref. 13).

plastic stress wave propagation. He suggested that, when the impact velocity is such that the stress wave that travels through the system has an amplitude equal to the dynamic yield stress of the material, ductility increases. At this stress level the plastic deformation takes place at the wavefront and the effective strain rate undergoes a significant increase over the one that one would expect if deformation were uniformly distributed over the specimen.

As one can see from the above, it is very difficult to predict *a priori* the medium of deformation under the imposed conditions. Additional complicating factors are introduced by the strain-rate dependence of the substructure morphology. Dislocations, deformation twins, phase transformations are all affected by the strain rate and strain state. They are discussed in the previous chapter. An effect that is being increasingly recognized as important is *adiabatic shearing*. This effect is simple to explain in general terms; however, the specific mechanisms are only partially understood. Zener and Hollomon[15] were the first to describe it. Recently it has received a great deal of attention.[16-21] Olson et al.[16] defined an adiabatic shear band as a strain-localization phenomenon that is generally attributed to a plastic instability arising from a thermal softening effect during adiabatic or near-adiabatic plastic deformation. In other words, as the strain rate is increased the deformation tends more and more towards an adiabatic state. Hence, if for a certain reason one has localized deformation, the temperature will increase at that region, above the average temperature. Since the flow stress decreases with temperature, the hotter region will undergo more and more deformation. This localization of deformation produces bands. One example of shear band formation is shown in Fig. 3.5. Depending on the temperature achieved, the rate of cooling (after deformation), and the composition of the alloy, the adiabatic shear band can present different characteristics. Moss[20] found that the shear strain across a band could be as high as 572. The shear strain rate inside a band was found to go up to $9.4 \times 10^7 \, \text{s}^{-1}$. Rogers and Shastry[19] discussed the phase transformation occurring within the shear bands when either the strain concentration or strain rate are sufficiently large. The deformation mechanism responsible for adiabatic shear bands is still not very well understood, but it is doubtful that dislocations play a preponderant role. Indeed, Olson *et al.*[16] have discussed an alternative mechanism, involving the formation and reclosure of shear cracks in the material. This mechanism seems to have a great potential, since nothing would impede crack reclosure, under the high stresses and temperatures, and since no oxidation of the fracture surfaces would be allowed.

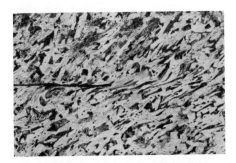

FIG. 3.5 Adiabatic shear band revealed by optical micrograph of steel. Tip of shear fracture preceded by shear zone (from Pearson and Finnegan in source cited in Ref. 10, p. 205).

At strain rates up to 10^4, one considers the stress as uniformly distributed over the length of the specimen (workpiece). At these velocities of deformation the inertia of the system becomes more and more important, and elastic and plastic waves propagating through the specimen might obscure the results. The upper limit for the strain rate is given by shock loading. In this situation, the strain is imparted by the passage of a shock wave throughout the specimen. Hence, one has a non-uniform distribution of stresses and strains during the deformation process.

As demonstrated in the previous chapter, the passage of a shock wave in a solid crystalline material can be fairly rigorously described in a phenomenological and mechanical way. However, its interaction with any specific structure is sometimes complicated by very subtle features which promote specific residual microstructures.

The propagation of a shock wave in a metal or alloy is shown schematically in Fig. 3.6(a). The deformation induced in a solid metal or alloy by a shock wave having a shape as shown in Fig. 3.6(a) can be separated into three prominent parts or regions: compression of the solid by the shock front (I), a region of constant peak pressure where no overall volume work is done, and which defines the shock pulse (II), and a rarefaction (relief) portion (III). Since $dV=0$ in region II, the major contributions to shock deformation producing permanent, residual microstructural phenomena must come from the compression (I) and relief portions (III) of the shock wave. One can estimate the strain rate at the shock front in an approximate way. A pressure of 20 GPa generates a shear strain of 0.10 for a metal such as copper. The rise-time of the shock front has not been accurately determined as yet, but can be estimated to be equal to 0.1 μs (or less). The strain rate obtained by dividing these two

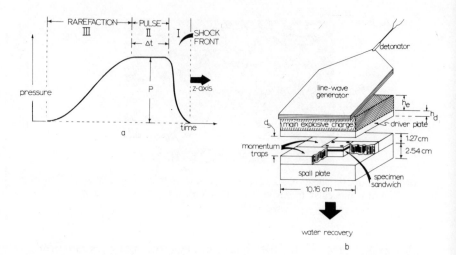

FIG. 3.6 (a) Idealized view of a shock pulse traveling through a solid metal. The pulse duration is denoted Δt. (b) Schematic view of an arrangement for shock loading sheet materials with plane compression waves, of known peak pressure.

values is 10^6 s^{-1}. At the rarefaction part of the wave, the strain rate is lower by about an order of magnitude. While the internal energy change across the shock front is calculated assuming an adiabatic process, the work done on the solid during rarefaction is calculated assuming an isentropic relief process,[2] according to the hydrodynamic theory, that assumes that the solid acts like a fluid. Consequently, the total change in internal energy after shock compression is expressed as:[22]

$$(E_f - E_0) = [(P_i + P_0)(V_0 - V_i)]/2 - \int_{V_i}^{V_f} P_s(V) \, dV \quad (3.4)$$

where E_f and V_f are the final values and P_i and V_i are intermediate values associated with the peak pressure, and $P_s(V)$ is the isentropic relief path.

Grace[23] has expressed the elastic energy stored by means of defects as a function of the pressure, but eqn. (3.4) cannot be applied because it does not take into account the additional processes taking place in crystalline solids. Indeed, Hsu et al.[24] have shown that the hydrodynamic model does not apply to nickel and that one has additional heating (above that predicted by the hydrodynamic model), dislocations and point defects; consequently, one has to add a term taking into account these additional processes

$$(E_f - E_0) = [(P_i + P_0)(V_0 - V_i)]/2 - \int_{V_i}^{V_f} P_s(V) dV + E_s \quad (3.5)$$

where E_s is the energy stored as defects or additional heating.

Nevertheless, one can illustrate how shock waves having specific peak pressures can alter the metallurgical (microstructural, physical, and mechanical) properties of metals by correlating the energy stored in the material with the elastic strain energy due to defects. Grace used the following equation:

$$\sigma = \sigma_0 + 2\alpha G |\mathbf{b}|(W^E)^{1/2} \quad (3.6)$$

where α is a constant, G is the shear modulus, $\mathbf{b}$ is the lattice dislocation Burgers vector, σ_0 is the initial yield stress (prior to shock-wave propagation). This expression is intended as an illustration of the overall response of a metal or alloy to the stored energy induced by shock-wave passage. The specific form will change depending upon the particular types of defects contributing to the stored energy.

In eqn. (3.5), one can decompose E_s into E_{s1}, the energy due to defects, and E_{s2}, the heating over and above that predicted by the hydrodynamic theory. Hence:

$$E_{s1} = (E_f - E_0 - E_{s2}) - [(P_i + P_0)V_0 - V_i)]/2$$
$$+ \int_{V_i}^{V_f} P_s(V) dV; (E_f - E_0 - E_{s2}) = C_R \Delta T \quad (3.7)$$

But E_{s1} is equal to W^E, and one has on substituting eqn. (3.7) into eqn. (3.6):

$$\sigma = \sigma_0 + 2\alpha G |\mathbf{b}| \left\{ C_R \Delta T - [(P_i + P_0)(V_0 - V_i)]/2 + \int_{V_i}^{V_f} P_s(V) dV \right\}^{1/2} \quad (3.8)$$

To a large extent, eqn. (3.8) is indicative of the effects of peak shock pressure on the residual mechanical properties, viz. yield strength, ultimate tensile strength, hardness, ductility, etc. However, these effects occur because of pressure-induced microstructural changes. Equation (3.8) is a modified version of Grace's[23] expression.

3.2. PERMANENT CHANGES: RESIDUAL MICROSTRUCTURE–MECHANICAL PROPERTY RELATIONSHIPS

This section will deal exclusively with shock waves since this is the regime for which substantial information is available in the literature. The strain rates in the range $1-10^6 \, s^{-1}$ deserve greater attention and have not been explored in sufficient depth, from a metallurgical point of view. There is an almost complete lack of substructural characterization by transmission electron microscopy. An up-to-date collection of contributions with detailed information on the metallurgical effects of shock waves is the book edited by the authors.[25]

Although transient effects are important in the generation and propagation of shock waves in metals and other materials, these effects are often difficult to monitor accurately, and are of little consequence with respect to their effect on the residual physical and mechanical properties of materials, which are dependent upon the more-or-less stable microstructures created in the shock front. Graham[26] has recently discussed transient phenomena in some detail, and we will not pursue this in any detail here because we will be more concerned with shock residual microstructures and properties which are of a much more practical consequence in evaluating the performance of explosively hardened, welded, formed, or consolidated materials.

It is well known that microstructures of all types have some influence on the physical and mechanical properties of materials. These include grain boundaries in polycrystalline materials, precipitates and other inclusions and dispersoids, and other types of point, line, planar, and volume defects. The passage of a shock wave through a metal can have three prominent effects; existing microstructure can be altered, new substructural features can be created, or both can occur simultaneously. Alterations can include the annihilation of crystal defects through shock-induced annealing which can result by shock heating at very high peak pressures (60 GPa in steels, for example). These effects are almost without exception pressure dependent, and as a consequence both microstructure and related properties are quantitatively if not qualitatively related to the peak pressure, P.

Difficulties have been encountered in attempts to relate residual microstructural phenomena to residual mechanical properties through peak shock pressure because in many cases the shock wave was not accurately defined and errors occurred in measuring both the peak pressure and the stress conditions. In order to quantitatively or even qualitatively assess the metallurgical effects of shock waves in metals, it is necessary to accurately define the stress (or strain) state, the pressure, and pulse duration. This can be achieved by utilizing experimental arrangements similar to that shown schematically in Fig. 3.6(b), and described in detail in the previous chapter and elsewhere.[25] In such designs, the shock wave can be very accurately described as a compressive wave, and the tensile relief wave is eliminated from the experimental sandwich by spallation in the spall plate. Consequently, thin sheets of metal or alloy material can be subjected to plane compressive shock waves producing microstructures which can be directly observed by transmission electron microscopy.

3.2.1. Grain Size Effects

It is rare that grain size is altered with shock wave passage in a metal or alloy except where a thermal transient promotes annealing or recrystallization and grain growth. However, it has now been demonstrated that the grain size prior to shock deformation has a very significant influence on residual properties, particularly hardness and yield strength. These features are illustrated in Fig. 3.7. which shows a Hall–Petch type response for residual yield stress and hardness at various shock pressures:

$$\sigma = \sigma_0 + KD^{-1/2} \tag{3.9}$$

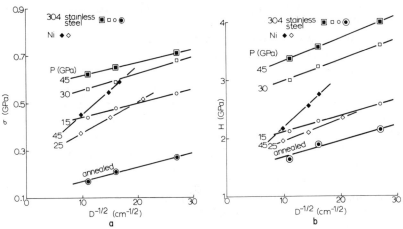

FIG. 3.7 Effect of initial grain size on the residual yield stress (a) and hardness (b) of shock-loaded nickel and stainless steel (from data in Ref. 27).

and

$$H = H_0 + K'D^{-1/2} \qquad (3.10)$$

where K and K' are constants and D is the average grain size (grain diameter). The relationship between eqn. (3.10) and eqn. (3.9) is certainly obvious.

Moin and Murr,[28] Greulich and Murr,[29] Kestenbach and Meyers,[30] Wongwiwat and Murr,[31] and Murr[27] have discussed the effects of grain size on the development of microstructure and have shown that variations in grain size can have a significant effect on the production of deformation twins, martensitic transformation, and even dislocations.

3.2.2. Shock-Induced Microstructures

Dislocations are generated in the shock front without exception. If the wave profile has a shape shown ideally in Fig. 3.6(a), the dislocations generated will remain as a relatively stable microstructure behind the shock pulse as described in the preceding chapter, although some rearrangement, multiplication, or annihilation can occur in the relief portion of the pulse. As shown in eqn. (3.9), the generation of dislocations can be influenced by the grain size, but there is also an important effect of slip systems available and the ability of dislocations generated in the shock front to cross-slip, glide, or climb. While glide and climb may be influenced by temperature and other parameters, cross-slip will depend primarily on the stacking-fault free energy. This is particularly true in face-centered cubic metals where slip is restricted to the $\{111\}$ planes, and dislocation motion is almost completely governed by the ability of dislocations to extend, forming partial dislocations separated by a region of stacking fault which increases with decreasing stacking-fault free energy.

Figure 3.8 illustrates these features for a number of common face-centered cubic metals and alloys with stacking-fault free energies noted,[32] and shock loaded at a similar peak pressure and at a constant pulse duration utilizing the experimental design concept illustrated in Fig. 3.6(b). Figure 3.8 shows that at high stacking-fault free energies (≥ 70 mJ/m^2) dislocations cannot extend appreciably and cross-slip is predominant. With sufficient time available in the shock pulse,[33] this produces dislocation arrays which form cell-like structures particularly prominent in shock-loaded nickel. As the stacking-fault free energy is reduced by alloying, cross-slip is discouraged or impossible, and dislocations form planar arrays which include extended stacking faults in

Open: 10.00 am to 1.00 pm and ...

Norris Green Library.
Open: 9.30 am to 1.00 pm and 2.00 pm to 4.30 pm.
Every Thursday.

Spellow Library.
Open: 9.30 am to 1.00 pm and 2.00 pm to 4.30 pm.

Old Swan Library.
Open: 9.30 am to 1.00 pm and 2.00 pm to 4.30 pm.

Issue No. **406**

Printed by
J & C Moores Ltd

CITY OF
Liverpool

FIG. 3.8 Examples of shock-induced microstructures in face-centered cubic metals having a range of stacking-fault free energies shown. (a) Ni(15 GPa), (b) Cu (15 GPa), (c) Fe–34% Ni (10 GPa), (d) $Ni_{80}Cr_{20}$ (8 GPa), (e) Inconel 600, (f) 304 stainless steel.

the {111} planes. In many materials, regular or periodic arrays of stacking faults can produce new phase regimes or twins. Indeed when intrinsic stacking faults form on every {111} plane, the region encompassed is a twin of the unfaulted region. During shock loading, these regions form irregularly, creating faulted twins or twin-faults. Figure 3.9 illustrates this phenomenon in type AISI 304 stainless steel subjected to explosively generated shock waves having various peak pressures indicated.

It is apparent from Fig. 3.9 that the microstructural 'density' is increasing with increasing peak pressure. Indeed, as the number of dislocations or stacking faults increases, the volume fraction of deformation twin-faults increases, their size increases and their spacing decreases (the inter-twin spacing decreases). Similarly, in the case of high stacking-fault free energy face-centered cubic materials, increasing dislo-

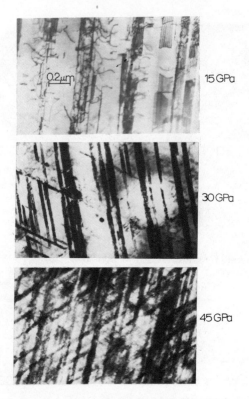

FIG. 3.9 Creation of twin-faults in AISI 304 stainless steel.

cation density with increasing peak shock pressure will cause an increase in the dislocation cell density, and this can only occur by decreasing the mean cell size or by increasing the number or density of dislocations composing the cell walls. This feature is illustrated in Fig. 3.10.

There are exceptions[34] to the rule of increase in flow stress with shock loading. Figure 3.11 shows a flow stress decrease found in RMI 38644 titanium alloy up to pressures of 10 GPa (100 kbar). This decrease is due to a pressure-induced phase transformation; the alloy is in the metastable beta state; shock loading produces the reversion to the omega phase, which has a lower yield stress and UTS. At lower pressures the transformation effect is more important than the one due to dislocations generated, resulting in an overall strength decrease.

It is apparent on comparing Figs. 3.9 and 3.10 that in addition to changes in dislocation density with shock pressure, other microstructural

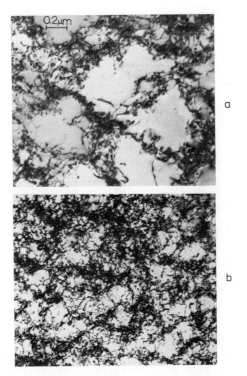

FIG. 3.10 Changes in dislocation density in high-stacking-fault free energy face-centered cubic materials.

geometries or dimensions change. These changes will influence the residual mechanical properties as generally described by Ashby:[35]

$$\sigma = \sigma_0 + K(\lambda^G)^m \tag{3.11}$$

where K is a constant which can be made similar or identical to that in eqn. (3.9), and λ^G is a wavelength equal to the spacing between specific phases or microstructural features. The specific nature of these regimes will determine the value of m, which will be equal to or less than 0·5 for grain boundaries, and increase for other, lower-energy interfaces, becoming unity for dislocation cells. In effect, gradients of plastic deformation are imposed by the microstructure, and λ^G can be regarded as the grain size, D, the dislocation cell size, d, or the inter-twin spacing, Δ. When microstructures become mixed, such as the intermixing of dislocation cells and deformation twins, they will each contribute to the resulting

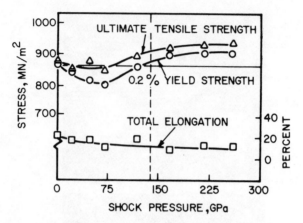

FIG. 3.11 Relationship between the flow stress and shock pressure in RMI 38644 titanium alloy.

strengthening, and this can be illustrated by an equation of the form

$$\sigma = \sigma_0 + K\rho^{1/2} + K'V_c d^{-1} + K''V_T \Delta^{-1/2} + K_0 D^{-1/2} \quad (3.12)$$

where the Ks are associated constants, and V_c and V_T represent the approximate volume fractions of dislocation cells and deformation twin-faults respectively. Certainly eqn. (3.12) must be regarded more as a qualitative guide than an empirically accurate description of microstructural strengthening, but a comparison of eqn. (3.12) with eqn. (3.9) will confirm the fact that peak pressure will have a profound effect upon both the residual microstructure and the associated physical and mechanical properties in shock-loaded metals and alloys.

Figures 3.12 and 3.13 illustrate the features implicit in eqn. (3.12) for shock-loaded nickel where dislocation cell structures form as shown in Fig. 3.10 with increasing pressure of roughly 30 GPa, at which point deformation twin-faults occur with increasing frequency (and decreasing inter-twin spacing, Δ) at increasing peak shock pressure ($P > 30$ GPa).

Figure 3.14(b) summarizes the variations in residual hardness with peak shock pressure for a large number of metals and alloys. The peak pressures do not exceed the range of systematic microstructural change, and are below the range of higher pressures which produce shock-thermal effects by shock heating.[11,20,21] However, Fig. 3.14(b) shows these effects for hardness changes at very high peak shock pressures, and Fig. 3.15 shows, by comparison with Fig. 3.9, the unique microstructural variations which occur in type AISI 304 stainless steel at very high

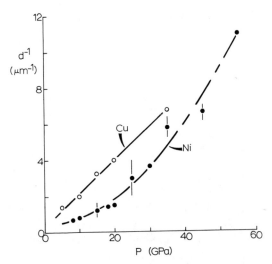

FIG. 3.12 The effect of pressure in shock-loaded nickel (illustrating eqn. (3.12)).

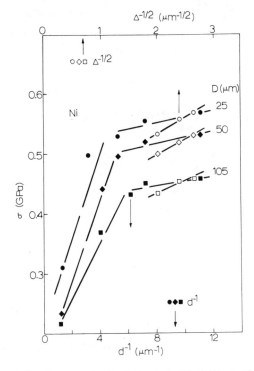

FIG. 3.13 Variation in stress in shock-loaded nickel (illustrating eqn. (3.12)).

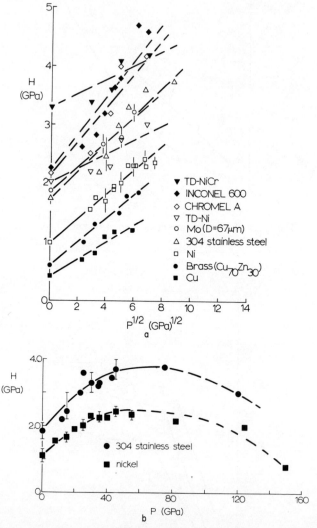

FIG. 3.14 (a) Summary of shock strengthening in a variety of metals and alloys (2 μs shock pulse duration) variations with grain sizes, are shown by error bars (from Ref. 27). (b) Hardness versus high shock pressure for nickel and type 304 stainless steel. Error bars correspond to variations in grain size (from Murr[27]).

pressures where shock heating effects are imposed upon the microstructural development. The sub-grain cell structures shown in Fig. 3.15 are to be recognized as distinct from the dislocation cell-type microstructures shown for shock-loaded nickel in Fig. 3.10.

FIG. 3.15 High-pressure (120 GPa) sub-grain microstructure in 304 stainless steel resulting from shock-thermal effects (compare with twin-fault microstructures shown in Fig. 3.9).

The shock heating effects illustrated in Figs. 3.15 and 3.16 must be recognized to be very different from those frictional heating effects which are important in explosive compaction, inducing particle melting and melt-controlled consolidation. However, at sufficiently high pressures,

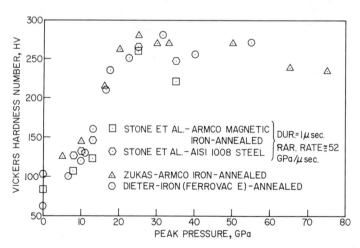

FIG. 3.16 Effect of shock-hardening pressure on the hardness of iron and steel (adapted from Ref. 68).

the heat generated by the shock transient can contribute to these effects, and vice versa.

Of particular significance from both a fundamental and applied point of view is the pressure-induced phase transformation undergone by iron and steel.[64,65] At 13 GPa the BCC (α) phase transforms into the HCP(ε) phase. The kinetics of this transformation are rapid enough for it to be produced by the shock wave. This phase transformation has been studied in detail by many investigators (e.g., refs. 66 and 67). The high-pressure phase retransforms to α in the release (or rarefaction) part of the way, but profuse remnants of the transformation are left in the microstructure, they have been described by Stone et al.[68] as 'shear plates', and are somewhat similar to elongated sub-grains. This additional increase in defects reflects itself in a significant increase in hardness. Figure 3.16 shows the experimental results found for shock-loaded iron and low-carbon steel by Dieter,[67] Zukas[66] and Stone et al.;[68] they agree fairly well. At 13 GPa one has an increase in hardness of approximately HV 100 (from HV 250 to HV 350). Above this pressure there is no substantial gain in hardness; indeed, shock recovery effects discussed in the preceding paragraph start above 50 GPa. The microstructural changes produced by this phase transformation can be seen by optical microscopy. Figure 3.17 shows the change introduced by the transformation.

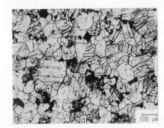

FIG. 3.17 Optical micrographs of AISI 1008 steel shock loaded below (a), and above (b) 13 GPa. The transformation debris can be clearly seen in (b) (from Ref. 69).

Deformation twins can be seen in Fig. 3.17(a), typical of a region shocked below 13 GPa. The dark regions that are highly etched are due to the transformation debris, in Fig. 3.17(b).

Figure 3.18 shows the deformation undergone by rapidly-solidified Mar M-200 powders during shock-wave consolidation.[70] The hardness is significantly increased by the passage of the shock wave: HV 400 to HV

FIG. 3.18 Shock-consolidated MAR M-200 superalloy (courtesy of B. P. B. Gupta). (a) TEM image from a region in (a) showing planar dislocations in arrays (b).

720. In addition to shock-hardening *per se*, one has high strain-rate deformation at the external regions of the particles and heating due to friction between the particles. The dendritic structure produced by the RSP[70] process can be clearly seen in Fig. 3.18(a) and the deformation at the boundaries of the particles is reflected by the distortion of the dendritic pattern, that was initially regular. The clear islands between particles represent melted (or recrystallized) regions; they are highly localized, and does not affect the regions adjacent to it because of the high cooling rates.

Figure 3.18(b) shows the deformation substructure in the center of one of the shock-consolidated Mar M-200 particles. It is characteristic of shock loading. The dislocations are organized in planar arrays and their density is extremely high. This high dislocation density is responsible for the significant hardness increase in dynamic consolidation. The substructure close to the interfaces where bonding occurred is different, because the uniaxial strain state condition required for shock waves is not obeyed.

3.2.3. Shock Deformation versus Conventional Deformation (Cold Reduction)

Although it is unnecessary to draw extensive comparisons between shock loading and more conventional deformation (characterized by more complex stress states and lower strain rates), it must be recognized that shock loading is unique as a result of the significant hardening and strengthening which arises from shock wave propagation, while the residual strains are small or even negligible. Although there are significant differences in a cold-reduction mode of deformation as compared to well-controlled shock compression, it is of interest to note that the corresponding microstructures attendant to the production of specific mechanical responses are normally produced in shock deformation at significantly lower true strains when compared to the through-thickness true strains in cold reduction. This feature is illustrated in the experimental data shown plotted for comparison in Fig. 3.19. It is particularly

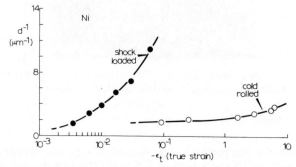

FIG. 3.19 Reciprocal dislocation cell size versus true strain for shock-loaded and cold roll reduced nickel sheet. (Shock loading data from Ref. 27; cold reduction data courtesy of Dana Rohr, Los Alamos National Laboratory.)

significant to note that, as implicit in Fig. 3.13, there is a particular strength (yield stress) or hardness associated with a particular dislocation cell size in nickel, and this is achieved at significantly lower and even negligible residual strains in shock deformation as compared to other forms of deformation. This is certainly one of the very practical and important aspects of shock hardening. Since the residual strain is very small, one would expect an anisotropy of strengthening; this is indeed observed. One would expect that no specific texture is introduced by the shock wave, in contrast to most conventional deformation processes. Indeed, Higgins,[71] Trueb,[72] and Dhere et al.[73] have found no change in texture upon shock loading copper, nickel and aluminum, respectively.

3.2.4. Effects of Shock Pulse Duration in Shock Loading

Murr and Kuhlmann-Wilsdorf[33] have discussed the fact that while the quantitative features of the microstructure, particularly dislocation density, are controlled almost exclusively by the peak pressure, the qualitative features such as the definition of dislocation cell structures, dislocation cell wall definition, and the like, depend upon the time the pulse is applied, i.e. the shock pulse duration (Δt in Fig. 3.6(a)). These features have recently been reviewed in considerable detail,[35] and will not be pursued here. However, it must be pointed out that as result of these effects, there is usually no significant variation in mechanical properties such as yield strength and especially hardness over a fairly wide range of shock pulse durations above about 1 μs. These effects are summarized in the hardness data plotted in Fig. 3.20.

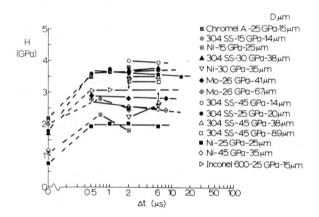

FIG. 3.20 Residual hardness (H) versus shock pulse duration (Δt) for a number of metals and alloys over a range of peak pressures. The initial grain sizes (D) are indicated (from Murr[36]).

It is important to realize that the pulse durations noted in Fig. 3.20 and in the other figure descriptions presented in this chapter were carefully controlled by adjusting the driver-plate thickness in experiments characterized explicitly by Fig. 3.6(b). In these experiments, the shock pulse duration, Δt, can be fairly accurately determined from.[37,38]

$$\Delta t \cong 2h_d/U_s, \quad (U_s = \text{shock velocity}) \quad (3.13)$$

where at high pressures (>5 GPa) we can write

$$\Delta t \cong 2h_d U_p \rho_0/P \quad (U_p = \text{particle velocity}) \quad (3.14)$$

It can be observed, somewhat qualitatively, that in order to maintain a constant shock pulse duration with increasing peak shock pressure, the flyer-plate thickness must be increased.

3.2.5. Effects of Point Defects, Precipitates and Other Second-Phase Particles

Because of the rapid movement of dislocations generated in the shock front,[39] intersections forming jogs favor vacancy formation in large quantities. These features were demonstrated experimentally by Kressel and Brown,[40] and Murr, et al.[41,42] have also shown that vacancies and vacancy clusters make a direct contribution to residual shock hardening. In addition, vacancies present in shock-loaded metals contribute significantly to thermal recovery phenomena, and hardness recovery has been demonstrated to occur even catastrophically for shock-hardened metals and alloys.[38] A high concentration of point defects, particularly vacancies, is therefore a somewhat unique feature of the residual shock microstructure in explosively loaded metals and alloys.

Although it might be expected that shock-induced point defects might lead to the formation of precipitates or other similar inclusions, evidence for this phenomenon is meager.[39,43] On the other hand, precipitates and other second-phase particles (both coherent and non-coherent) can have a profound effect on shock-loaded metals and alloys. Initially, of course, particle distributions strengthen and harden a matrix by providing for deformation gradients having a wavelength equal to the mean particle (or interparticle) spacing (eqn. (3.11)). Following shock loading, dislocations generated in the shock front add additional, significant strength which can be described by an equation of the form of eqn. (3.12). Furthermore, such particles or precipitates can not only block the initial (shock) formation of crystal defects,[44] but can also block their annihilation by shock heating or post-shock annealing or heat treatment. This feature is illustrated in Fig. 3.21. Figure 3.21 also shows the very rapid and even catastrophic recovery alluded to above. However, this phenomenon is not universal, and Inconel 718 shock loaded to a pressure of 51 GPa and cold rolled (20% reduction in the thickness) exhibited about the same recovery response, when subjected to isochronal anneals (1 hour). This is evidenced by Fig. 3.22. The shocked and rolled conditions show the same hardness decrease.

Because of the density differences imposed by precipitates and other inclusions, impedance mismatch can lead to local spallation at the particle–matrix interface or other aberrations in the wave profile can

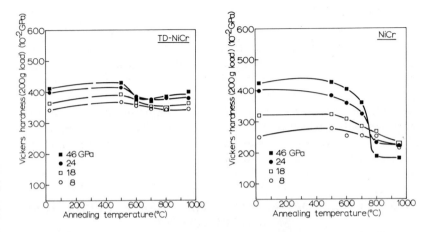

FIG. 3.21 Comparison of hardness and recovery in shock-loaded $Ni_{80}Cr_{20}$ and TD–Ni Cr (the same alloy with 2 volume percent ThO_2 included as a dispersed particle distribution) for a constant annealing time of 1 h.

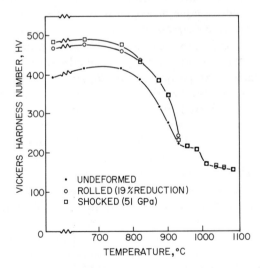

FIG. 3.22 Isochronal (1 h) curves for undeformed, rolled (20% reduction) and shocked (52 GPa) Inconel 718 superalloy having been pre-aged at $1400°F/4\,h$ (from Ref. 45).

occur through the mechanism proposed by Meyers[46] for the production of a wavy-wave in a polycrystalline metal or alloy. Moreover, Murr and Foltz[47] have demonstrated experimentally that coherent precipitates can become non-coherent following shock loading at sufficiently high peak pressures, and that with the loss of coherency, the precipitates become engulfed by dislocations. It is also certain that many precipitates and other inclusions serve as dislocation sources during the shock-loading of metals and alloys which contain them, when they have different compressibilities. Geometrically-necessary dislocations are generated around the interface to accommodate the strains.

3.3. RESPONSE OF METALS TO THERMOMECHANICAL SHOCK TREATMENT

In the foregoing treatment in Section 2 above, we have attempted to provide an overview of structure–property relationships (mainly simple mechanical relationships), associated with explosively shock-deformed metals and alloys. It should be apparent from this treatment that there are some unique features of shock-induced microstructures when compared with more conventional deformation microstructures, and there are correspondingly unique mechanical properties. Figure 3.22 is a particularly convincing example of the unique features of shock hardening. Overall, the metallurgical effects of shock wave propagation are not generally unlike those of other deformation modes where microstructures are created which evoke particular physical and mechanical responses.

Cooperative strengthening effects between mechanical and thermal treatments can be achieved by thermomechanical processing (TMP) or thermomechanical treatments (TMT). In such treatments, microstructures created by mechanical (deformation) treatment can be altered through thermal (annealing, quenching, etc.) treatments, and vice versa. Central to the effective utilization of this concept is of course the ability to control both composition and microstructure, and to carefully monitor these changes, in consonance with alterations in specific properties.

The phase transformations that deformation and heat treatments can create in steels make them particularly amenable to TMT, and there are numerous examples of such applications.[48,49] There are, by comparison, currently few applications of shock TMT or thermomechanical shock treatment (TMST), although the concept of TMST has been discussed in a recent review by Meyers and Orava.[50]

The results shown in Fig. 3.22 suggest that the creation of second phases by aging and other treatments followed by shock loading, or similar TMST schedules could provide unique metallurgical properties, and indeed this has been demonstrated for example in nickel-base superalloys such as Udimet 700[51] and Inconel 718.[45,52] Figures 3.23 shows constant-load creep curves for Udimet 700 and Inconel 718 at 649 °C. The benefits arise mainly by the development of a high volume fraction of finely-dispersed precipitates (γ'), and a finely-dispersed, thermally-stabilized dislocation substructure. Aging treatments of 100 h at 789 °C gave rise to extensive δ formation along the {111} planes in the Inconel 718 matrix, while it was only occasionally observed in the cold-rolled material and absent in the undeformed material for the same heat treatment. Some of these effects arise from the high vacancy concentrations associated with the shock deformation as briefly described above and in the previous chapter. Some evidence for the effect of vacancies on precipitation is also suggested in the work by Greenhut, *et al*.[53] involving an aluminum alloy.

Because of the ability of high-pressure shock waves to produce particular microstructural phenomena, it is of course futile to assess the shock TMP response of a specific material on the basis of room temperature tensile properties alone, because, as shown in Fig. 3.23, there are frequently very particular metallurgical effects which are recognized only under very specific conditions. Certainly Fig. 3.23 illustrates a condition where shock TMP is superior to conventional TMP.

3.3.1. Shock–Mechanical Treatment, Stress-Cycling and Repeated Shock Loading

As we noted previously (Fig. 3.14(b)), most metals and alloys, depending upon their particular melting points and other properties, will begin to heat up appreciably both during and after shock wave propagation at sufficiently high peak pressure. This limitation can be circumvented to some extent by conventionally deforming a material prior to shock deformation, by shocking repeatedly, or by some other stress-cycling involving shock deformation. In any of these treatments, which may not always specifically involve TMP (although many processes begin with some conventional annealing treatment), the initial deformation process (either conventional or shock loading) creates a microstructure which in effect work hardens the material. The second deformation process then involves an interaction with this microstructure creating modifications which can lead to an increase (or decrease) in defect density, a change in the type of defect, or both.

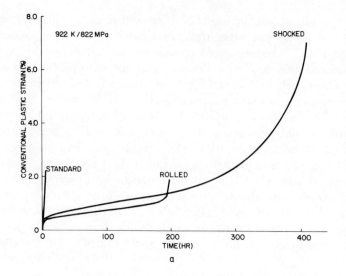
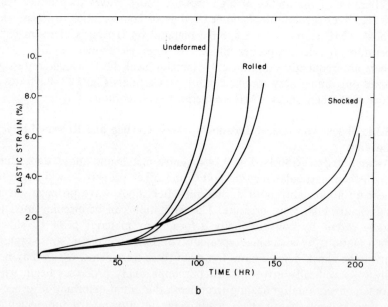

FIG. 3.23 Constant-load creep curves. (a) Udimet 700 at 649 °C; TMP schedule: 1177 °C, 4 h + cold work + 843 °C, 24 h + 760 °C, 16 h; $\sigma = 0.8$ GPa. (b) Inconel 718 at 649 °C; shocked and rolled: 954 °C, 1 h + 704 °C, 4 h + cold work + 677 °C, 8 h F.C. to 621 °C, total 18 h; undeformed, $\sigma = 0.8$ GPa.

Staudhammer and Murr[54] have described experiments involving schedules of cold-reduction and shock cycling of stainless steels which clearly illustrate a sensitivity of residual microhardness of sheet samples to the particular deformation schedules. This is shown in Fig. 3.24 for

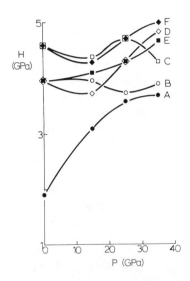

FIG. 3.24 Residual hardness (H) versus shock pressure (P) for 304 stainless steel sheets subjected to deformation schedules shown. —·—. A: the annealed sheet with no deformation prior to or after shock loading; —○—○— B; shocked +15% cold reduction by rolling; □—□— C; shock loading; ■—■— E: 15% cold reduction + shock loading 15% cold reduction; ◆—◆— F: 30% cold reduction + shock loading. (Data from Ref. 54.)

deformed type 304 stainless steel. Although it was originally demonstrated by Staudhammer and Murr[54] that the relative amounts of residual α' martensite changed with the deformation cycling, it has only recently been shown that this feature, along with the hardness responses shown in Fig. 3.24 probably occur by one deformation mode (cold reduction) being more conducive to martensite formation because nucleation and growth is retarded by adiabatic heating associated with shock loading at room temperature.[55] This occurs because deformation of AISI 304 stainless steel produces twin-fault intersections which increase with increasing deformation. α' martensite nucleates at these intersections as originally described by Olson and Cohen,[56] but at high strain rates, adiabatic heating suppresses this process. Consequently, shock loading can produce additional intersections but not necessarily additional martensite, whereas cold reduction will cause more efficient transformation at the intersections formed during deformation.

When the number of such shock-induced intersections in AISI 304 stainless steel is increased substantially, large amounts of α' martensite can form irregular morphologies by coalescence, and this can be achieved by repeated shock loading as described by Kazmi and Murr.[57] However, while dramatic changes in α' martensite content are observed as shown in Fig. 3.24, there is very little difference in net, residual yield stress increase, while the hardness increment increases substantially with the production of large volume fractions of martensite. This is illustrated in Fig. 3.25 for AISI 304 stainless steel and similar results were observed on doubly-shocked Inconel 718 in earlier work by Meyers.[52] However, Meyers[52] concluded from TMP experiments involving doubly shocked Inconel 718 that multiple shocking will yield attendant strength improve-

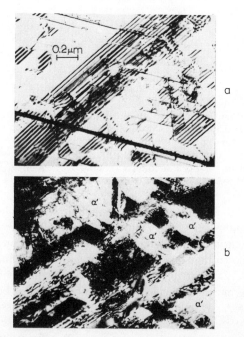

FIG. 3.25 Production of α' martensite by repeated shock loading of 304 stainless steel. (a) single shock event: 15 GPa 2 µs; (b) event in (a) repeated three times (shock loading three times at 15 GPa, 2 µs). The volume fraction of α' martensite: (a) about 2% (b) 26%.

ments. This feature is demonstrated in Fig. 3.26, but it is imperative, as pointed out earlier, to recognize that specific conditions (residual mechanical properties) may not always be able to provide a generalized concept of the repeated shock loading effects. The response may be very different at different shock mechanical or TMP treatments:

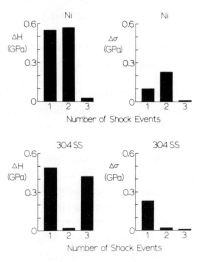

FIG. 3.26 Residual hardness increase, ΔH, and yield stress increase, $\Delta\sigma$, after repeated shock loading of nickel and 304 stainless steel (304 SS). The unshocked hardnesses: stainless steel 1.8 GPa, nickel 1.0 GPa. The original yield stresses were stainless steel 0.25 GPa, nickel 0.14 GPa respectively. The increment is taken as the 0.2% offset yield stress before and after each event; $\Delta\sigma = \sigma_f - \sigma_0$. Each shock event was characterized by a peak shock pressure of 15 GPa of 2 μs duration (data from Ref. 57).

It should also be recognized, as is perhaps implicit in Fig. 3.25, that repeated shock loading is not an additive process in terms of residual mechanical properties. That is, shock loading twice at a constant pressure, P, is not generally equivalent to shock loading once at twice the pressure, i.e. $2P_1 \neq P_1 + P_1$, etc. This occurs because each wave establishes a different kind of interaction with the structure of the material. In the case of the 304 stainless steel in Fig. 3.25, the second event does not produce much additional martensite and the twin-faults as shown in Fig. 3.25(a) probably impede additional twin-fault development, however after the third shock event, the change in microstructure causes an attendant change in hardness, while the yield stress becomes saturated.

In the case of nickel, the formation of dislocation cells as shown in Fig. 3.10(a) typifies the first shock event. Successive events simply decrease the cell size and increase the dislocation density. The second event causes a marked reduction in the dislocation size as well as a densification of the cell wall dislocation density. The third event, while decreasing the dislocation cell size and increasing the dislocation density, does not cause a concomitant increase in the hardness and yield stress.

3.3.2. Microstructural Stability and Thermal Stabilization of Substructure
There is some evidence that certain shock-induced microstructures are mechanically unstable.[50,59] This can arise because dislocation arrays can be easily rearranged when deformed after shock loading. This is particularly true in metals and alloys which, because of their stacking-fault energy (FCC materials) or available slip systems (or multi-slip systems) (BCC materials), form poorly-structured dislocation cells or other tangled arrays of dislocations (see Figs. 3.8 and 3.10). Indeed such work-softening, which results when materials with unstable microstructures are deformed in uniaxial tension, has been observed in nickel[59] (which forms dislocation cell structures in response to shock loading as shown in Figs. 3.8 and 3.10), and in a magnetic ingot iron (which forms dislocation cell-like tangles similar to those shown in Fig. 3.8(c)). This is apparently true only in the case where the microstructures are poorly developed.

In addition, work softening can be reduced or eliminated by thermal stabilization of the microstructure. This treatment simply involves an anneal following shock loading which allows the dislocations composing the microstructure to rearrange themselves into a more stable (equilibrium) sub-structure. As illustrated in Fig. 3.27, work softening can be eliminated with little loss in yield strength by annealing shock loaded metals under appropriate conditions. Such thermal stabilization treatments do not, strictly speaking, qualify as being described as thermomechanical processing. They are really mechanical–thermal treatments (MTT), and as such are strictly a substructural control process. Such treatments can be useful in stabilizing other microstructures because they can affect impurity segregation, and solubility, and as a consequence alter impurity pinning effects on dislocations and other recovery microstructures. The general subject of mechanical–thermal treatments of microstructures has been treated in a review by McElroy and Szkopiak.[60] The shock-loaded nickel specimen of Fig. 3.27 (25 GPa, 2 μs) exhibits a significant work-hardening rate after being annealed at 400 °C for one hour, without significant loss in the yield stress.

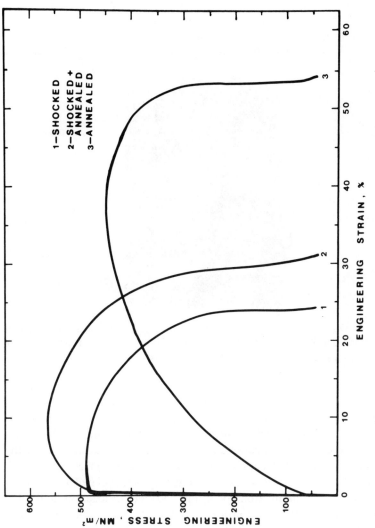

FIG. 3.27 Elimination of work-softening in shock-loaded nickel (25 GPa; 2 μs) by thermal stabilization of the substructure (from ref. 61).

On a substructural level, work softening can be explained by the replacement of the unstable substructure generated by shock loading by the one stable under the imposed conditions. Longo and Reed-Hill[62] found work softening in a number of high S.F.E. FCC alloys and attributed it to dynamic recovery during the tensile test. The same explanation applies to the case of shock loading, as evidenced by Fig. 3.28. This TEM specimen was cut from the region in the tensile specimen

FIG. 3.28 Dislocation substructure generated by conventional deformation of shock-loaded nickel (25 GPa, 2 μs) in a tensile test. The elongated cells are associated with work softening.

where work softening was occurring (neck region). An elongated cell many times larger than the shock-induced cells is clearly seen; its walls exhibit a much higher dislocation density than the original walls. It seems that the applied stress during tensile testing promotes the movement of the existing dislocations into these new arrays, rather than the generation of new dislocations (as in the case of conventional work hardening). So, the process of work softening is associated with dislocation cross-slip, annihilation, or climb, but no dislocation multiplication. Indeed, Kuhlmann-Wilsdorf and Laird[63] have analyzed similar elongated cells in fatigue. They are of 'near-minimum energy configurations of the stored dislocation content'. Fatigue tends to generate dislocation configurations that are more stable than conventional tensile testing. Prestrained material (by uniaxial tensile extension, for instance) can undergo cyclic softening by fatigue, which is accompanied by cell formation. This lends support for the explanation advanced by Meyers[59] and Hsu[61] for work softening.

3.4. SUMMARY AND CONCLUSIONS

Beginning with a development of a simple expression for the influence of shock or pressure waves on the residual yield strength of metals and alloys (eqn. 3.16), we have demonstrated experimentally observed microstructural development in response to uniaxial shock loading, and the corresponding influence on residual hardness and strength of shock-loaded metals and alloys. We have, to a large extent, limited our attention to an exposition of the conditions which are likely to be encountered, perhaps in a more complex and synergistic way, in explosive compaction and consolidation of powders. However, this development is particularly applicable to the response of metals and alloys in explosive fabrication hardening, and other forms of explosive metal working.

We have shown in Section 3.2, that shock-induced microstructures, while frequently similar to those characteristic of more conventional deformation, are sometimes unique, and form at significantly lower equivalent or true strains than in conventional deformation. However, as we demonstrate in Section 3.3, certain instabilities can occur for particular shock-induced microstructures. While these microstructures can be modified by heat treatment, there can be significant mechanical property consequences. Indeed, shock thermomechanical treatments or processing can also have significant effects on the residual mechanical properties of shock-loaded metals and alloys. More importantly, these effects can be very specific to a particular response such as creep life, fatigue life, tensile strength, ductility, etc. Furthermore, shock-wave interactions with existing microstructure as in the case of the repeated shock loading of a particular material or solid volume can also result in substructures and related mechanical properties which are unique to that particular mode of deformation.

It must be recognized that, in the case of explosive consolidation of metal powders for example, that the shock wave can produce effects within individual particle volumes which have been described here in terms of metallurgical effects. These effects most certainly must include thermal effects not only associated with the shock transient heating, but also with the frictional contact which characterizes the compaction and consolidation process and on the high-velocity deformation taking place in the external layers of the particles. As a consequence, the properties of explosively consolidated products will depend to a large extent upon this process synergism. To some extent, these consolidated masses will function as complex composites of phases having different stoichiometry, crystallography, and microstructure which can be altered by altering the shock or pressure-wave parameters (peak pressure and pulse duration for example).

ACKNOWLEDGEMENTS

Many of the observations presented here, and the research described in connection with our own explosive shock loading programs over the past decade and more, have been supported by the Department of Energy through contracts with Sandia Laboratories (L.E.M.), and by the National Science Foundation and the Instituto Militar de Engenharia (M.A.M.). Portions of this work were also supported by the Research and Development Division of New Mexico Tech.

REFERENCES

1. ARGON, A. S., ed. *Constitutive Equations in Plasticity*, MIT Press, Cambridge, Mass, 1975.
2. HART, E. W. *Trans. A.S.M.E.*, **98** (1976), 193.
3. SWEARENGEN, J. C., ROHDE, R. W., and HICKS, D. L. *Acta Met.*, **24** (1976), 969.
4. CAMPBELL, J. D. *Dynamic Plasticity of Metals*, Springer-Verlag, Vienna Austria, 1972.
5. MAIDEN, C. J., and GREEN, S. V., *Trans. A.S.M.E., J. Appl. Mech.*, **33** (1966), 496.
6. LINDHOLM, U. S. In *Techniques of Metals Research*, Vol. 5, ed. R. F. Bunshah, Interscience, NY, 1974, p. 199.
7. LINDHOLM, U. S., and YEARLEY, J. *Exptl. Mech.*, **7** (1967), 1.
8. CULVER, R. S., *Exptl. Mech.*, **12** (1972).
9. CULVER, R. S. In *Metallurgical Effects at High Strain Rates*, Eds. R. W. Rohde, B. M. Butcher, J. R. Holland, and C. H. Karnes, Plenum, N. Y., 1973, p. 519.
10. WARNES, R. H. DUFFEY, T. A. KARPP, R. R., and CARDEN, A. E., in *Shock Waves and High-Rate Phenomena in Metals: Concepts and Applications*, eds. M. A. Meyers and L. E. Murr, Plenum, N. Y., 1981, p. 23.

11. KOLSKY, H. *Proc. Phys. Soc. (London)* **B62** (1949), 676.
12. WASLEY, R. J. *Stress Wave Propagation in Solids*, M. Dekker, New York, 1973.
13. ORAVA, R. N., and WITTMAN, R. H. In *Advances in Deformation Processing*, eds. V. Weiss and J. Burke, Plenum, N.Y. 1978.
14. ORAVA, R. N. In *Proceedings of the First International Conference of the Center for High Energy Forming*, ed. A. A. Ezra, U. of Denver, Colo., 1978, p. 7.5.1.
15. ZENER, C., and HOLLOMON, J. H. *J. Appl. Phys.*, **15** (1944), 22.
16. OLSON, G. B., MESCALL, J. F., and AZRIN, M. In *Shock Waves and High-Strain-Rate Phenomena in Metals: Concepts and Applications*, eds. M. A. Meyers and L. E. Murr, Plenum, N. Y. 1981, Chapter 14, p. 221.
17. SHOCKEY, D. A., and ERLICH, D. C. Source cited in ref. 16, chapter 15, p. 249.
18. DORAIVELU, S. M., GOPINATHAN, V., and VENKATESH, V. C. Source cited in ref. 16, chapter 16, p. 263.
19. ROGERS, H. C., and SHASTRY, C. V. Sources cited in ref. 16, Chapter 18, p. 285.
20. MOSS, G. L. Source cited in ref. 16, chapter 19, p. 299.
21. National Materials Advisory Board, *Materials Response to Ultra-High Loading Rates*. Report NMAB 356, 1980, Chapter 8.
22. WALSH, J. M., and CHRISTIAN, R. H. *Phys. Rev.* **97** (1955), 1544.
23. GRACE, F. I. *J. Appl. Phys.*, **40** (1969), 2649.
24. HSU, C. Y., HSU, K. C., MURR, L. E., and MEYERS, M. A. Source cited in ref. 16, p. 433.
25. MEYERS, M. A., and MURR, L. E. (eds.) *Shock Waves and High-Strain-Rate Phenomena in Metals: Concepts and Applications*, 1981, Plenum Press, New York.
26. GRAHAM, R. A. In *Shock Waves and High-Strain-Rate Phenomena in Metals: Concepts and Applications*, Chap. 23, Meyers, M. A. and Murr, L. E. (eds) 1981, Plenum Press, New York.
27. MURR, L. E. In *Shock Waves and High-Strain-Rate Phenomena in Metals: Concepts and Applications*, Chap. 37, Meyers, M. A., and Murr, L. E. (eds.) 1981, Plenum Press, New York.
28. MOIN, E., and MURR, L. E. *Mater. Sci. Engr.*, **37** (1979), 249.
29. GREULICH, F., and MURR, L. E. *Mater. Sci. Engr.*, **39** (1979), 81.
30. KESTENBACH, H.-J., and MEYERS, M. A. *Met. Trans.*, **7A** (1976).
31. WONGWIWAT, K., and MURR, L. E. *Mater. Sci., Engr.*, **35** (1978), 273.
32. MURR, L. E. *Interfacial Phenomena in Metals and Alloys*, 1975, Addison-Wesley, Reading, Mass.
33. MURR, L. E., and KUHLMANN-WILSDORF, D. *Acta Met.*, **26** (1978), 847.
34. RACK, H. J. *Met. Trans.*, **7A** (1976), 1571.
35. ASHBY, M. F. *Phil. Mag.*, **21** (1970), 399.
36. MURR, L. E. In *Shock Waves and High-Strain-Rate Phenomena in Metals: Concepts and Applications*, Chap. 42, Meyers, M. A. and Murr, L. E. (eds.), 1981, Plenum Press, New York.
37. MCQUEEN, R. G., and MARSH, S. P. *J. Appl. Phys.*, **31** (1960), 1253.

38. MURR, L. E., and ROSE, M. F. *Phil Mag.*, **18** (1968), 281.
39. MEYERS, M. A., and MURR, L. E. In *Shock Waves and High-Strain-Rate Phenomena in Metals: Concepts and Applications*, Chap. 30, Meyers, M. A., and Murr, L. E., (eds.), 1981, Plenum Press, New York.
40. KRESSEL, H., and BROWN, N. *Appl. Phys.*, **38** (1967), 1618.
41. MURR, L. E., INAL, O. T., and MORALES, A. A., *Acta Met.*, **24** (1976), 261.
42. MURR, L. E., INAL, O. T., and MORALES, A. A. *Appl. Letters*, **28** (1976), 432.
43. STEIN, C. *Scripta Met.*, **9** (1975), 67.
44. MURR, L. E., VYDYANATH, H. R., and FOLTZ, J. V. *Met. Trans.*, **A1** (1970), 3215.
45. MEYERS, M. A. Thermomechanical Processing of a Nickel-Base Superalloy by Cold-Rolling and Shock-Wave Deformation, Ph.D. Dissertation, 1974, Univ. of Denver, Denver, Colorado.
46. MEYERS, M. A. *Mater. Sci. Engr.*, **30** (1977), 99.
47. MURR, L. E., and FOLTZ, J. V., *J. Appl. Phys.*, **40** (1969), 3796.
48. DELAEY, L., *Zeitschriff Metallk.*, **63** (1972), 531.
49. ZACKAY, V. F., *Mater. Sci. Engr.*, **25** (1976), 247.
50. MEYERS, M. A., and ORAVA, R. N. In *Shock Waves and High-Strain-Rate Phenomena in Metals: Concepts and Applications*, Chap. 45, Meyers, M. A., and Murr, L. E. (eds.), 1981, Plenum Press, New York.
51. ORAVA, R. N. *Mater. Sci. Engr.*, **11** (1973), 177.
52. MEYERS, M. A., and ORAVA, R. N. *Met. Trans.*, **7A** (1976), 179.
53. GREENHUT, V. A., CHEN, M. G., BANKS, R., and GOLASKI, S. Long-range diffusion of vacancies and substitutional atoms during high-strain-rate deformation of aluminum alloys, *Proc. ICM II*, Boston, Mass., 1975.
54. STAUDHAMMER, K. P. and MURR, L. E. *Mater. Sci. Engr.*, **44** (1980), 97.
55. STAUDHAMMER, K. P., FRANTZ, C. E., HECKER, S. S., and MURR, L. E. In *Shock Waves and High-Strain-Rate Phenomena in Metals: Concepts and Applications*, Chap. 7, Meyers, M. A. and Murr, L. E. (eds.), 1981, Plenum Press, New York.
56. OLSON, G. B., and COHEN, M. *J. Less Common Metals*, **28** (1972), 107.
57. KAZMI, B., and MURR, L. E. In *Shock Waves and High-Strain-Rate Phenomena in Metals: Concepts and Applications*, Chap. 41, Meyers, M. A., and Murr, L. E. (eds.), Plenum Press, New York.
58. STAUDHAMMER, K. P. Effect of Shock Stress Amplitude, Shock Stress Duration, and Prior Deformation on the Residual Microstructure of Explosively Deformed Stainless Steels, Ph.D. Dissertation, 1975, New Mexico Institute of Mining and Technology, Socorro, New Mexico 87801.
59. MEYERS, M. A. *Met. Trans.*, **8A** (1977), 1581.
60. MCELROY, R. J., and SZKOPIAK, F. C. *Intl. Metall. Reviews*, **17** (1972), 174.
61. HSU, K. C., M.Sc. Thesis, 1981, New Mexico Institute of Mining and Technology, Socorro, New Mexico 87801.
62. LONGO, W. P., and REED-HILL, R. E. *Metallography*, **4** (1974), 181.
63. KUHLMANN-WILSDORF, D., and LAIRD, C. *Mater. Sci. Eng.*, **27** (1977), 137.
64. DUVALL, G. E., and GRAHAM, R. A. *Reviews of Modern Phys.*, **49** (1977) 523.
65. BARKERS, L., and HOLLENBACH, R. E. *JAP*, **45** (1974), 4872.
66. ZUKAS, E. G. *Metals Eng. Quart.*, **6** (1966), 1, May.

67. DIETER, G. E. In *Response of Metals to High Velocity Deformation*, Shewmon, P. G., and Zackay, V. F., (eds.) Interscience, New York, 1961, p. 419.
68. STONE, G. A., ORAVA, R. N., GRAY, G. T. and PELTON, A. R. An Investigation of the Influence of Shock-Wave Profile on the Mechanical and Thermal Responses of Polycrystalline Iron, U.S. Army Research Office Final Report, Grant No. DAAG29-76-G-0180, September 1978, *Report No. SMT-1-78*.
69. MEYERS, M. A., SARZETO, C. and HSU, C. Y. *Met. Trans.*, **11A** (1980), 1737.
70. MEYERS, M. A., GUPTA, B. B., and MURR, L. E. *J. Metals*, **33** (1981), 21.
71. HIGGINS, G. T. *Met. Trans.*, **2** (1971), 1277.
72. TRUEB, L. F. *JAP*, **40** (1967), 2976.
73. DHERE, A. G., KESTENBACH, H.-J., and MEYERS, M. A. *Mater. Sci. Engr.*, **54** (1982) 113.

Chapter 4

HIGH-RATE STRAINING AND MECHANICAL PROPERTIES OF MATERIALS

J. HARDING

Department of Engineering Science,
University of Oxford, UK

4.1. INTRODUCTION

In many conventional metal working or forming processes, such as machining, rolling or wire drawing, very large plastic strains may be applied to the workpiece in very short intervals of time, of the order of 10^{-4} s. Estimates of the resulting strain rate vary from 10^2 to 10^4 s^{-1},[1] and may be as high as 10^5 s^{-1},[2] i.e. some 6 to 9 orders of magnitude greater than the strain rates encountered in conventional tension, compression or torsion testing. The development of high energy rate fabrication methods, including the use of explosives, has meant that even such processes as sheet or plate forming can now be carried out at relatively high strain rates, in the range 10 to 10^3 s^{-1},[3] while in processes where the explosive is in direct contact with the workpiece the pressures generated may be so great that the metal behaves as if it were a fluid, i.e. its response is characterised by a viscosity rather than a flow stress.[4]

It is now well established that many metals and alloys show a significant increase in flow stress with increase in strain rate[5] and there is considerable evidence to suggest that this rate dependence becomes even more marked at strain rates greater than about 10^3 s^{-1},[6,7] i.e. at rates attained in many high energy fabrication processes. If such processes are to be understood, therefore, and the plastic deformation involved is to be analysed in more than a superficial manner, it is clear that account must be taken of the effect of strain rate on the mechanical response of the workpiece material.

Over the last two to three decades extensive studies have been made of the mechanical behaviour of metals and alloys when deformed plastically at high rates of strain. This chapter describes some of the experimental techniques which have been employed and attempts to review the various types of behaviour which have been observed. In seeking to relate such data to high-rate fabrication processes it should be noted that explosive forming is normally a 'stand-off' operation, i.e. an operation where the explosive shock is transmitted to the workpiece through an intermediate (transfer) medium, e.g. water, leading to strain rates typically of the order of $100\,\text{s}^{-1}$ whereas explosive welding and cladding are 'contact' explosive operations where the strain rates are several orders of magnitude greater than this. Strain rate effects have also been observed in dynamic powder compaction processes[8] where, under explosive loading, volumetric compaction rates of the order of $10^3\,\text{s}^{-1}$ have been achieved. The effect of strain rate on the mechanical behaviour of metal powders has received little attention but some evidence has been obtained[9] which shows a very strong rate dependence of the shear flow stress for copper powder.

4.2. TESTING TECHNIQUES AT HIGH RATES OF STRAIN

Some of the earliest work on the impact loading of materials was that due to J. Hopkinson[10] and his son, B. Hopkinson.[11] They showed that steel wires could withstand transient stresses as high as twice the static yield stress without permanent deformation. Later G. I. Taylor[12] and von Kármán and Duwez[13] independently conducted plastic wave propagation experiments and developed a theory of material response based on the assumption of a unique stress–strain curve,

$$\sigma = f(\varepsilon), \tag{4.1}$$

i.e. on the assumption of *rate-independent* material behaviour. According to this theory the velocity of longitudinal plastic waves, c_p, is given by

$$c_p = \sqrt{((1/\rho)(d\sigma/d\varepsilon))} \tag{4.2}$$

where ρ is the density of the material in which the wave is propagating.

The corresponding equation for the velocity of longitudinal elastic waves,

$$c_0 = \sqrt{(E/\rho)}, \tag{4.3}$$

relates to the special case where $d\sigma/d\varepsilon = E$.

Subsequently Bell[14] and Sternglass and Stuart[15] demonstrated that incremental loading waves, superimposed on specimens already strained into the plastic region, travelled at a velocity approximately equal to that for elastic waves, c_0, rather than that for plastic waves, c_p, where $c_0 \gg c_p$. This result, which was later confirmed by Campbell and Dowling[16] for torsional waves, raises doubts about the validity of using a rate-independent equation. Efron and Malvern,[17] however, were able to show that eqn. (4.1) still gave an adequate description of the observed wave propagation provided a raised, or dynamic, stress–strain curve was used in the determination of c_p. In effect, a mechanical equation of state involving stress, strain and plastic strain rate, of the form

$$\sigma = f(\varepsilon, \dot{\varepsilon}) \tag{4.4}$$

is assumed, with the additional assumption that during the wave propagation experiment the strain rate may be taken as constant but of a significantly higher value than the quasi-static. Nevertheless, it is clear that wave propagation experiments only provide an indirect indication of the effect of strain rate on the material mechanical response.

An alternative approach may be followed in which wave effects are eliminated by performing experiments under high-rate steady-state plastic flow conditions as found, for example, in such metal working processes as machining. However, strain and strain rate variations across the workpiece can be very large and temperature can rise by several hundred degrees so that it remains very difficult to deduce the mechanical response of the workpiece material from such tests. Nevertheless, Stevenson and Oxley[18] have successfully used a machining test to estimate material response at strain rates from 10^3 to $10^5\,\text{s}^{-1}$, obtaining results consistent with those from a more direct type of experiment to be described in a later section.

In general, therefore, to determine the mechanical behaviour of materials at high rates of strain it is necessary to use test techniques which have been specifically designed for the purpose. These techniques may be conveniently divided into two categories, those for testing at intermediate (or rapid) strain rates, where it is not necessary to take account of wave propagation effects in the testing machine, see Section 4.2.1, and those for testing at impact rates of strain, see Section 4.2.2, where careful design to either eliminate or allow for wave propagation effects in the testing machine is required. Some attempts to perform tests at even higher strain rates will be described in Section 4.2.3.

4.2.1. Testing Techniques at Intermediate Rates of Strain

Intermediate strain rates may be considered to fall in the approximate range from 1 to $100\,\text{s}^{-1}$. At $100\,\text{s}^{-1}$ a strain of 0·5 is attained in a time of 5 ms. This is likely, therefore, to be the lower bound for the duration of such a test. The longitudinal elastic wave velocity is typically of the order of 5000 m/s so that an elastic wavefront will travel some 25 m in a time of 5 ms. In a load cell of length 0·25 m there will be at least 100 transits of the wavefront during the course of such a test, sufficient for it to be assumed that stress equilibrium is established in the testing machine as well as in the specimen so that wave propagation effects need not be taken into account.

Intermediate rate testing machines may operate either under direct mechanical loading or by means of fluid pressure, i.e. the load is applied pneumatically or hydraulically. Mechanical testing techniques subdivide into those which involve direct physical impact and those which do not. For reasons of experimental simplicity one of the most popular forms of intermediate strain rate test has used the Charpy or Izod pendulum loading system, the hammer being constrained to impact an anvil at the free end of a short tensile specimen.[19] Provided that only low impact velocities are employed, so that the rise time of the loading wave is no faster than about 1 ms, oscillatory signals arising from elastic wave propagation in the dynamometer should not be encountered.[20] There is, however, a constant temptation to increase the impact velocity in an effort to achieve higher effective strain rates while ignoring oscillatory effects due to wave propagation.

Other workers[21,22] have replaced the pendulum by a rotating flywheel. This is accelerated to the required angular velocity and then spring-loaded ratchets on the periphery are released and brought into contact with an anvil on the free end of the specimen. Although strain rates as high as $10^3\,\text{s}^{-1}$ have been claimed,[22] as noted above this was at the price of ignoring wave propagation effects.

A flywheel was also used in the high-speed torsion test developed by Vinh et al.[23] This is effectively an impact test, in that a sliding dog clutch between the flywheel and the specimen is suddenly engaged to initiate the test, and shear strain rates up to $1050\,\text{s}^{-1}$ are again claimed. However, wave propagation effects do not appear to have been taken into account and no experimental torque–time records are presented. Direct compression–impact tests using a drop-hammer have been employed by Suzuki et al.[24] and by Samanta[25] at strain rates up to about $600\,\text{s}^{-1}$, again ignoring wave propagation effects.

Of the techniques not involving direct physical impact, and hence limited to speeds where the problems associated with wave propagation are normally avoided, probably the cam plastometer[26] has been most widely used. This also utilises the energy stored in a flywheel to which a cam with a logarithmic profile is attached. When the flywheel is at speed the cam is brought into contact with a platen supporting the short compression specimen and, provided the flywheel has sufficient energy, the specimen is compressed between the platen and the load cell at a nearly constant true strain rate which may be as high as $200\,\text{s}^{-1}$.[27] This technique has been extensively used to determine the high strain compressive strength of many metals and alloys at intermediate strain rates and over a wide range of temperature in deformation processing studies.[28,29] Similar tests have been performed in torsion, also using a flywheel to store and transmit the energy, by Bailey et al.[30] They obtained shear stress–strain curves for aluminium and two aluminium alloys at constant effective strain rates from 10^{-2} to $100\,\text{s}^{-1}$ at temperatures from 200 to 600°C.

Perhaps the most popular of the intermediate strain rate machines are those in which the load is applied pneumatically or hydraulically. Early examples of this technique were the designs of Clark and Wood[31] and Campbell and Marsh.[32] The flexibility of the loading system is well illustrated by the variety of machines developed since then. These include a universal tension/compression machine,[33] a machine in which the uniaxial strain rate could be suddenly increased during the course of the test,[34] used in studies of strain rate history effects, and two biaxial loading tension/torsion machines.[35,36]

The principle of operation is illustrated in Fig. 4.1, a schematic view of the machine developed by Cooper and Campbell. The specimen S is inserted between the fixed crosshead C, to which is attached a set of strain gauges, and the moving crosshead Y, which is integral with the piston P in the smaller cylinder. Valves V1, V2 and V3 are opened and the oil pressure in the reservoir R and above and below piston P, i.e. in regions U and L, is raised to the required level. For a tension test the oil pressure is suddenly released from region L through valve V6 and the dump valve D. The strain rate is determined by the variable orifice size in valve V6. For a compression test valve V5 is used and the oil pressure

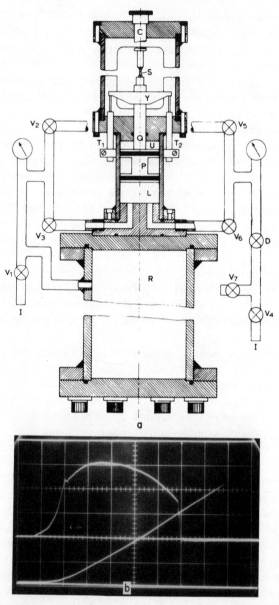

Fig. 4.1 Intermediate-rate hydraulically-operated testing machine. (a) schematic arrangement; (b) typical oscilloscope traces for tensile test on steel specimen (time base, 1 ms/cm; mean strain rate, $50\,s^{-1}$).

released from region U. Linear variable differential transformers (LVDTs) T1 and T2 measure the crosshead velocity and their integrated output gives the crosshead displacement. In a more recent modification[34] a second dump valve and variable orifice valve, in parallel with D and either V5 or V6, allows a sudden increase in strain rate to be applied during the course of the test by the delayed operation of the second dump valve.

Shown in Fig. 4.1(b) is a typical set of oscilloscope traces for a tensile test on an alloy steel specimen at 250 °C. The upper trace follows the variation of load with time and the lower trace that of displacement with time. The sweep speed was 1 ms/cm and the average strain rate $50 \, s^{-1}$. Using a transient recorder the load and displacement signals may be displayed directly as a load–displacement curve on an X–Y recorder.

A schematic view of a hydraulically-operated biaxial tension/torsion machine[36] is shown in Fig. 4.2. The specimen is fixed between the tension load cell, attached to the tension actuator, and the torsion load cell, attached to the torsion actuator. The two actuators operate independently through individual variable orifice valves and strain rates in each mode, depending on specimen dimensions, from 0.01 to $100 \, s^{-1}$ may be applied. Load and displacement traces and torque and rotation

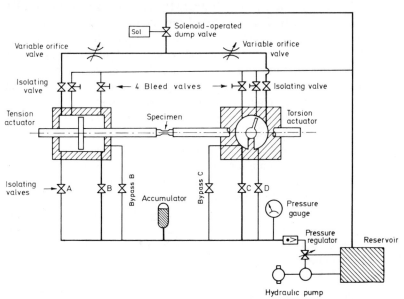

FIG. 4.2 Schematic arrangement of intermediate rate biaxial testing machine.

traces, the latter obtained from an angular velocity transducer specifically developed for the purpose,[37] are displayed separately on two double-beam oscilloscopes. A proportional straining path is followed for various combinations of tension and torsion and effective stress–strain curves may be derived and the dependence of the yield locus on strain rate determined.

4.2.2. Testing Techniques at Impact Rates of Strain

At strain rates of the order of $1000\,\text{s}^{-1}$, where a strain of 0.5 is attained in only 500 µs, elastic wave propagation in the testing machine becomes important. Wave propagation effects may also lead to a non-uniform stress state in the specimen if its dimensions are too large. Testing techniques have to be devised, therefore, which either eliminate these effects or make it possible for them to be taken into account.

In the first of these approaches a number of investigators have used an expanding ring technique. Dynamic expansion is achieved either by detonation of an explosive charge on the axis of the test ring,[38] the resulting pressure being transmitted through an intermediate driver ring, or by electromagnetic repulsion[39] resulting from the discharge of a condenser bank through a suitably designed coil, the same principle as is used in electromagnetic forming. Initially the test ring is accelerated to a high radial velocity. It then separates from the driver ring and is left in free flight during which it decelerates entirely under the action of the internal hoop stress. Analysis of the test requires a double-differentiation of the radial displacement–time record in order to determine the stress, an intrinsically inaccurate process. In a recent improvement of the technique[40] the radial velocity, rather than displacement, is measured, so that stress is determined by means of a single differentiation and displacement by a single integration, thereby considerably improving the experimental accuracy. Using this technique a strain rate of $23\,000\,\text{s}^{-1}$ [41] was achieved in tests on aluminium alloys. However, the other principal disadvantages of the technique remain, namely that no meaningful data can be obtained until after the test ring separates from the driver ring, which at the higher strain rates may be at a significant plastic strain, that following separation the strain rate decreases very rapidly with increasing strain and that considerable care is required to obtain a rotationally symmetric loading.

Because of these difficulties most workers have preferred to opt for the second approach, to recognise the presence of stress waves and to design the test so as to take them into account. This approach was pioneered by

Kolsky[42] in 1949. He devised the first split Hopkinson's pressure bar (SHPB or Kolsky bar), a technique for determining the impact properties of materials which is now widely accepted and used very much as a standard. Because of its importance this technique will be described in some detail. It involves the use of stress waves to apply the load to the specimen. Several methods have been used to generate these stress waves. The most common is by direct physical impact[43] but other methods include the rapid release of a stored elastic strain energy,[44] the detonation of an explosive charge[45] or the discharge of a condenser bank.[46]

In its earliest, and simplest, form the SHPB was used to perform dynamic compression tests.[47] The specimen is aligned between two long thin coaxial loading bars, as shown schematically in Fig. 4.3(c). A calibration (elastic) test is first performed with the specimen removed, as shown in Fig. 4.3(a). The corresponding Lagrange (wave propagation) diagram, also illustrated in Fig. 4.3(a), shows the advancing (compressive) elastic wavefront, $dx/dt = c_0$, which is reflected at the free end of the output bar as a receding (unloading) wavefront, $dx/dt = -c_0$. Assuming ideal alignment at the interface between the two bars the signals from the strain gauges on each bar should be of identical magnitude, equal to a strain of V/c_0 or a stress of $\rho c_0 V$, where V is the impact velocity. The duration of the input bar signal, $t_1 = 2l_1/c_0$, should exceed that of the output bar, $t_2 = 2l_2/c_0$, where l_1 and l_2 are defined in Fig. 4.3. Typical oscilloscope traces for such a test are shown in Fig. 4.3(b), the upper trace deriving from the input bar gauges and the lower trace from those on the output bar. Timing pips are at $5 \mu s$ intervals.

When a specimen, of softer material than the bars, is inserted between them, see Fig. 4.3(c), the incident wave is partially reflected at the input bar/specimen interface and partially transmitted through the specimen. The transmitted wave, which is recorded on the output bar strain gauges, is proportional to the load supported by the specimen. The reflected wave travels back up the input bar as an unloading wave, modifying the signal recorded on the input bar gauges. Stress reflections occur between the specimen end faces. Specimen dimensions, therefore, must be small so as to allow many such reflections and the early attainment of stress equilibrium across the specimen.

Typical oscilloscope traces (for a test at an impact velocity of 5m/s on a glass reinforced plastic specimen) are shown in Fig. 4.3(d). The upper trace, obtained from the input bar strain gauges, initially follows the elastic response for an impact at 5m/s until unloaded by the returning

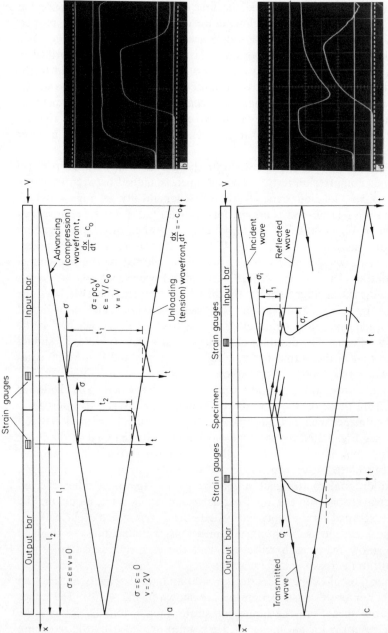

FIG. 4.3 Impact compression testing with the SHPB. (a) Test arrangement and Lagrange diagram for elastic test; (b) typical oscilloscope traces for elastic test (timing pips at 5 μs intervals); (c) test arrangement and Lagrange diagram for specimen test; (d) typical oscilloscope traces for specimen test (timing pips at 5 μs intervals).

reflected wave after time T_1, the time for an elastic wave to travel from the input bar strain gauges to the specimen interface and back. Once stress equilibrium across the specimen is achieved the stress on the specimen/output bar interface determined from the transmitted stress wave, and the stress on the specimen/input bar interface determined from the elastic and input bar strain gauge signals, should be the same. Figure 4.4 shows computer plots of oscilloscope traces for a test similar to

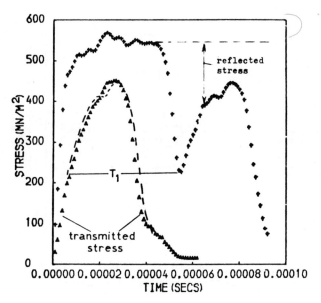

FIG. 4.4 Computer plots of oscilloscope traces showing stress difference across specimen.

that of Fig. 4.3(d). A good correlation is apparent between the stress–time signals for the two faces of the specimen. From traces such as those of Fig. 4.4 the variation of stress and particle velocity with time at each end of the specimen may be determined and the dynamic stress–strain curve obtained.

The SHPB apparatus has been modified in a variety of ways to allow testing under different stress systems. Probably the two most important of these are the tensile and the torsional loading versions. Lindholm and Yeakley[48] adapted the compression SHPB to perform tensile tests using a hollow cylindrical output bar and a complicated 'top-hat' design of specimen divided into four equal tensile segments, see Fig. 4.5(a). More recently Nicholas[49] has developed an alternative modification to the compression SHPB, see Fig. 4.5(b), involving a split shoulder, or collar, around the specimen which carries, in effect, the entire compression pulse but is unable to support any of the reflected tensile loading wave. Strain

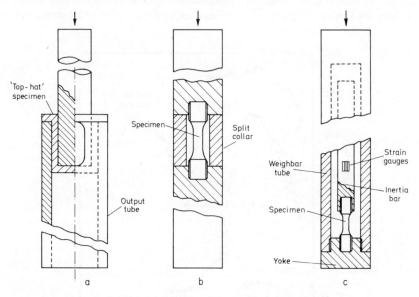

FIG. 4.5 Tensile versions of SHPB. (a) 'top-hat' specimen;[48] (b) split-collar design;[49] (c) tubular weighbar design.[51]

rates up to about $1000 \, s^{-1}$ were achieved but the magnitude of the initial compressive impact is limited by the need to prevent the specimen from yielding first in compression. Difficulties are also experienced in eliminating uneven loading of the specimen due to play in the threaded joints under first compressive and subsequently tensile loading. Direct tensile loading in the SHPB has been achieved both by using a magnetic loading technique,[46] limited in that particular design to only small specimens, and by the use of explosives,[50] where the application was to the determination of dynamic fracture toughness.

Probably the most successful tensile version of the SHPB, however, is that using a tubular input bar (weighbar) within which the output bar (inertia bar) slides freely,[51] see Fig. 4.5(c). Careful alignment at the weighbar/yoke interface allows the *compressive* loading wave in the weighbar to be reflected as a *tensile* loading wave in the specimen and inertia bar without the development of bending stresses. To determine the input loading wave an elastic test at the same impact velocity is required. Stress equilibrium is assumed across the specimen. A typical oscilloscope trace for the transmitted stress wave, recorded on the inertia bar strain gauges, is shown in Fig. 4.6 for an impact at about 15 m/s on a

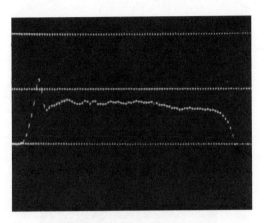

FIG. 4.6 Typical oscilloscope trace for impact tension test using tubular weighbar SHPB (Pure iron; timing pips at 2 μs intervals).

pure iron specimen.[52] The trace, which is proportional to the stress supported by the specimen, shows a large drop in load at yield, very little work-hardening in the post-yield region and fracture after about 160 μs. The trace is interrupted at 2 μs intervals.

A torsional version of the SHPB was developed by Baker and Yew[53] and Campbell and Lewis.[54] The general arrangement is shown in Fig. 4.7. The input bar is clamped at a point near the strain gauges and a static torque T_0 is applied at the input end. The test is initiated by the fracture of a brittle bolt and the sudden release of the clamp, causing a loading wave, $T_1 = T_0/2$, to travel towards the specimen and an unloading wave, $-T_1$, to travel towards the input end. The transmitted torque, T_2, is recorded on the output bar strain gauges.

Typical oscilloscope traces for a test on a copper specimen at room temperature at a strain rate of about $900 \, \text{s}^{-1}$ are shown in Fig. 4.7(b). The upper trace follows the transmitted torque, proportional to the shear stress supported by the specimen, while the lower trace follows the input torque T_1 and its subsequent unloading by the wave reflected from the

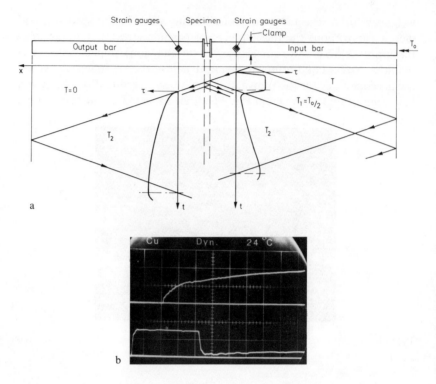

FIG. 4.7 Torsional version of SHPB. (a) test arrangement and Lagrange diagram; (b) typical oscilloscope traces for a specimen test.

specimen. The principal difficulty of the 'stored torque' technique lies in the design of a clamp which will release rapidly without introducing either axial or bending waves. The main advantages are the non-dispersive nature of torsional waves, allowing, in principle, faster rise times than with axial loading, and the much smaller effect of stress concentrations at the shoulders of the torsional specimen, allowing very short gauge lengths to be used and hence the early attainment of stress equilibrium across the specimen. It is also possible to reach significantly higher strains when testing in pure shear, an aspect of the technique which has been exploited in the determination of stress–strain data applicable to metal-working processes.[55]

In an interesting modification to the torsional SHPB Clyens and Johnson[9] devised a technique for determining the effect of shearing rate on the dynamic flow stress of powdered materials. Shear strain rates of about $2500\,\text{s}^{-1}$ were achieved, of the same order as the volumetric compaction rates in explosive powder compaction processes. A very significant raising of the flow stress at these rates over that observed in quasi-static tests was reported.

4.2.3. Attempts to Reach Higher Strain Rates

The maximum strain rate which may be reached in tests of the type described so far is of the order of $3000\,\text{s}^{-1}$. As discussed in the introduction, however, strain rates one or two orders of magnitude greater than this may be expected in some explosive metal-working processes. Attempts have been made, therefore, to modify the SHPB for use at strain rates in this range. The most common approach involves a significant reduction in the effective gauge length of the specimen. This is achieved in both the punch-loading,[7] see Fig. 4.8(a), and the double-

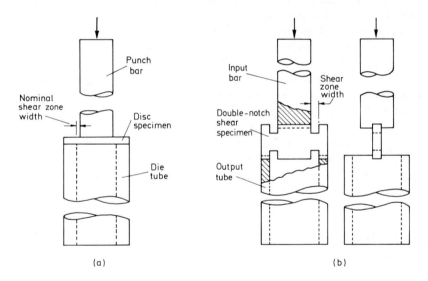

FIG. 4.8 Shear loading versions of SHPB. (a) punching test; (b) double-notch shear test.

notch shear loading,[6] see Fig. 4.8(b), versions of the SHPB. In both the output bar is replaced by a tube, the specimen being either a flat plate in which a circular hole is punched or a thin strip into which two pairs of notches have been cut. The two tests differ in that the effective shear zone width in the punch test is considerably greater than the clearance between the punch bar and the die tube, about 0.025 mm, and cannot therefore be clearly defined, so only approximate estimates of shear strain and shear strain rate may be made. In the double-notch shear test, however, the notch width, 0.84 mm, defines the size of the shear zone (i.e. the effective gauge length) and more accurate estimates of shear strain rate may be made. In both tests strain rates in excess of $10^4 \, \text{s}^{-1}$ have been reported, with rates as high as $40\,000 \, \text{s}^{-1}$ in the double-notch shear test. The principal disadvantage of the double-notch shear specimen is that at shear strains greater than about 20% it ceases to deform in pure shear.[56] Since some 20 µs is required before a relatively constant strain rate is established, reliable data are only obtained over a limited range of shear strains, from ~ 10 to $\sim 20\%$.

In an alternative approach to the problem of raising the maximum strain rate attainable in the SHPB several investigators have developed miniaturised versions of the apparatus. Lindholm,[57] for example, was able to approach compressive strain rates of $10^5 \, \text{s}^{-1}$ by using ultra high-strength loading bars, allowing a tenfold increase in particle velocity (and hence in impact velocity) and by reducing the specimen dimensions, to 4 mm diameter by 4 mm long. To minimise dispersion of the large amplitude stress pulses it was also necessary to reduce the diameter of the loading bars to 4.7 mm. Otherwise the test procedure was similar to that for a standard SHPB test.

More recently Gorham[58] has developed a modified Hopkinson bar system in which a projectile, accelerated by a gas gun, impacts directly on a miniature specimen, 1 mm in diameter by 0.46 mm long, at velocities up to 100 m/s. Strain is monitored with a high-speed camera and the stress is determined from strain gauges on a miniature output bar, 3 mm in diameter by about 150 mm long. A Fourier transform technique is used to account for, as far as possible, the effects of dispersion on the rising characteristic of the stress wave in the output bar. The apparatus was used to test strong metals at strain rates of the order of $50\,000 \, \text{s}^{-1}$.

A very different technique, which has the potential for reaching even higher strain rates under controlled conditions, is based on the high velocity impact of flat parallel plates. High intensity shock waves are generated in the impacting plates and a uniaxial state of strain is developed, for which the governing wave theory is one-dimensional. These shock waves are often characterised by an elastic precursor, followed by a plastic shock which travels at a speed dependent on the impact velocity. The amplitude of this elastic precursor, which may greatly exceed the quasi-static elastic limit of the material, decays as the wave propagates, the rate of decay being related to the plastic strain rate. Measurements are made of the shock wave transit time through the target plate and the free surface velocity of its back face. The measured wave profile, i.e. the variation of the free surface particle velocity with time, is compared with that predicted theoretically based on an assumed constitutive relationship and hence the validity of the assumed constitutive relationship is assessed. This technique suffers the disadvantage discussed in Section 4.1, namely that it cannot test the fundamental assumptions on which the particular form of constitutive relationship is based but merely assesses which version of this form is most appropriate. Nevertheless it has the potential for reaching higher strain rates than can be attained by any other method so far discussed. Two versions of plate impact test for determining material properties have been reported. The first is a direct uniaxial strain version of the SHPB[59] where a thin plate specimen is compressed between two hard, high-impedance elastic plates giving strain rates of the order of 10^5 s^{-1}. Experimental difficulties with this arrangement have led to the development of an alternative version, a 'pressure-shear' test[59,60] in which the impacting plates, while remaining parallel, are inclined to the direction of impact. In the pressure shear test a shear stress-strain curve at plastic strain rates of $\sim 10^5$ s^{-1} is obtained directly from measurements of the normal and transverse components of particle velocity at the back surface of the target plate.

4.3. MECHANICAL PROPERTIES OF MATERIALS AT HIGH RATES OF STRAIN

Extensive reviews of data available at the time on the rate-sensitivity of a wide range of metals and alloys have been published by Lindholm and Bessey,[61] in 1969, Eleiche,[62] in 1972, and Campbell,[5] in 1973. In this present section we only have space to highlight the main features and to include some reference to more recent work. It is clear that most metals and alloys exhibit some increase in strength with increasing strain rate. Typical stress–strain curves are shown in Fig. 4.9 for fcc aluminium[63] and cph titanium,[64] obtained in the torsional SHPB apparatus, and for bcc En2A mild steel[65] and orthorhombic α-uranium,[66] obtained in the tensile SHPB apparatus, at strain rates from $\sim 10^{-4}$ to ~ 2000 s^{-1}. For each of these widely differing materials a significant raising of the

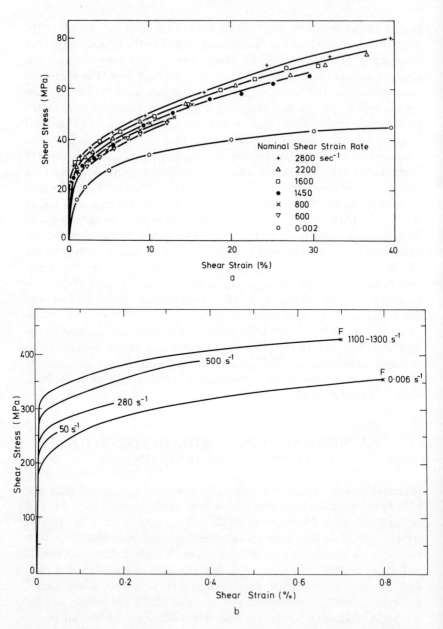

FIG. 4.9 Effect of strain rate on stress–strain curves. (a) aluminium;[63] (b) titanium;[64] (c) mild steel;[65] (d) uranium.[66]

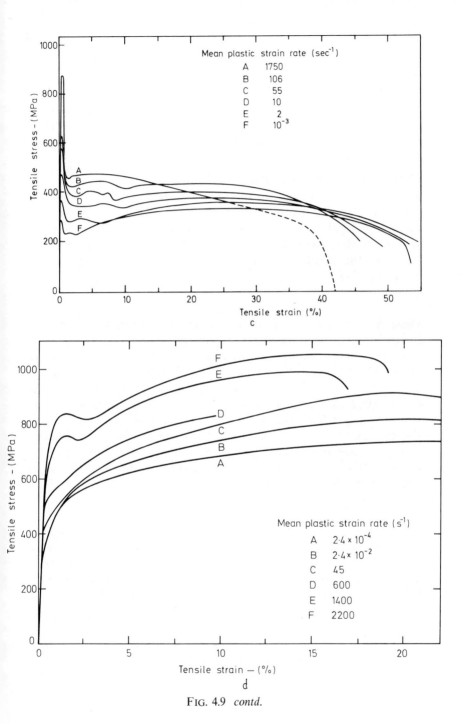

Fig. 4.9 contd.

stress–strain curve with increasing strain rate is apparent. The effect of strain rate on the upper yielding stress of the mild steel specimens is particularly marked while the development of a yield drop in the tensile impact test on α-uranium at the highest strain rates should also be noted. A similar effect was observed in tensile impact tests on α-titanium, particularly at low temperatures.[67]

4.3.1. Theoretical Considerations

It is now well established that thermally-activated processes control the yield and flow of many metals and alloys over a wide range of temperature and strain rate. Thermal energy is able to assist in overcoming short-range barriers to flow and the rate of straining will be related to the absolute temperature according to an Arrhenius type exponential law,

$$\dot{\varepsilon}_p = \dot{\varepsilon}_0 \exp[-\Delta G_{(\tau^*,T)}/kT] \qquad (4.5)$$

where ΔG, the free energy of activation, is a function of the local (thermal) stress τ^* and the absolute temperature T, and $\dot{\varepsilon}_0$ is a frequency factor (or nominal limiting strain rate) which depends on the mobile dislocation density, i.e. on the structural state of the material. A dependence on stress is implicit, both through the parameter $\Delta G_{(\tau^*,T)}$ and also, possibly, through a stress dependence of $\dot{\varepsilon}_0$. It is usual, however, to assume that $\dot{\varepsilon}_0$ is independent of strain rate and temperature (and hence of stress) and to take ΔG to be a unique function of stress. Assuming that a single thermally-activated mechanism controls flow and that the corresponding force–displacement relationship is rectangular, ΔG may be expressed as a linear function of stress of the form

$$\Delta G = \Delta G_0 - V(\tau - \tau_a) \qquad (4.6)$$

where ΔG_0 is the free energy of activation in the absence of stress, V the activation volume, τ the total applied stress and τ_a the athermal component of stress, i.e. that associated with long-range stress fields. Other barrier shapes have been considered by Davidson and Lindholm.[68] At high rates of strain eqns. (4.5) and (4.6) may be combined to give

$$\tau = \tau_a + \Delta G_0/V + (kT/V)\ln(\dot{\varepsilon}_p/\dot{\varepsilon}_0) \qquad (4.7)$$

Taking V to be independent of stress, as is implied by the rectangular barrier shape, and assuming that $\sigma = \sqrt{(3\tau)}$, a logarithmic strain-rate-sensitivity parameter, λ, may be defined as

$$\lambda = (\partial \sigma/\partial \log \dot{\varepsilon}_p)_{T,\varepsilon_p} \simeq 4kT/V \qquad (4.8)$$

4.3.2. Strain-rate Dependence of fcc Materials

Experimental results for the variation of flow stress with the logarithm of the plastic strain rate over a strain rate range from $\sim 10^{-4}\,\text{s}^{-1}$ to $> 1000\,\text{s}^{-1}$ in tests on four widely differing materials are shown in Fig. 4.10. For fcc 1100.0 aluminium,[69] see Fig. 4.10(a), at strain rates greater than about $10^{-3}\,\text{s}^{-1}$ a constant value for λ of about 5 MPa is obtained at constant temperature and plastic strain, as predicted by eqn. (4.8). A similar response is shown in tests on fcc copper and lead specimens.[69] There is a slight increase in λ, corresponding to a decrease in V, with increasing plastic strain. An athermal, i.e. rate independent, response is observed at strain rates below $10^{-3}\,\text{s}^{-1}$. The threshold strain rate, below which athermal behaviour is observed, varies with the material being tested[70] and for heat-treated aluminium alloys 6061–T6 and 7075–T6 essentially athermal behaviour is observed at strain rates up to $1000\,\text{s}^{-1}$. Equation (4.8) also predicts a linear dependence of λ on temperature. There is some evidence to support this prediction[71] but conflicting results have been obtained in other investigations.[5,72]

4.3.3. Strain-rate Dependence of hcp and Orthorhombic Materials

Less information is available on the response of hcp metals and alloys. Lawson and Nicholas,[73] in shear tests on α-titanium at strain rates from 10^{-4} to $5000\,\text{s}^{-1}$, report a constant value of λ at room temperature and 10% plastic strain of about 50 MPa, an order of magnitude greater than that for fcc materials. Other investigators have reported a more complex behaviour[64,67] with two regions of mechanical response, as shown in Fig. 4.10(b) where, at room temperature and 5% plastic strain, λ approximately equals 20 MPa at tensile strain rates below about $10\,\text{s}^{-1}$, rising to approximately 75 MPa at strain rates above about $100\,\text{s}^{-1}$. In the high rate-sensitivity region λ is relatively independent of temperature but at lower strain rates λ decreases with temperature, but in only an approximately linear manner. The corresponding activation volume is about $24\,b^3$, where b is the Burgers vector. Other hcp materials, for example beryllium[61] and Ti6A14V alloy,[74] do not show this increase in λ at strain rates above about $100\,\text{s}^{-1}$. However, in Ti6A14V alloy at temperatures above room temperature, there is some evidence[75] for a continuous increase in λ with strain rate at all rates above $10^{-4}\,\text{s}^{-1}$.

As an example of the very wide range of materials in which significant strain rate effects are observed results are also included for α-uranium[66] see Fig. 4.10(c), a material with a complex orthorhombic structure. As for α-titanium, two regions of rate-sensitivity are observed, correspond-

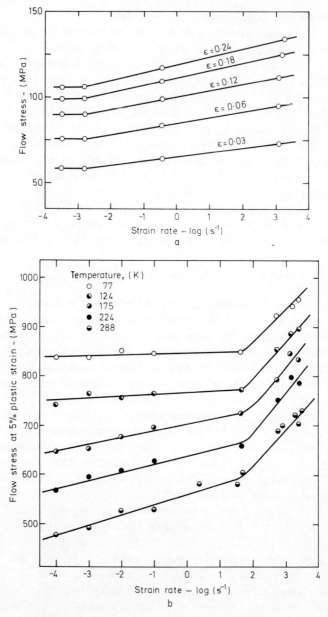

FIG. 4.10 Variation of flow stress with strain rate. (a) aluminium;[69] (b) titanium;[67] (c) uranium;[66] (d) alloy steel.[79]

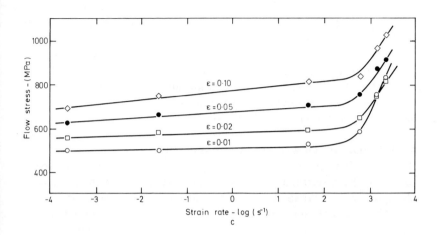

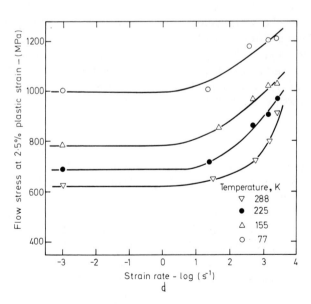

Fig. 4.10 contd.

ing to $\lambda \simeq 12$ MPa at strain rates below about $100 \, \text{s}^{-1}$ increasing to $\lambda \simeq 300$ MPa at a strain rate of $1000 \, \text{s}^{-1}$. However, in tests on other grades of α-uranium[76] this region of greatly increased rate-sensitivity was either absent or less marked, corresponding to λ values of ~ 160 MPa or ~ 80 MPa. It was concluded that at rates above about $100 \, \text{s}^{-1}$ the rate dependence of α-uranium is very sensitive to details of fabrication history, impurity content and heat treatment.

4.3.4. Strain-rate Dependence of bcc Materials

Because of their generally high rate-sensitivity bcc materials, and steels in particular, have received probably the most attention. In the early work of Manjoine,[77] who tested mild steel in tension at strain rates up to $\sim 1000 \, \text{s}^{-1}$, λ was found to increase continuously with strain rate to a maximum value in room temperature tests of ~ 90 MPa. Similar behaviour was observed in tensile tests on polycrystalline high-purity iron[52] where the rate dependence of the lower yield stress was found to be related almost entirely to the friction stress in the Petch equation. In these tests the maximum value of λ for the rate dependence of the friction stress was found to be ~ 100 MPa, again at a strain rate of the order of $1000 \, \text{s}^{-1}$. However, in other work on annealed high-purity iron[78] λ for the rate-dependence of both the proportional limit and the 1% flow stress was found to be nearly constant at a value of ~ 36 MPa over the entire range of strain rates from 10^{-4} to over $1000 \, \text{s}^{-1}$. Constant values of λ for the lower yield stress were also obtained in the thermally-activated region, i.e. region II of Fig. 4.11(b), in double-notch shear tests on mild steel[6] where, against the prediction of eqn. (4.8), λ was found to decrease with increasing temperature, from 61 MPa at 195 K to 30 MPa at 493 K.

For alloy steels in tension,[79] see Fig. 4.10(d), there is some evidence that the flow stress at room temperature and below is rate (but not temperature) independent at strain rates up to $\sim 10 \, \text{s}^{-1}$. Above $10 \, \text{s}^{-1}$ λ increases continuously, reaching, at a strain rate of $1000 \, \text{s}^{-1}$, values of ~ 170 MPa in room temperature tests and ~ 110 MPa in tests at 77 K. A similar response was observed in room temperature compression tests on 4340 steel.[80] In other steels, however, rate-dependent behaviour is found at all strain rates above $\sim 10^{-3} \, \text{s}^{-1}$.[81] In general, the higher the quasi-static strength of the steel the higher the threshold strain rate below which rate-independent behaviour is obtained. Above this threshold strain rate, however, a very rapid increase in rate-sensitivity may be observed irrespective of the static strength of the steel.

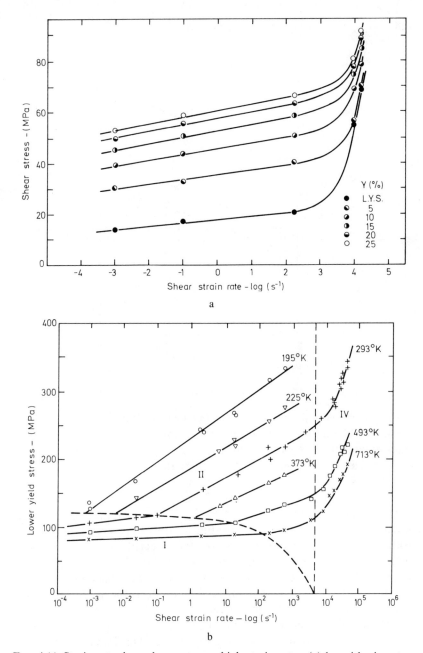

FIG. 4.11 Strain-rate dependence at very high strain rates. (a) logarithmic rate-sensitivity of aluminium;[7] (b) logarithmic rate-sensitivity of mild steel;[6] (c) (overleaf) linear rate-sensitivity of mild steels at strain rates $> 5000^{-1}$ s.[6]

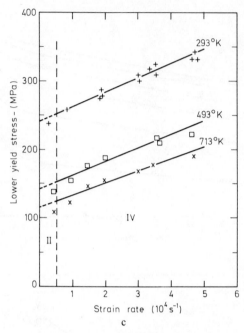

FIG. 4.11. contd.

4.3.5. Mechanical Response at Very High Strain Rates

Following the development of techniques for testing materials at strain rates greater than $\sim 5000\,\mathrm{s}^{-1}$ considerable evidence has accumulated to suggest that, at these very high strain rates, there is a much more marked increase of flow stress with strain rate. Behaviour of this type has been reported in single crystals of zinc,[82] aluminium[83] and copper,[84] in polycrystalline aluminium,[7,85,86,87] see Fig. 4.11(a), copper[7] and titanium[88] and in mild steel,[6,7] see Fig. 4.11(b), alloy steel[89] and brass.[7] Although the maximum values of the logarithmic rate-sensitivity parameter, λ, in Figs. 4.11(a) and 4.11(b), 100 MPa and 300 MPa respectively, are significantly greater than the maximum values previously quoted for aluminium and mild steel, the more important difference is that at shear strain rates greater than $\sim 10^4\,\mathrm{s}^{-1}$ a linear variation of flow stress with strain rate, rather than with the logarithm of the strain rate, is often observed, see Fig. 4.11(c). This implies that flow is viscous in nature. At room temperature the corresponding macroscopic viscosity coefficients, defined as $\eta = (\partial \tau / \partial \dot{\gamma})$, were found to fall in the range 1 to $10\,\mathrm{kNs/m^2}$,[5] although in some materials two regions of linear response have been observed, the macroscopic viscosity coefficients at the higher rates being $< 1 \text{ kNs/m}^2$.[88,89] It is of interest, therefore, to note that in an analysis of the explosive welding process Kowalick and Hay[90] related the periodic wave formation at the interface to values of Reynolds number based on a typical viscosity of 2 kNs/m^2.

It is usual to associate this change in mechanical response at strain rates above $\sim 5000 \text{ s}^{-1}$ with a change from thermal activation to viscous drag as the rate-controlling mechanism. However, in recent work on 1100.0 aluminium and 99.999% copper, Lindholm[57] has suggested that in aluminium the transition is to a second thermally-activated mechanism. In copper, despite some experimental scatter, his results for strain rates up to nearly 10^5 s^{-1} fail to show any dramatic increase in strain-rate sensitivity and are adequately represented by a constant value of λ. He concludes that at strain rates greater than 10^4 s^{-1} the validity of Hopkinson bar methods must come under careful scrutiny and further work is required before these inconsistencies can be resolved.

Doubts about the transition to a viscous drag mechanism have also been raised by Clifton.[59] Observations of the very rapid relaxation of stress behind the elastic front in his plate-impact experiments lead him to conclude that for 3003 aluminium a transition to a viscous-drag controlled mechanism does not occur at strain rates less than 10^5 s^{-1}. More recent results, however, using the pressure-shear test,[60] suggest that a transition in rate-controlling mechanism may occur between strain rates of 10^3 and 10^5 s^{-1}.

4.3.6. Effect of Changing Strain Rate (strain rate history)

In the work described so far the logarithmic rate-sensitivity parameter, λ, has been determined from a comparison of stress–strain curves obtained at different nominally-constant strain rates. The stress increment is taken between two such curves at the same plastic strain. It is implicitly assumed that the structural state of the material, i.e. the density and distribution of dislocations, is determined by the current plastic strain and is independent of the path by which this plastic strain was reached. This assumption may be checked by performing tests in which the strain rate is suddenly changed at some predetermined plastic strain and the resulting stress increment at that strain compared with the stress increment obtained from a comparison of constant-rate tests at the same strain. If the structural state of the material is uniquely defined by the current plastic strain, i.e. if it is independent of the previous strain rate history, then the two measurements of rate-sensitivity, λ, determined

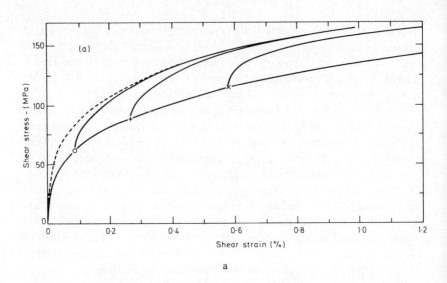

a

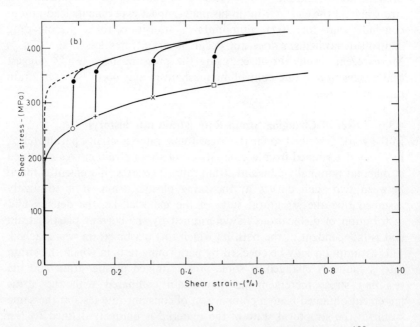

b

FIG. 4.12 Stress–strain curves for constant-rate and rate-jump tests.[100] (a) copper; (b) titanium; (c) mild steel.

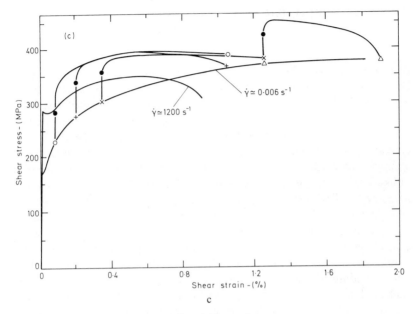

FIG. 4.12 contd.

from constant rate tests (now referred to as the 'apparent' rate-sensitivity parameter) and λ', determined from rate-jump tests (referred to as the 'intrinsic' rate-sensitivity parameter) should be the same.

Over the last few years considerable evidence has been accumulated which shows that, in general, λ and λ' are not the same. At low and intermediate rates in compression on niobium and molybdenum[34] λ' may be either greater than or less than λ depending on the base strain rate from which the jump test is performed. At higher rates the torsional SHPB has been used by several investigators to perform jump tests from quasi-static rates, of about $10^{-3}\,\text{s}^{-1}$, to impact rates, of about $1000\,\text{s}^{-1}$. Such tests have been performed on single crystals of LiF[91] and on polycrystalline fcc aluminium,[92,93,94,95] lead[96] and copper,[94,95,97,98] hcp magnesium,[95] zinc[95] and titanium,[98] and various steels.[98,99] The different types of behaviour observed are illustrated in Fig. 4.12, for room temperature tests on copper, where $\lambda' < \lambda$, titanium, where $\lambda' \simeq \lambda$, and mild steel, where $\lambda' > \lambda$. In tests at other temperatures on titanium[100] greater differences are apparent between λ and λ'. Clearly at high strain rates there is a very significant, and by no means simple, effect of strain rate history on the mechanical response.

4.4 MECHANICAL EQUATIONS OF STATE AT HIGH RATES OF STRAIN

It is apparent from the preceding section that many problems remain to be resolved before the mechanisms governing the rate-sensitivity of metals and alloys are fully understood. Conflicting evidence has been obtained regarding the postulated change from a thermally-activated to a viscous damping mechanism at high strain rates. In the thermally-activated region the frequency factor $\dot{\varepsilon}_0$ has been taken as a constant and a single rate-controlling mechanism has been assumed to operate. It is unlikely that both assumptions are always true. More importantly, it is necessary to find a way of characterising the effect of strain rate on the changing material structure, i.e. to define a structure sensitive parameter.

Nevertheless, for the practising engineer some general overall description of the effect of strain rate on the mechanical response of materials would be of real value, despite this lack of a detailed fundamental understanding of the processes involved. In an early attempt to provide such a description for the response of steel specimens Rosenfield and Hahn[101] collated a great deal of the experimental data available at the time and introduced the concept of four regions of material response, athermal (region I), thermally-activated (region II), reduced rate-sensitivity at low temperatures, possibly due to twinning (region III) and viscous damping (region IV). This approach was later extended by Lindholm[57] in the light of Ashby's deformation maps.[102] More difficult, but potentially more rewarding, in practical terms, have been the attempts to develop mechanical equations of state for these four regions, including, where appropriate, the effect of strain rate.

In its most usual form the mechanical equation of state,

$$\sigma = f(\varepsilon_p, \dot{\varepsilon}_p, T) \qquad (4.9)$$

related flow stress to current values of plastic strain, strain rate and temperature in a deforming metal or alloy. It is implicit in eqn. (4.9) that strain rate history effects can be ignored, i.e. that the structural state is taken to be uniquely defined by the plastic strain. Although evidence reviewed in Section 4.3.6 strongly suggests that this is not the case, nevertheless, for many practical purposes, a suitable form of eqn. (4.9) may provide an adequate description of the mechanical behaviour.

Based on experimental data obtained in torsion, Vinh et al.[23] proposed an empirical form for eqn. (4.9),

$$\tau = F\gamma^n(\dot{\gamma}/\dot{\gamma}_0)^m \exp(W/T). \qquad (4.10)$$

F, n, m and W are parameters which have to be determined for each material. It was found that eqn. (4.10) fitted the experimental data over a strain rate range from 10^{-3} to $1000\,\text{s}^{-1}$ for duralumin, copper and mild steel. For mild steel, however, the parameters F, n and W were required to be functions of strain rate, effectively introducing yet more disposable parameters. All tests were performed at ambient temperature but the effect of adiabatic heating was taken into account.

A more fundamental approach, see Section 4.3.1, leads to the alternative form given in eqn. (4.7) which, for practical purposes, adequately describes the general stress, strain rate, temperature dependence of fcc materials, although details of the mechanical response remain unresolved. Equation (4.7) also leaves unspecified the work-hardening behaviour of the material. Other evidence, from high rate torsional tests on copper,[98] supports the use of a power law, $\tau \alpha \gamma^n$, with n independent of strain rate for tests at 10^{-3} and $900\,\text{s}^{-1}$. Temperature effects were not considered and the full constitutive relationship had the form

$$\tau = A\gamma^n \left[1 + m \ln (1 + \dot{\gamma}/B) \right] \qquad (4.11)$$

where A, B, m and n are constants and, for $\dot{\gamma} < B/10$, $\lambda \to 0$ so that B represents the strain rate below which the response is athermal.

For bcc and hcp materials eqn. (4.7) is no longer found to give an adequate description of the behaviour in region II. Recently, however, experimental data for several alloy steels[81] and for Ti6Al4V alloy[103] have been shown[75] to follow more closely a relationship of the form

$$\tau = \tau_a + C(\dot{\varepsilon}_p/\varepsilon_0)^{T/D} \qquad (4.12)$$

where C and D are constants which depend on the material. Equation (4.12) corresponds to a stress dependence of the free energy of activation of the form

$$\Delta G = E\{1 - \ln(\tau/\tau_a - 1)\} \qquad (4.13)$$

where E is also constant. It follows that the activation volume also varies with stress and becomes very large as $\tau \to \tau_a$. Because of the change in shape of the shear stress–strain curve at high rates in torsion[23,98] and the early onset of plastic instability at high rates in tension,[79,81] a simple function to describe the work-hardening behaviour of bcc materials at different strain rates is not easy to find. The development of adiabatic shear bands at very high strain rates[104,105] and the problems associated with predicting plastic instability under combined dynamic stresses add further complications to the search for a suitable form of constitutive relation for use in analysing high energy rate processes when applied to bcc materials.

4.5. SUMMARY

Reliable and well-established techniques are available for testing materials in simple tension, compression or shear at strain rates up to about $5000\,\text{s}^{-1}$, and under biaxial stress (tension/torsion) at rates up to about $100\,\text{s}^{-1}$, i.e. at rates applicable to most 'stand-off' explosive operations. Some success has been achieved in adapting the SHPB technique for performing tests at rates in excess of $5000\,\text{s}^{-1}$, although the validity of the data obtained is still open to some doubt and a need remains for reliable results at strain rates of the order of $10^5\,\text{s}^{-1}$ in connection with contact explosive operations. Plate impact tests provide a potentially powerful alternative technique for obtaining data at these very high rates.

Over recent years a great deal of experimental evidence on the effect of strain rate on the mechanical behaviour of metals and alloys has been accumulated. Despite considerable gaps in our understanding of the details of the fundamental processes involved it has, nevertheless, been possible to derive empirical, or semi-empirical, equations of state which, for many practical purposes, provide an adequate description of the mechanical behaviour at rates of strain up to about $5000\,\text{s}^{-1}$. At rates above this conflicting results have been reported and more work needs to be done.

REFERENCES

1. ATKINS, A. G. *J. Inst. Metals*, **97** (1969), 289–298.
2. JOHNSON, W. Mechanical properties at high rates of strain, *Inst. of Phys. Conf. Ser. No. 47, 1979*, Institute of Physics, London, 337–359.
3. CRUVER, R. W., LIEBERMAN, I., PEARSON, J., PHILLIPS, J. L. and ZERNOW, L. *High Velocity Forming of Metals*, 1964, Prentice-Hall, Englewood Cliffs, New Jersey, 39–76.
4. VON KLEIN, W. *Z. Metallkde.*, **62** (1971), 589–92.
5. CAMPBELL, J. D. *J. Mat. Sci. and Eng'g.*, **12** (1973), 3–21.
6. CAMPBELL, J. D. and FERGUSON, W. G. *Phil. Mag.*, **21** (1970), 63–82.
7. DOWLING, A. R., HARDING, J. and CAMPBELL, J. D. *J. Inst. Metals*, **98** (1970), 215–24.
8. CLYENS, S. and JOHNSON, W. *J. Mat. Sci. and Eng'g.*, **30** (1977), 121–39.

9. CLYENS, S. and JOHNSON, W. *Int. J. Mech. Sci.*, **19** (1977), 745–52.
10. HOPKINSON, J., *Original Papers by J. Hopkinson*, 1901, Cambridge University Press, Cambridge, 316.
11. HOPKINSON, B. *Proc. Roy. Soc.*, **A74** (1905), 987.
12. TAYLOR, G. I. *J. Inst. Civil Eng'rs.*, **26** (1946), 486.
13. VON KÁRMÁN, T. and DUWEZ, P. *J. Appl. Phys.*, **21**, (1950), 9.
14. BELL, J. F. *Tech. Report No. 5*, Dep't. Mech. Eng'g., The Johns Hopkins University, Baltimore, 1951.
15. STERNGLASS, E. J. and STUART, D. A. *J. Appl. Mech.* **20** (1953), 427–34.
16. CAMPBELL, J. D. and DOWLING, A. R. *J. Mech. Phys. Solids*, **18** (1970), 43–63.
17. EFRON, L. and MALVERN, L. E. *Exp. Mech.*, **9** (1969), 255.
18. STEVENSON, M. G. and OXLEY, P. L. B. *Proc. Inst. Mech. Eng'rs.*, London **185** (1970–71), 741–54.
19. CLARK, D. S. and DATWYLER, G. *Proc ASTM*, **38** (part 2), (1938), 98.
20. BROWN, A. F. C. and EDMONDS, R. *Proc. Inst. Mech. Eng'rs.*, London, **159** (1948), 11–14.
21. MANN, H. C. *Proc. ASTM*, **36** (part 2), (1936), 85.
22. NADAI, A. and MANJOINE, M. J. *J. Appl. Mech.*, **8** (1941), A77–A91.
23. VINH, T., AFZALI, M. and ROCHE, A. *Mechanical Behaviour of Materials*, ICM3, (1979), Pergamon Press, Oxford, **2** 633–42.
24. SUZUKI, H., HASHIZUME, S. YABUKI, Y., NAKAJIMA, S. and KENMOCKI, K. *Report of the Institute of Industrial Science*, University of Tokyo, March 1968, **18** No. 3, Serial No. 117.
25. SAMANTA, S. K. *Int. J. Mech. Sci.*, **11** (1969), 433–53.
26. ALDER, J. F. and PHILLIPS, V. A. *J. Inst. Metals*, **83** (1954–5), 80–86.
27. HOCKETT, J. E. *Trans. AIME Met. Soc.*, **239** (1967), 969–76.
28. COOK, P. M. *Proc. Conf. Prop. Mat. at High Rates of Strain*, 1957, Inst. Mech. Eng'rs., London, Session 3, Paper 2.
29. ARNOLD, R. R. and PARKER, R. J. *J. Inst. Metals*, **88** (1959–60), 255–59.
30. BAILEY, J. A., HAAS, S. L. and SHAH, M. K. *Int. J. Mech. Sci.*, **14** (1972), 735–54.
31. CLARK, D. S. and WOOD, D. S. *Proc ASTM*, **49** (1949), 717.
32. CAMPBELL, J. D. and MARSH, K. J. *Phil. Mag.*, **7** (1962), 933–52.
33. COOPER, R. H. and CAMPBELL, J. D. *J. Mech. Eng'ng. Sci.*, **9** (1967), 278–84.
34. CAMPBELL, J. D. and BRIGGS, T. L. *J. Less Common Metals*, **40** (1974), 235–50.
35. LINDHOLM, U. S. and YEAKLEY, L. M., *Exp. Mech.*, **7** (1967), 1–7.
36. RANDALL, M. R. D., D.Phil. Thesis, 1972, University of Oxford.
37. BRIGGS, T. L., CAMPBELL, J. D. and TASSICKER, O. J. *J. Phys. D., Sci. Instruments*, **4** (1971), 240–42.
38. HOGGAT, C. R., ORR, W. R. and RECHT, R. F. *First Int. Conf. of Center for High Energy Forming*, 1967, University of Denver, **2**, 7.4.1–7.4.25.
39. NIORDSON, F. *Exp. Mech.*, **5** (1965), 29–32.
40. WARNES, R. H., DUFFEY, T. A., KARPP, R. R. and CARDEN, A. E. *Int. Conf. Metallurgical Effects on High Strain-Rate Deformation and Fabrication*, Albuquerque, New Mexico, June 22–26, 1980.

41. CARDEN, A. E., WILLIAMS, P. E. and KARPP, R. R. *Int. Conf. Metallurgical Effects on High Strain-Rate Deformation and Fabrication*, Albuquerque, New Mexico, June 22–26, 1980.
42. KOLSKY, H. *Proc. Phys. Soc. (London)*, **B62** (1949), 676–700.
43. CAMPBELL, J. D. and DUBY, J., *Proc. Roy. Soc.*, **A236** (1956), 24–40.
44. CAMPBELL, J. D. and LEWIS, J. L., Oxford University Engineering Laboratory, 1969, *Report No. 1080/69.*
45. DUFFY, J., CAMPBELL, J. D. and HAWLEY, R. H. *J. Appl. Mech.*, **38** (1971), 83–91.
46. HARDING, J. *J. Mech. Eng'g. Sci.*, **7** (1965), 163–76.
47. BACK, P. A. A. and CAMPBELL, J. D. *Proc, Conf. Prop. Mat. at High Rates of Strain*, 1957, Inst. Mech. Eng'rs., London, Session 6, Paper 2.
48. LINDHOLM, U. S. and YEAKLEY, L. M. *Exp. Mech.*, **8** (1968), 1–9.
49. NICHOLAS, T. *Exp. Mech.*, **21** (1981), 177–85.
50. COSTIN, L. S., DUFFY, J. and FREUND, L. B. *ASTM STP No. 627* (1977), 301–18.
51. HARDING, J., WOOD, E. O. and CAMPBELL, J. D. *J. Mech. Eng'g. Sci.*, **2** (1960), 88–96.
52. HARDING, J. *Acta Met.*, **17** (1969), 949–58.
53. BAKER, W. E. and YEW, C. H. *J. Appl. Mech.*, **33** (1966), 917–23.
54. CAMPBELL, J. D. and LEWIS, J. L. *Exp. Mech.*, **12** (1972), 520–24.
55. STEVENSON, M. G. Mechanical properties at high rates of strain, *Inst. of Phys. Conf. Ser. No. 21, 1975*, Institute of Physics, London, 393–403.
56. HARDING, J. and HUDDART, J. Mechanical properties at high rates of strain, *Inst. of Phys. Conf. Ser. No. 47, 1979*, Institute of Physics, London, 49–61.
57. LINDHOLM, U. S. High velocity deformation of solids, 1977, *IUTAM Symposium*, University of Tokyo, 26–35.
58. GORHAM, D. A. Mechanical properties at high rates of strain, *Inst. of Phys. Conf. Ser. No. 47, 1979*, Institute of Physics, London, 16–24.
59. CLIFTON, R. J. Mechanical properties at high rates of strain, *Inst. of Phys. Conf. Ser. No. 47, 1979*, Institute of Physics, London, 174–86.
60. LI, C. H. and CLIFTON, R. J. To appear in *Proc. of the American Physical Soc., Topical Conf. on Shock Waves in Condensed Matter*, Stanford Research Institute, Menlo Park, California, June 23–25, 1981.
61. LINDHOLM, U. S. and BESSEY, R. L. Air Force Materials Laboratory, Wright-Patterson Air Force Base, 1969, *Tech. Report No. AFML-TR-69-119.*
62. ELEICHE, A. M. Air Force Materials Laboratory, Wright-Patterson Air Force Base, 1972, *Tech. Report No. AFML-TR-72-125.*
63. TSAO, M. C. C. and CAMPBELL, J. D. Oxford University Engineering Laboratory 1973, *Report No. 1055/73.*
64. ELEICHE, A. M. and CAMPBELL, J. D. Oxford University Engineering Laboratory 1974, *Report No. 1106/74.*
65. DOWLING, A. R. and HARDING, J. *First Int. Con. of Center for High Energy Forming*, 1967, University of Denver, **2**, 7.3.1–7.3.26.
66. HUDDART, J., HARDING, J. and BLEASDALE, P. A. *J. Nucl. Mat.*, **89** (1980), 316–30.

67. HARDING, J. Arch. Mech. Stosowanej, **27** (1975), 715–32.
68. DAVIDSON, D. L. and LINDHOLM, U. S. Mechanical properties at high rates of strain, *Inst. of Phys. Conf. Ser. No. 21, 1975*, Institute of Physics, London, 124–37.
69. LINDHOLM, U. S. J. Mech. Phys. Solids, **12** (1964), 317–35.
70. HOLT, D. L., BABCOCK, S. G., GREEN, S. J. and MAIDEN, C. J. Trans. ASM, **60** (1967), 152–59.
71. LINDHOLM, U. S. *Mechanical Behaviour of Materials under Dynamic Loads*, 1968, Springer, New York, 77–95.
72. SAMANTA, S. K. J. Mech. Phys. Solids, **19** (1971), 117–35.
73. LAWSON, J. E. and NICHOLAS, T. J. Mech. Phys. Solids, **20** (1972), 65–76.
74. MAIDEN, C. J. and GREEN, S. J. G. M. Defense Research Labs., 1965, *Report No. DASA-1716*.
75. HARDING, J. *Seventh Int. Conf. on High Energy Rate Fabrication*, University of Leeds, September 15–18, 1981.
76. HUDDART, J. and HARDING, J. Oxford University Engineering Laboratory, 1979, *Report No. 1304/79*.
77. MANJOINE, M. J. J. Appl. Mech., **11** (1944), A211–18.
78. DAVIDSON, D. L., LINDHOLM, U. S. and YEAKLEY, L. M. Acta Met., **14** (1966), 703–10.
79. HARDING, J. Metals Technology, **4** (1977), 6–16.
80. LINDHOLM, U. S., YEAKLEY, L. M. and BESSEY, R. L. Air Force Materials Laboratory, Wright-Patterson Air Force Base, 1968, *Tech. Report No. AFML-TR-68-194*.
81. HARDING, J. Oxford University Engineering Laboratory, 1978, *Report No. 1266/78*.
82. FERGUSON, W. G., HAUSER, F. E. and DORN, J. E. Brit. J. Appl. Phys., **18** (1967), 411–17.
83. FERGUSON, W. G., KUMAR, A. and DORN, J. E. J. Appl. Phys., **38** (1967), 1863–69.
84. EDINGTON, J. W., Phil. Mag., **17** (1969), 1189–1206.
85. HAUSER, F. E., SIMMONS, J. A. and DORN, J. E. *Response of Metals to High Velocity Deformation*, 1961, Interscience, New York, 93–114.
86. YOSHIDA, S. and NAGATA, N. Trans. Japan Inst. Metals, **7** (1966), 273–79.
87. DHARAN, C. K. H. and HAUSER, F. E. Exp. Mech., **10** (1970), 9.
88. WULF, G. L. Int. J. Mech. Sci., **21** (1979), 713–18.
89. WULF, G. L. Int. J. Mech. Sci., **20** (1978), 843–48.
90. KOWALICK, J. F. and HAY, D. R. Metall. Trans., **2** (1971), 1953–58.
91. CHIEM, C. Y. and DUFFY, J. J. Mat. Sci. and Eng'g., **48** (1981), 207–22.
92. DUFFY, J. Mechanical properties at high rates of strain, *Inst. of Phys. Conf. Ser. No. 21, 1975*, Institute of Physics, London, 72–80.
93. FRANTZ, R. A. and DUFFY, J. J. Appl. Mech., **39** (1972), 939–45.
94. KLEPACZKO, J. J. Mat. Sci. and Eng'g., **18** (1975), 121–35.
95. SENSENY, P. S., DUFFY, J. and HAWLEY, R. H. J. Appl. Mech., **45** (1978), 60–66.
96. FRANTZ, R. A. and DUFFY, J. Brown University, *Report No. NSF GK-26002X*.
97. KLEPACZKO, J., FRANTZ, R. A. and DUFFY, J. *Polska Akademia Nauk*

Instytut Podstawowych Problemow Techniki, Eng'g. Trans., **25** (1977), 3–22.
98. CAMPBELL, J. D., ELEICHE, A. M. and TSAO, M. C. C. *Fundamental Aspects of Structural Alloy Design*, 1977, Plenum Publishing Corp., New York, 545–63.
99. WILSON, M. L., HAWLEY, R. H. and DUFFY, J. Brown University, 1979, *Report No. NSF ENG 75–18532/8*.
100. ELEICHE, A. M. and CAMPBELL, J. D. Air Force Materials Laboratory, Wright-Patterson Air Force Base, 1976, *Tech. Report No. AFML-TR-76-90*.
101. ROSENFIELD, A. R. and HAHN, G. T. *Trans. ASM*, **59** (1966), 962–80.
102. ASHBY, M. F., *Acta Met.*, **20** (1972), 887–897.
103. HARDING, J. Unpublished work.
104. BEDFORD, A. J., WINGROVE, A. L. and THOMPSON, K. R. L. *J. Aust. Inst. Metals*, **19** (1974), 61–73.
105. COSTIN, L. S., CRISMAN, E. E., HAWLEY, R. H. and DUFFY, J. Mechanical properties at high rates of strain, *Inst. of Phys. Conf. Ser. No. 47, 1979*, Institute of Physics, London, 90–100.

Chapter 5

BASIC CONSIDERATION FOR COMMERCIAL PROCESSES

D. B. CLELAND

Nobel's Explosives Co. Ltd, Stevenston, Ayrshire, UK

5.1. EXPLOSIVE CLADDING

5.1.1. Introduction

Clad plates have been commercially available since the 1930s initially by the hot rolling of stainless steels and nickel alloys to carbon steels. Copper alloys have also been offered by some mills. Explosive cladding was first introduced by Du Pont of USA in the early 1960s and subsequently by other firms in various parts of the world, several of whom are licensees of Du Pont. The advent of explosive cladding opened up the range of metal combinations available in clad form to include virtually all the ductile engineering alloys, and has set higher standards of bond strength and continuity of bond. Guaranteed bond strength 50% greater than the levels offered for roll bonded clads became available. However, the explosion bond market complements rather than replaces the rolled product, the latter being more economical in large area thin gauge clads, whereas explosive clads supply the demand for combinations outside the roll bonded range or where the standard of bond quality required, or the dimension and quantities do not suit the roll bonded route. In practice several explosive cladding organisations work in close co-operation with mills producing rolled clad plates and it is common for slabs to be explosively bonded and then rolled out and marketed by the mill in preference to their traditional methods. The worldwide production of explosively clad plate is several tens of thousands of square metres per annum.

Whichever the production method the advantage of clad plate is to permit the designer and manufacturer of pressure vessels, heat exchangers and other equipment to minimise his costs without sacrifice of reliability or operating life. Higher operating pressures and temperatures and increasing corrosion problems have led to accelerated use of more expensive metals and alloys. Whereas stainless steel, and nickel alloys such as Monel, have long been used in fabrication, increasing attention is being given to such materials as titanium or Hastelloy where cost per tonne can be, say, 40 times that of a boiler quality steel. The rarer metals such as tantalum cost substantially more.

By use of clad plate in construction the fabricator can have the benefit of the expensive corrosion resistant metal on the working surfaces combined with the strength of a low cost carbon steel or low alloy backer metal. A certain penalty has to be paid in the use of clad plate in that the welding of composite materials requires special techniques, the most complex being that used with plates clad with reactive metals and involving inserts and capping strips as shown in Fig. 5.1.

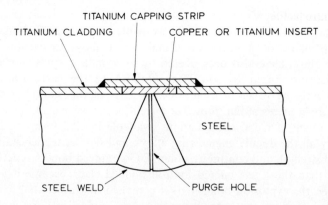

FIG. 5.1 Joining of titanium clad plates.

By its nature explosive bonding is applicable only to flat surfaces such as plates or disc forgings, or to internal or external cylindrical surfaces, and cannot be used to line shaped components. However clad plates can be formed and machined or hot pressed and cold spun as readily as homogeneous metal apart from certain reservations on temperatures. Table 5.1 lists limiting temperatures for several composites. For some clads, excessive heating can result in the development of intermetallics at the interface with loss of bond strength. Figure 5.2 shows results for titanium clad to carbon steel. The enormous range of metal specifications, thicknesses and sizes of plate which are possible, requires almost all explosively clad plates to be produced to specific requirements. The variations in product and the problem of working with large explosive charges gives restricted scope for mechanisation of the process and results in a significant part of the cost of cladding to be related to workmanship and so not to vary greatly with the value of the metals being clad.

TABLE 5.1
LIMITING TEMPERATURES FOR CLAD PLATE

Combination	Maximum temperature°C	
	Fabrication (Hot forming)†	Service
Stainless/CS	1 100	
Monel/CS	1 100	
Hastelloy/CS	1 200	
Cu Ni/CS	1 000	
Copper/CS	925	540
Titanium/CS	800	400
Copper/Aluminium	260	150
Aluminium/CS	315	260

† Minimum time at temperature.

The point at which explosively clad plate produced in final gauge thickness becomes economically attractive will thus depend on the thickness ratio, the value of the metals and the size of the clad plate. As examples, tube plates of typical size clad with titanium 6–12 mm thick on carbon steel become less expensive than solid with base thicknesses of over 16 to 20 mm. Large shell plates clad with 3 mm titanium are economic with any base thickness which can be processed say 12 mm or over depending on size. At the other end of the range 18/8 stainless tube plates require a backer thickness of 40 mm or more to be attractive and shell plates 17 mm. In the above examples no allowance has been made for additional fabrication costs which will increase slightly the break-even thickness but the tube plate costings include extra cladding thick-

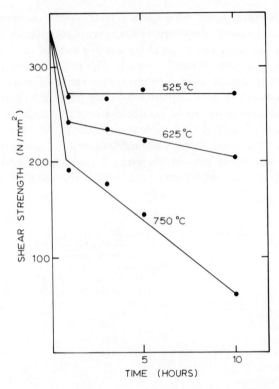

FIG. 5.2 Titanium/carbon steel clad shear strength after heating.

ness for face machining. Explosively clad slabs which are subsequently hot rolled to thinner gauges will extend downward the break-even thicknesses of these combinations.

Advantages may result in the use of clad plate other than savings in material costs. For instance a vessel can have external heating jackets or other fitments more readily attached to the steel base metal of a clad than to say the equivalent solid stainless steel vessel. Transition joints which are sections of clad plates of metals which cannot be welded to each other by conventional means, are used to fabricate all-welded assemblies for structural or electrical conduction applications.

5.1.2. Cladding Sites and Facilities

Commercial operations involve the use of very large explosive charges and the noise problem is of first importance in selecting a cladding site. Fabricators are always seeking larger clad plates to minimise their costs and clad slabs for onward rolling are more economic in higher tonnage. Charge weights on a single shot in excess of 1 tonne are common.

Firing can be carried out in an open site, or in a tunnel which can be specially constructed or adapted from disused mine workings. A further alternative is a mounded-over area with controlled venting. Tunnel operations are effective in reducing noise, particularly if the ends are sealed off when firing. The capital cost of a specially constructed tunnel is very high and a better solution is the adaptation of an old mine with sufficient internal volume to reduce noise level without sealing the entrance. Such facilities are not readily available and also give rise to problems of lighting and the risk of rock fall from the tunnel roof. The majority of cladding sites are situated in the open and so have to be located at a considerable distance from built-up areas to avoid disturbance or damage. A distance of 7 km from the nearest residential zone is a reasonable minimum figure. During firings, an area of about $\frac{1}{2}$ km radius must be clear of all persons except the cladding operators.

A road must exist to the cladding site or be constructed and be of sufficient carrying capacity to accommodate heavy lorries transporting metals to and from the site. Power supplies and communication links have also to be provided.

In designing the cladding facility a decision has to be made as to how much of the operation is to be carried out at the firing site. In the one extreme it can be used purely for firing with the prepared clads and explosives brought in, fired and taken away. All preparation work before firing and subsequent finishing work is carried out at a separate plant. On the other hand workshops can be set up adjacent to the firing area so that metal storage and preparation, assembly of the clads, explosive preparation and finishing work can take place on the one site. The advantage of an integrated site is that the various stages of the operation are carried out by the same team with continuity of supervision and quality control. Even so some operations such as hydraulic press levelling of clads and heavy machining of plates will have to take place at other plants.

A typical open cladding site can therefore comprise the following:

1. A firing point about $\frac{1}{2}$ km from the remainder of the plant, preferably screened by the natural lie of the ground, and with protective shelters for the shotfirers.
2. A workshop for the preparation and assembly of clads with facilities for cutting, machining, welding and abrading of metals.

3. Storage facilities for metals and other materials, fuels, gases, etc.
4. Explosive preparation buildings.
5. Magazines for explosives storage. The magazines and the preparation building must comply with regulations as to construction and safe distances.
6. Offices and amenity buildings.

For handling the incoming metals and the clad plates, mobile cranes or heavy duty fork lift trucks are used. The main workshop is constructed so that cranes can operate inside. An onsite transporter is available for the movement of heavy clads to and from the firing point and a specially constructed van for carrying explosive. A bulldozer is also required for restoring the firing area after cladding. Provision has to be made for operating in winter with snow clearing equipment and the use of FWD vehicles for transport of the cladding team to and from the site.

Clads are fired on a bed of sand or similar material with a depth of up to 1.5 m. The base plate is laid directly on the sand unless an anvil plate or insulating layer is required. More complex arrangements using reinforced concrete firing plinths or massive permanent steel anvils or water beds have been tried without worthwhile effect. It might be expected that such devices would absorb the shock waves in the base plate and minimise the reflected tension waves which could affect the interface. In practice the simple sand bed has proved the most satisfactory arrangement. On firing, the clad is driven down into the sand sometimes to a considerable depth and the resultant cavity has to be filled in and levelled off for the next shot.

5.1.3. Range of Products
Routine production of clads falls into the following categories:

(a) Shell plates for the construction of process vessels for use in the chemical or petrochemical industries, and for shells, channels and covers for heat exchangers.
(b) Tube plates for heat exchangers.
(c) Clad slabs for hot rolling to thinner gauges for the production of large area shell plates.
(d) Internal cladding of nozzles and other fitments for pressure vessels.
(e) Internal or external cladding of tubular components for direct use or for subsequent drawing to smaller diameter.
(f) Transition joints.

BASIC CONSIDERATION FOR COMMERCIAL PROCESSES 165

Other specialised applications are met as required but the foregoing list represents the bulk of commercial work. Metal combinations cover the spectrum of engineering alloys, the limitations being thòse of sufficient ductility and adequate impact strength. A rough guide is that cladding metals with an elongation of 20% can be readily handled, and lower elongation can in many instances be successfully bonded. Base steels must have adequate impact strength at ambient temperatures to avoid risk of cracking due to the explosive forces. Table 5.2 gives a list of the more common cladding metals.

TABLE 5.2
STANDARD CLADDING COMBINATIONS

Cladding metals	Base metals
Stainless steels	
Copper and copper alloys	
Nickel and nickel alloys	Carbon and low alloy plates and forgings.
Hastelloy	
Titanium and zirconium	Stainless steel.
Tantalum	
Aluminium	
Copper	Aluminium

5.1.4. Bonding Parameters

Other contributors to this volume have discussed the principles of explosive welding and the methods of determining the optimun parameters for achieving a good bond. The purpose of this chapter is to detail how the parameters, once established from theoretical considerations or from experimental data, are applied to large scale operations. Obviously working methods will vary between differing operators but the information herein represents a typical approach to the problems of commercial production.

With the parallel gap system, which is used for virtually all large shots, only three variables have to be determined for a particular clad. (a) Mass of explosives per unit area of plate. (b) Velocity of detonation of the explosive. (c) Dimensions of the gap between the plates.

For routine work these variables are well established and tabulated so the operation can proceed with confidence. For new combinations the parameters are estimated from basic principles and optimised by test firings. The aim is to develop systems which can be relied upon to give clads of acceptable quality on a production basis and the variables are chosen to give good bond quality within practical processing tolerances. Plate velocities and collision angles will therefore tend to be somewhat higher than theoretical minimum.

A wavy interface is always aimed for and its presence is taken as an essential feature of a satisfactory clad. Typical wave pitches lie between 0.5 mm and 1.2 mm with amplitudes of a quarter to a third of the pitch. Exceptions occur with combinations such as aluminium to steel where some intermetallics are inevitably produced at the interface. A strongly developed waveform concentrates the intermetallics at the vortices leaving the remainder of the interface free of their weakening effects.

5.2. DESIGN OF CLAD ASSEMBLIES

5.2.1 General

For given dimensions of guaranteed sound bond, specified by the end user, the cladder must add extra material to allow for edge effects. Areas of peripheral weak bond or non-bond can occur on clads due partly to the reduction in effective explosive loading at the edges of the charge and partly to the collision angle of the cladding plate becoming progressively smaller as the detonation point approaches the extreme edges.

The increase in dimensions required varies with the thickness of the cladding plate and the particular alloy concerned but, typically, is 50 to 150 mm extra on length and breadth. Where a section of the clad plate has to be removed for destructive tests a further addition to the sizes will have to be made.

For some combinations, e.g. titanium clad to carbon steel, best results are obtained by reducing the base size, i.e. that the cladding plate overhang the base plate by 25 mm all round. The excess cladding metal sheared off by the explosive forces, to give a clean edge and the improved edge bond quality, allows the minimum of extra size of clad for given finished dimensions. Where the cladding metal is particularly soft and ductile such as pure copper, edge shearing is less effective and such combinations are fired size for size, as are some other metals in the thicker cladding gauges.

The maximum size of clad which can be fired depends on a number of factors. Assuming that the cladding or base thickness is not a process limitation, the main restriction is often the size of cladding metal plates which can be purchased. For example thin titanium cannot be readily obtained in width over 2 to 2.5 m and some other metals are even more restricted. One solution is to preweld two or more sheets to form a single piece. The problem then is to obtain a welding procedure which gives weld of acceptably high quality with virtually no distortion of the welded plate. As always flatness of the cladding plate is of prime importance. Electron beam welding has been shown to give the best quality and flatness, but the chambers available are restricted in the size of plate they can handle. Plasma welding has also given good results, and is not restricted to size. TIG welding tends to give excessive distortion unless special precautions are taken. Prewelding can add significantly to the cost of a clad.

The cladding of two adjacent cladding plates on to a single base plate, either singly or simultaneously has been proposed as a method of eliminating prewelding. The cladding plates are then connected by a welding operation after cladding. Of the two routes only simultaneous cladding has proved practical but excessive non-bond areas along the line of abutment make it unattractive except in special circumstances. For example, clads of Hastelloy on carbon steel were manufactured using two unwelded cladding plates. The design of the vessel to be fabricated had large openings along the centre line of the plates which meant that only part of the butting edges remained to be joined by welding. Cladding each half of the base in two separate shots creates the problem of uneven distortion; the base tending to bend into a vee configuration which is difficult to rectify. The same difficulty arises where the customer wishes only part of the base plate clad, the remainder being unclad. Unless the base plate is very thick and rigid such part-area clads are to be avoided.

Other limitations on size of clad plate are obviously the maximum weight of plate which can be handled and the maximum permitted weight of explosive which can be fired. Cladding sites and facilities are chosen so that these limits rarely restrict operations.

Minimum sizes are governed in practice by economic factors when the edge trim allowances and handling costs become proportionately uneconomic.

5.2.2 Shell Plates

The simplest arrangement of a cladding shot is a rectangle with the base and cladding plate of equal size. For large area plates in normal commercial production the cladding thickness can be 3 mm and upwards. Thinner cladding plates are possible but on restricted areas. Base thickness should be at least three times the cladding thickness to provide sufficient mass to give optimum bond quality. Lower ratios are possible using an additional anvil plate under the backer to give extra mass but with the penalty of additional cost. Anvils can be flattened and reused but the life of an anvil is usually three or four shots before suffering permanent damage. Anvils are also used in large area shots where the backer is thin, even if the thickness ratio is adequate, so as to give rigidity in handling and reduced deformation in firing. For a clad say 6 m × 2 m a base thickness below 25 mm will require extra support and for a clad 3 m × 1 m below 12 mm thickness.

To ensure that the final clad is not under gauge in thickness an extra 1 mm can be added to the thickness for surface abrading of steel—more if it is to be machined clean—and cladding plates specified with positive manufacturing tolerances.

5.2.3. Tube Plates

In the case of tube plates for heat exchangers it is standard practice for the clad plates to be face machined before drilling and additional metal thickness must be provided to ensure that the faces can be machined clean and flat. All clad plates distort in firing and it is routine for the plates to be levelled in a hydraulic press. Some residual out of flatness will remain and so, based on a knowledge of the press capability with plates of varying size and thickness, an extra machining allowance is added to the cladding thickness, and slightly more to the base thickness. This ensures that the tube plate can be fully machined and still maintain the specified minimum thickness of cladding as in Fig. 5.3. It is worth noting that the minimum cladding thickness and minimum base thickness do not add up to the total thickness.

Tube plates which are normally circular in form and relatively thick, both in cladding and base metal, are designed rather differently from shell plates. The base plates are cut in disc form with the appropriate extra allowance on diameter. If the steel is a forging it will be purchased in disc form. The cladding plate is used in the square format rather than circular—square blanks generally cost no more than discs. The overhanging arrangement is used even with soft or ductile metals and the cladding plate corners are sheared off.

To ensure a clean shear, a stress raising groove is cut on the surfaces of the plate at a diameter corresponding to the base diameter.

Departures from the standard arrangement occur when part of the clad is required for destructive testing. A square plate is then clad from

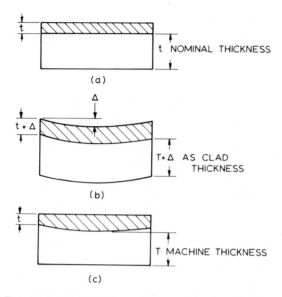

FIG. 5.3 Machining allowances on tube-plate thickness.

which the disc is cut leaving corner areas for test. Alternatively two or more discs can be combined on a rectangular plate which has the advantage that one set of destructive tests applies to the several discs cut out.

Where only limited destructive testing is called for, such as one shear test, it is possible to retain the square cladding on circular base configuration by including an extention of the base plate under one of the cladding plate corners, of sufficient area to yield a test coupon.

Forged bases are supplied in disc form, so the provision of test material is difficult. One solution is to clad an oversize disc and cut the test coupon from the periphery, but the extra cost is high, typically 30% extra to carry out a shear test. Alternatively a separate test plate can be fired along with the production tube plates. If several different destructive tests are required a separate test plate is often the only effective way of providing sufficient material, and if one plate is used to prove several production clads the cost is minimised.

5.2.4. Metal Requirements

In most cases plates are intended for the construction of pressure vessels or heat exchangers which are fabricated to comply with established codes such as ASME. It is essential therefore that all metals are purchased to the necessary specifications, inspected and certificated at source in accordance with the code and the customer's requirements. In addition the cladder may add further requirements such as maximum yield strength, impact quality and freedom from inclusions and laminations to ensure the best quality of clad plate. Base steels, which are normally to a boiler plate specification, are purchased in the normalised condition, and gas cut to the required size. Steel cut to size by shearing is liable to develop cracks from the sheared edges and some steels, e.g. ASTM A515, must be renormalised after gas cutting to eliminate the risk of cracking from the edge heat affected zone. Mill finish of plate steel is acceptable as surface preparation is part of the cladding process but surfaces are preferably shot blasted to remove scale. Care must be taken to avoid excessive pitting or other surface flaws which result in undue preparation work and possible adverse effect on the bond quality. Steel plates are purchased with the best available flatness, e.g. Euronorm 29–69 close tolerance. Forged discs are supplied proof machined to size.

Cladding plates are specified in the annealed condition and can be flame cut or sheared to size except for brittle grades such as ferrite stainless steel which must be sawn or machined. Flatness is again of great importance and the supplier must be held to the tightest standard which is practical.

5.2.5. Extension Bars

Improved edge bond quality and freedom from edge cracking of the cladding plate can often be obtained by fitting flat steel bars butting against the cladding plate edge. The compressive wave in the cladding plate is transmitted to the steel bar, but the reflected tension wave from the steel edge does not return through the butted contact into the cladding plate. Extension bars can be fitted to critical parts of the clad such as corners and sides remote from the initiator, or for convenience in assembly, all round the cladding plate. Extension bars are not normally used for thicknesses below 5 mm. The explosive charge is extended over the bar surface.

5.2.6. Temperature of Metals

Certain steels have brittle transition temperatures which are above ambient temperatures, particularly in winter. To avoid the risk of fracture, such steels are heated before cladding using gas heaters or electric pads. The temperature is raised to a level such that allowing for cooling in transit to the firing point the temperature of the steel at firing is above the critical figure. Cladding metals in normal use are clad cold but some exceptions such as ferritic stainless plates must be warmed before firing. Care has to be taken that the temperature does not affect the explosive and conversely to avoid chilling the cladding plate, the explosive should also be warmed before use. Clads with heated bases can be set on insulating layers at the firing point especially in wet conditions.

5.3. ASSEMBLY OF CLADS

5.3.1 Metal Preparation

The mating surfaces of the two metals require to be cleaned and smooth to a finish of about 2.54 μm. Coarser surface finishes can be successfully clad but as a general working rule the finer the finish the better the quality of the bond. Very fine finishes have been found to be necessary for collision velocity and angles which approach minimum values, but a 2.54 μm finish is satisfactory for routine work. Some metals such as aluminium or titanium sheets are supplied by the manufacturers with smooth surfaces and need only light buffing. Others such as steel plate require substantial preparation either by abrading with special purpose machines or in the case of forgings or very thick plate steel bases by machining in a lathe or planer. Local defects such as pitting are removed by hand tools after machine abrading. Plates with prepared surfaces which are not to be set up and fired immediately must be protected from contamination and corrosion, and before use given a final buffing to restore the finish.

5.3.2. Assembly

Clads may be assembled *in situ* in the firing area or may be preassembled in a workshop and transported to the firing area. The latter has the advantage that the work is carried out in good conditions and is not affected by weather in an open firing area or poor visibility in the case of tunnel operations. In practice both systems are used with an open firing area. Simple straightforward clads are assembled in the firing position, and workshop assembly is used for complex shots or when the weather is poor. Transporting of assembled shots must be done with great care to avoid disruption of the components and the settings rechecked before loading on the explosive.

The main item to be controlled in assembly is the gap between the cladding plate and the base. Gap dimensions are commonly in the range 50% to 100% of the cladding plate thickness, over which distance the plate will develop a substantial proportion of its terminal velocity. However, other considerations may require the use of different gaps in practice. Tolerance on gap is around $-25\% + 50\%$ for straightforward clads but less for critical systems. If both plates are perfectly flat and the cladding plate is sufficiently thick to be rigid the procedure is to place small pegs of the correct height at the edges of the base plate. Such spacers are ejected when the plate is fired and in any case are in edge trim areas so that the pay area of the clad is unaffected. However, if the cladding plate is not perfectly flat it may be necessary to introduce spacers at areas inboard of the edge. Obviously any such internal spacers must not significantly affect the bonding action so only very light components can be used. Forms of spacer which have been employed include foil ribbons, wire helices, thin tubes and polystyrene foam pellets or blocks. Further control can be exercised by light clamps round the edges of the assembly. A cladding plate which has a slight out of flatness would best be set up with the concave face upwards and held level with edge clamps in conjunction with central spacing. The amount of correction which can be made on thicker cladding plates is limited before collapse of the light spacers and the real solution is to ensure that the cladding plate is flat before starting, if necessary having it specially levelled.

For large area clads with thin cladding plates, i.e. shell plates where the cladding thickness is typically 3 or 4 mm, the situation is somewhat different as the plates are flexible and have to be supported by a pattern of spacers placed on the base plate. Also the weight of the explosive has a measurable effect. The number and distribution of the spacers is a matter of judgement depending on the size and thickness of plate and the amount of distortion originally present. One form of distortion which is difficult to control is an 'oil can' effect where an area of the plate springs from a concave to a convex form. Additional control can be obtained by attaching rigid steel sections to the plate edges which also can act as the explosive charge frame and by edge clamps. Downward loading of the plate surface to correct a convex area can be applied to a limited extent. The explosive weight will contribute and its effect can be judged by applying dummy weights in the critical region. Otherwise downward pressure at particular points can be effected by small diameter pegs passing through the explosive layer and loaded by sandbags or, in critical cases, by a gantry assembled over the clad plates. Provided the pegs are of small diameter their presence in the explosive has no adverse effect on the cladding quality. The gap is checked by inserting calibrated probes from the edges. As well as an overall tolerance on the gap the rate of change of gap is significant as a variation will cause fluctuation in the impact point velocity. Where variation in gap is unavoidable it is better to have an increasing gap rather than a diminishing gap along the line of detonation, and the initiation point can be chosen accordingly, or the spacer height can be slightly increased remote from the initiator so that gap changes are generally positive. Once the gap has been satisfactorily established extension bars are fitted, if required, and a light frame attached to the edge of the cladding plate to act as a container for the explosive. The clad is then ready for transportation to the firing area.

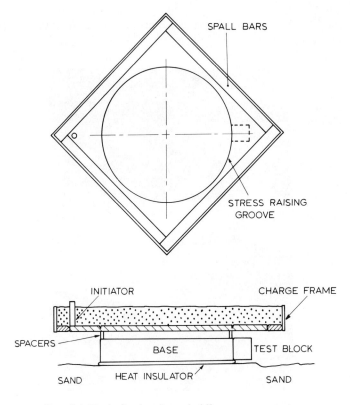

FIG. 5.4 Typical tube-plate cladding arrangements.

5.3.3. Protection of Cladding Plate Surface

Provided that entrapped air has been avoided the application of the explosive directly to the cladding plate surface has very little deleterious effect. Very soft metals can show a slight roughening and reactive metals such as titanium can be discoloured.

In such cases the effect can be eliminated by coating the top surface with a rubberised paint. Other types of paint have also been used. The ideal coating is one which gives protection but is sufficiently disrupted by the detonation to be easily removed. However, the majority of commercial clads are fired without any protection of the cladding metal. Buffer layers of substantial thickness between the explosive and the cladding plate are not used except where it is desired to increase the mass or rigidity of a thin sheet or foil. Contact between the buffer and sheet must be perfect to avoid damage to the metal.

Figure 5.4 shows a typical cladding set up illustrating some of the aspects discussed. Figures 5.5 and 5.6 are of a heavy stainless steel clad during setting up and after firing.

FIG. 5.5 Loading explosive on a stainless steel clad slab.

FIG. 5.6 Clad slab after firing.

5.4 EXPLOSIVES

5.4.1. Main Charge

The first requirement of an explosive used with the parallel gap arrangement is to give the appropriate detonation velocity which may be between 1.5 km/s and 3.5 km/s, the majority of combinations lying around the middle of the range. Other desirable features are that it can be readily spread in a uniform layer, is sufficiently sensitive to propagate in the minimum required thickness and preferably is non-toxic. Types of explosive which have been used are those based on ammonium nitrate–fuel oil, nitroglycerine based powders, TNT–ammonium nitrate and PETN–ammonium nitrate. All are used in grain or powder form to allow easy spreading in a uniform layer. Gelatinous explosives or sheet explosives are not suitable for large shots. It is essential to avoid any entrapped air in the explosive charge, otherwise damage to the cladding plate surface will result. Hence the use of free running compositions. Slurried explosives have been suggested and can be readily poured to a uniform layer. However, tests have shown that the risk of entrapped air pockets cannot be easily eliminated and there was no advantage in their use over powders except perhaps in an unusual application where the charge was fired in a vertical position. The slurry was allowed to gel before being turned vertical.

Precise control of velocity of detonation means that the explosives must be adapted or modified to suit the application by altering the composition or physical form of the constituents or by the use of velocity modifying additives. Varying metal combinations require different optimum detonation velocities and the velocity achieved with a given composition will also vary with depth of charge. To cover the full range, the cladder must either have a stock of precalibrated compositions from which to choose, or must mix or blend the explosive charge for each individual clad. Tolerance on velocity is about $\pm 10\%$ and less for critical combinations. Calibration shots are used to check the velocity of detonation particularly when batches are changed. The explosive is poured on to the cladding plate surface and spread out and screeded level. Charges are prepared on a weight basis rather than volume and the density is left in the as spread condition. However, in some critical shots the powder is applied in two increments with light tamping to give uniform density of the charge and care is taken that the explosive fully fills the corners and edges of the container.

The velocity of detonation of each production clad is measured and recorded and if necessary the composition of subsequent clads in a series is adjusted to optimise the velocity. Measurement of velocity is by electronic pick-up or by the Dautriche method using calibrated detonating cord. The accuracy of the latter is just sufficient to confirm that a charge is within tolerance, but the method is simple and reliable and is used for routine monitoring of shots.

5.4.2. Initiation

Initiation of the main charge is normally by a substantial booster of cast, high velocity explosive to ensure that the main charge immediately propagates at full velocity and to set up the collapse angle of the cladding plate. The metal below the booster is indented and bonding will not commence below a radius of about 25 mm. The initiator is therefore placed on trim material, at a corner with a square cladding plate or at the centre of an edge. Occasionally it is necessary to place the initiator within the finished area of the clad, e.g. with a recessed forging the initiator is located centrally. In such cases the non-bonded area is accepted or is rectified by trepanning out and replacing by weld deposit. Plane wave generators are used in cylindrical clads, nozzles or tubes to give uniform axial detonation but not with plate clads except in special applications.

5.5. DOUBLE SIDED CLADS

Plates clad on both faces are sometimes required, not necessarily with the same grade or thickness of cladding metal. As such clads require virtually double the work of a single-sided clad the economic break-even point is much higher, and they are usually in the form of thick tube plates or slabs for onward rolling to thinner gauge.

Two methods have been used. The base can be clad on each side in two separate shots with, if necessary, a stress relief and levelling between stages. Procedures are straightforward and standard with only extra care being needed in the second firing to avoid damage to the original cladding. The strength of the initial bond is sufficient to eliminate any risk of delamination by reflected shock waves, but surface damage from debris in the sand bed must be prevented.

Alternatively both sides of the clad can be fired together provided that the cladding and therefore the charges are identical. The plates are set up in the vertical position and both charges fired simultaneously and symmetrically. Practical difficulties arise in setting up the clad in the vertical orientation both in gap control and containment of the powder explosive and a fairly complex jig or framework has to be fabricated. The method would only apply to compact clads with rigid cladding plates and preferably to a sequence of identical shots allowing a standard framework to be designed and manufactured. One further possible disadvantage is that in the rare event of one charge misfiring the clad would be projected violently to one side and could suffer damage.

5.6. MULTILAYER CLADS

Standard examples of multilayered clads are aluminium to steel transition joints which have an interlayer of pure aluminium below aluminium alloy, tantalum clads which are sometimes required with a copper layer, and cryogenic aluminium to stainless clads which have a silver or tantalum interlayer.

The method adopted is usually to clad the two metals to the base in one shot, although cryogenic clads and other special requirements are fired as two separate shots.

For a single shot clad the parameters are chosen so that conditions at both interfaces are within the limits which give acceptable results. For example with the structural transition joints, the bond between the two grades of aluminium has a wider range of acceptable conditions and the aluminium to steel bond is the controlling factor.

Assembly of multilayer clads requires particular care to ensure that the two gaps are correct particularly if one of the cladding metals is thin and the spacing is small with low tolerances. Spacer locations should not be superimposed.

Simultaneous bonding of several layers of thin material has been proved experimentally but has not found application in production.

5.7. POST CLADDING OPERATIONS

5.7.1. Preliminary Examinations
After firing, the clad is removed from the firing area and washed down and examined for defects. The bond is checked by ultrasonic testing as confirmation that the process has been satisfactory. However, the final ultrasonic examination is not carried out until all finishing operations have been carried out and the clad is ready for shipment.

5.7.2. Stress Relief
Explosive cladding creates a work hardened layer at the interface which can be removed by stress relief. Some combinations such as titanium to steel are stress relieved as standard, but the majority of clads do not require heat treatment unless specifically requested by the customer. If part of the clad is to be used for destructive tests the cladder may opt to include a stress relief to ensure for example that the elongation of the steel has not been reduced below specification limits. Figure 5.7 shows an example of the effect of heat treatment on the interface hardness.

Stress relief is normally 625°C for 1 h per 25 mm total thickness, although for titanium clads 3 h per 25 mm thickness at 525°C is preferred, with a maximum treatment of 6 h. With titanium clads it is essential to ensure that there is no iron contamination of the titanium surface before heating and all such plates are checked with a Feroxyl solution before stress relief.

5.7.3. Levelling
All clad plates are distorted or dished by the explosive forces and most

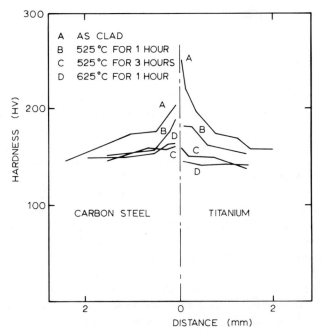

FIG. 5.7 Stress relief of clad interface.

require correction. Shell plates are levelled in rolls to tolerances equivalent to steel plate, e.g. Euronorm 29–69. Tube plates are flattened in a hydraulic press to tolerances specified to comply with the machining allowances included in the cladding plate thickness and are checked by a special gauge which measures the deviation from a true plane.

Clads which have bases of exceptional thickness may not need subsequent flattening but the majority require rectification. The degree of distortion resulting from cladding varies considerably and is influenced among other factors by the condition of the sand bed. There is some evidence that completely dry or alternatively very wet sand gives least distortion. A review of a large number of tube plates of normal proportions clad in disc form on a sand bed suggested an average out of flatness after firing given by:

$$\Delta = 16 - \frac{2T}{L}$$

where Δ = overall out of flatness in mm/m diameter, T = base thickness (mm) and t = cladding thickness (mm).

5.7.4. Cutting and Trimming

Clads produced to the customer's required sizes plus edge trim are more often than not supplied in the as-clad dimensions for the customer to trim or machine to his finished sizes. Where several pieces have been combined in one shot or where test coupons have to be removed the clad plate has to be cut. Plasma arc cutting is the preferred method although titanium clads and thinner stainless or nickel alloy clads can be cut with oxyacetylene although not so cleanly as with plasma. Alternatively, the cladding metal can be trepanned and the base steel gas cut. Transition joint materials which are supplied in cut strips or blocks are cut to size by a cold saw or abrasive discs.

5.7.5. Ultrasonic Testing

Ultrasonic testing is the standard method of evaluating the continuity of bond of the completed clad and is applied to all products. Procedures and standards can be to the cladder's own specifications or to those of recognised authorities such as ASTM A578. Probes of 10 to 25 mm diameter at frequencies of 2 to 5 MHz are used as appropriate. Shell plates are examined on a grid scan, but tube plates are given a 100% check. Acceptance standards vary according to the application but for tube plates it is normally 98% minimum continuity of bond with no individual non-bond area greater than 650 mm^2. Non-bond areas when they occur are usually at the edges of the clad and in practice the individual non-bond area criterion is the important factor. The minimum percentage continuity is rarely invoked.

In the majority of cases ultrasonic testing shows very clearly the distinction between bonded and non-bonded areas with a complete change in screen pattern. If the metals have similar acoustic impedances, such as stainless steel to carbon steel, there is no significant response from the interface with soundly bonded metals and only the back wall reflection is observed. With a non-bonded area the interface reflection appears at full height and back wall reflection is lost. Complete absence of back wall reflection is taken as indicative of non-bond and this criterion is standard in clad specifications. The change from the bonded to the unbonded pattern is abrupt with virtually no transition so the test cannot be used to distinguish clearly between a good bond of full strength and a continuous but weaker bond. Careful ultrasonic examination of sample pieces which included weak bonded areas and which were sectioned so that the ultrasonic response could be correlated with the interface conditions did show slight correspondence but not sufficient to be relied on in routine examination.

However, the belief that the ultrasonic test can indicate bond strength does exist in some quarters and one major international company specifies a minimum interface to back wall response ratio as a measure of bond strength. The level of interface response observed relative to the back wall reflection depends primarily on the acoustic mismatch of the metals and the transmission properties of the base. Titanium to steel clads always show some interface response with good bond and even more with aluminium clads. In extreme cases it may not be possible to achieve a satisfactory back wall response even with sound bond. An example is a clad product routinely produced as an intermediary in a hot rolling process in which an interlayer is explosively clad to the face of very thick stainless or nickel alloy slabs. The most satisfactory method of checking the bond is by observing changes in the interface response.

Another factor which can affect ultrasonic examination is the presence of inclusions or laminations in the base steel which can cause local obscuring of the back face reflection and preventing the latter from being used as the indicator of sound bond.

For transition joint clads which are supplied in sawn blocks or strips, the ultrasonic examination can be supplemented or replaced by a visual examination of the cut edges to confirm the presence of a satisfactory waveform.

5.8. DESTRUCTIVE TESTING

A number of standard tests for clad plates exist of which the ones most often called for are those described in specifications ASTM A263, A264, A265 and B432, which are broadly similar but deal with different cladding metals. They are also often called for with cladding metals not included in the specifications.

A full set of destructive tests comprises one shear test, one tensile test and two bend tests, with the cladding in compression and in tension. For overall thickness below 37 mm the tensile and bend specimens are full thickness. Thicker clads have the bend specimens reduced to 37 mm by removing both metals pro rata to their original thickness, and the tensile test piece is machined from base metal only. The specifications were originally drawn up for clads produced by hot roll bonding and are concerned as much with the properties of the hot rolled metal as with the bond quality. The bend testpieces for instance are bent round mandrels corresponding to the details of the base metal specification and the requirement is freedom from cracking. Failure of explosively bonded clads in bend and tensile tests are virtually unknown. As a full set of tests uses a significant amount of metal and increases the cost of the clad it is common practice among fabricators to restrict testing to a shear test only, and many accept clad plates without destructive testing, their experience having shown that the bond strength of clad plates can be relied upon.

Another standard for clad plate is the German specification ADW8 which requires a shear test similar to ASTM, a tensile test, impact testing of the steel and a side bend test. The side bend test takes the form of a 10 mm wide section cut from the thickness of the clad and bent round a 40 mm diameter mandrel with its axis vertical to the interface. No cracking or disruption of the bond is permitted. The impact specimens are taken from steel immediately adjacent to the interface and the results obtained can be affected by the work hardened zone, as is the performance in the side bend test. It is advisable to stress-relieve all clads subjected to ADW8 even if the combination is not one normally heat treated. The same does not apply to ASTM A263, etc. where the bend tests do not stress the interface significantly and are included primarily to confirm the bend properties of the base steel. Heat treatment would only be carried out if it was felt that the requirements of the tensile test, particularly minimum elongation, made it desirable.

Another routine test which is carried out is a ram test which measures the bend strength in tension, and is often used for aluminium clads in place of the more usual shear test. It has also found application in proving the bond strength of tube plates in which the interface is subject in service to high tensile load. The proportions of the ram tensile specimen can be chosen to correspond with the tube diameter and the ligament area of the actual tube plate so that the test gives a true indication of the tensile load which could be sustained by each tube in the bundle.

Other tests which are occasionally used are miniature tensile specimens machined from thick clads with the axis normal to the interface and flat sections notched to the interface through both metals to leave a section of the interface in shear.

A simple qualitative test of bond quality is to chisel the interface of a sample of the clad. A soundly bonded specimen will part in the softer metal rather than at the interface.

Figure 5.8 shows some destructive test specimens and Table 5.3 typical results for shear and ram tests.

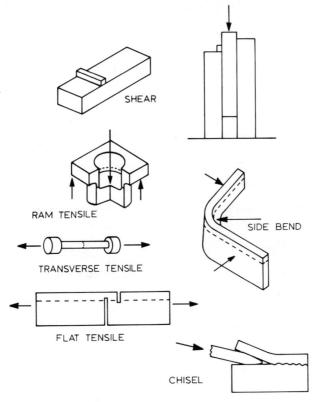

FIG. 5.8 Test specimens of clad plate.

TABLE 5.3
STRENGTH OF EXPLOSIVELY CLAD PLATES

	Shear test		Ram test	
	Min. strength guaranteed (N/mm^2)	Typical strength (N/mm^2)	Min. strength guaranteed (N/mm^2)	Typical strength (N/mm^2)
Aluminium	55	88	68	100
Brasses/Bronzes	140	280		
Copper	103	140		
Cupro-nickel 90/10	140	235		350
Cupro-nickel 70/30	140	216		
Nickel and nickel alloys	206	343		350
Stainless steel	206	363		
Tantalum	103	140		
Titanium	140	245	170	300

5.9. TUBULAR COMPONENTS

5.9.1. Nozzles

The bores of forged inlet and outlet connections for pressure vessels can be clad with stainless steels, nickel or copper alloys or titanium in thicknesses of 3 to 6 mm, and in diameters from 50 to 600 mm or more. The main controlling factor is the ability of the forging to withstand the detonation forces and Fig. 5.9 gives a guide to minimum wall thicknesses

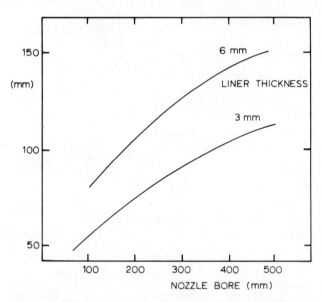

FIG. 5.9 Minimum nozzle wall thickness.

with forged steels having a minimum impact strength of 35J/cm^2 at ambient temperature. In many instances the finished profile of the nozzle has a minimum wall thickness below the critical value so the 'as-clad' blank has to be appropriately increased often conveniently to a simple cylinder with an outside diameter corresponding to the flange diameter, as Fig. 5.10.

The liner is formed from sheet material with a seam weld. Occasionally a standard tube can be used, but more often than not tubing of the precise dimensions is not available. A parallel gap between the liner and the forging base is provided similar to the corresponding plate clad. In calculating the sizes of the components, allowance is made for a slight

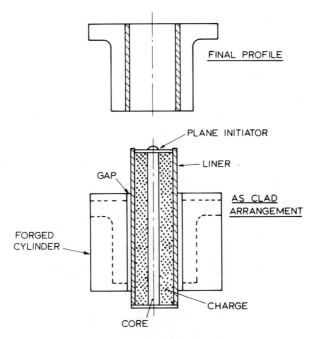

Fig. 5.10 Cladding of nozzles.

growth of the internal bore of the forging due to the explosive forces (Fig. 5.11) and also for the thinning of the liner as it expands through the radial gap. The latter is only significant with small diameter nozzles as the gap is not affected by diameter, and is for example about 10% for a 50 mm diameter nozzle.

The standard of bond offered is the same as for tube plates up to within 25 mm of the ends of the forging as clad, so extra length is required if the whole of the nozzle is required with guaranteed bond. There is also slight belling of the forging ends but within the guaranteed length a final inside diameter of better than ± 3 mm is achieved.

Extra length of the liner is necessary, particularly at the initiator end to allow for the axial initiation system and to make sure that the explosive is propagating uniformly. A 125 mm extension is provided at the initiator end for large diameter nozzles plus a further 25 mm extension at the other end.

The forging bore is smooth machined, and to give an accurate and uniform radial gap the liner must be fabricated to close tolerances with a maximum deviation from a true cylinder of ± 1 to 2 mm depending on

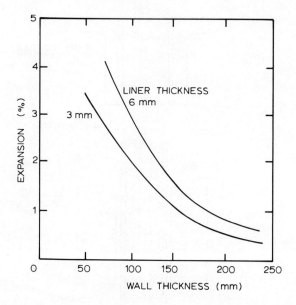

Fig. 5.11 Bore expansion for nozzles up to 250 mm diameter.

thickness. Spacing is controlled by attachments at the forging ends which can make slight correction in liner ovality.

The explosive is poured inside the metal liner with a central core to give the correct thickness of charge. The downward limit of diameter is determined by the point at which the core diameter becomes zero which in the case of a 3 mm liner is about 50 mm final inside diameter. The maximum length of nozzles handled is about 1 m. Longer lengths are possible provided the lining is accurately formed. Minimum length is usually determined by cost factors bearing in mind the extra length of the components. Nozzles 600 mm diameter and 200 mm length have been successfully produced.

5.9.2 Other Tubular Components

Items other than nozzles can be internally or externally clad, but in the majority of cases the thickness of the base tube is insufficient to withstand the detonation process and additional support must be provided. For internal cladding it will take the form of a heavy metal die of either split or unit construction. In the latter case the die is designed with a clearance of 12 to 25 mm into which is located a medium which can be subsequently removed to allow extraction of the components after firing.

Water is the simplest material but will not give adequate support at the ends of the die. As well as allowing over-expansion at the ends, extraction may be impossible. A heavy thixotropic slurry of barytes has been used successfully for the production of light gauge tubular clads and it shares with water the advantage of simple filling and removal. Heavier tubes have been clad using molten bitumen or lead as the intermediate support but require preheating of the die and tubular components together with careful filling to prevent voids. Removal also calls for the use of heat.

A split die does not give rise to such difficulties but is considerably more expensive to manufacture and would be used for a production run, whereas the unit die would be more economic for a small number. Some progressive growth of the die is inevitable with corresponding changes in the diameter of the clad components, but if they are subsequently to be drawn down to smaller gauge it is not a problem. However, the overall cost of producing clads with a die is inevitably high and has only been used for items for which there was no other possible route.

External cladding of tubes can be simpler in arrangement. A steel core with a bitumen intermediate has been used but water is satisfactory provided the end distortion can be discarded. Provided the thickness ratio is not extreme it is possible to achieve an internal clad by imploding the base on to the supported liner.

General consideration of explosive loading and gapping are as for nozzles with the additional problem that long tubes will require spacers other than at the ends. Metal spacers can be tack welded to the inner tube at intervals and care taken in inserting into the outer tube. With external cladding particularly using a light gauge cladding tube it is essential to achieve uniform axial detonation to avoid local gathering and creasing of the metal. External clads of 0.75 mm tantalum have been successfully produced in short lengths.

5.10. EXPLOSIVE HARDENING

The application of explosive shock waves to most metals results in some degree of work hardening but its use as an industrial process is limited to 13% manganese steel or Hadfield steel which has the characteristic of a very substantial increase in hardness from 200 Brinell as cast to 500 Brinell when work hardening is fully developed.

Manganese steel is used for the jaws of rock crushing machinery, digger teeth, conveyor components and such like applications where wear rates are high. It is also used for railway crossings subjected to heavy duty. Under ideal circumstances the metal develops a natural hardness due to the working conditions which maximise its operating life. Where abrasion is high the metal wears away before the surface has achieved its maximum hardness and the operating life is thereby reduced. By prehardening the surface and outer layers with explosive the initial wear rate can be reduced sufficiently for the natural work hardening to fully develop deeper in the metal and so achieve the maximum possible overall life.

The process consists simply of detonating a thin layer of high velocity explosive applied to the working surfaces of the component. A special explosive has been developed for the purpose in the form of a soft flexible sheet 3 mm thick which can be readily cut to size and attached to the metal either by its natural tackiness or by a coating of cellulose varnish. One layer of explosive will give a hardness of 400 HV and may be sufficient. Critical areas can have a double thickness of explosive or two shots in sequence to give maximum hardening.

The component to be treated is laid on a bed of sand. As only a few kg of explosive will be fired at one time the operation can be carried out at any convenient site. Often the item is associated with quarrying operations and hardening can be done within the quarry, possibly by the local operators after suitable instruction. Large components can be hardened in sections to minimise noise. The explosive propagates readily at high velocity even in thin layers and is detonator sensitive. The use of explosives will reveal inherent faults in the metal such as surface inclusion but will not damage the component other than possible spalling of the ends of thin sections such as crusher jaw teeth. Spalling can be prevented by continuing the explosive round the ends of the teeth to cancel out the reflected tension waves in the metal.

The economic benefit of the process depends largely on local conditions and the relationship between wear rates due to abrasion and the natural work hardening. Only a comparative test can establish the improvement in life to be gained in a particular case.

Chapter 6

MECHANICS OF EXPLOSIVE WELDING

H. EL-SOBKY

*Department of Mechanical Engineering,
University of Manchester Institute of Science and Technology,
Manchester, UK*

6.1. INTRODUCTION

The explosive welding process is one of the most useful and widely employed applications of the high energy rate methods to the fabrication of materials. Its major advantage lies in that it does not suffer from the limitations imposed on other welding processes by their specific characteristics. For example, in fusion welding melting of the two metals to be welded is essential at the interface, and in pressure welding large plastic deformation is required. Both processes are therefore limited to metals which have comparable melting points and plastic flow stresses respectively. Practical problems are also encountered in the case of metals and alloys with high melting points (e.g. tungsten) or with high mechanical strengths. Neither melting nor excessive plastic flow are essential for explosive welding to succeed though they may be observed under certain conditions, and hence no serious limitation is imposed by large differences in the corresponding metal properties. The use of explosives imposes nevertheless a limitation from the view point of safety regulations (see Chapter 5).

The process is a true example of multidisciplinary research as the phenomena associated with it fall under the various branches of engineering science. Although a large volume of research effort has been expended over the last twenty years or so and has resulted in a sound understanding of the process, a great deal of work in, and collaboration between various specialised fields are still necessary to achieve a comprehensive quantitative theory capable of giving an accurate description and prediction of the parameters and of the characteristic features of explosively welded components.

It is not possible, within the available space, to provide a comprehensive literature survey or an extensive theoretical analysis of the various aspects of the process. Instead, we will confine ourselves to the task of providing the reader with some working knowledge of the mechanics of this operation, of the determination of the welding parameters and the characteristic interfacial waves in as much detail as the state of the art and available space allow.

6.2. THE MECHANISM OF EXPLOSIVE WELDING

Figure 6.1 shows a typical set-up for explosive welding in which one component is driven by the detonation of an explosive charge to collide and weld with another.

Different mechanisms have been suggested to explain the process of the explosive welding in the early stages of development.[1] Some considered it to be essentially a fusion welding process (Zernow et al., Davenport, Duvall, and Philipchuck reviewed in ref. 2) which relies on the dissipation of the kinetic energy at the interface as a source of heat sufficient to cause bilateral melting across the interface and diffusion within the molten layers. Such fluid diffusion would lead to a gradual transition from one metal to another. In explosive welding it is, however, generally observed that the transition is quite a sharp one, even in those cases where pockets or layers of solidified melt—identified by the grain structure—appear locally. These form well defined boundaries with the parent metal materials. Other early investigators regarded it as a pressure weld operation which relies on large plastic deformation at the interface to allow clean surfaces to be formed and for a solid diffusion process to take place providing that a high pressure level is maintained for a sufficient length of time. This, however, is not the case in explosive welding where peak pressures are maintained for a few microseconds only and the coefficient of diffusion is small. In addition, we observe that the interfacial waves, vortices and/or melt pockets often present cannot

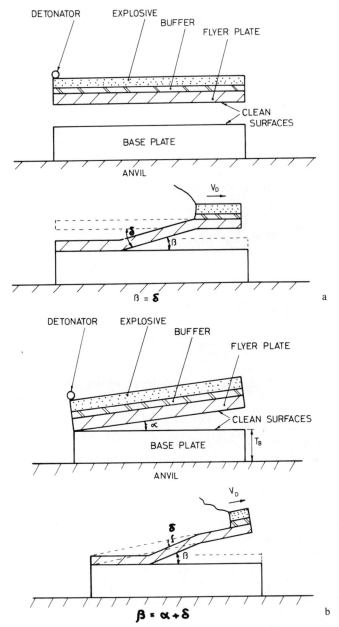

FIG. 6.1. Basic welding systems. (a) Arrangement for parallel plate welding (top, before detonation; bottom, during the detonation propagation), (b) arrangement for welding of plates inclined initially at angle α (top, before detonation; bottom, during the detonation propagation), (c) (overleaf) welding velocity diagrams.

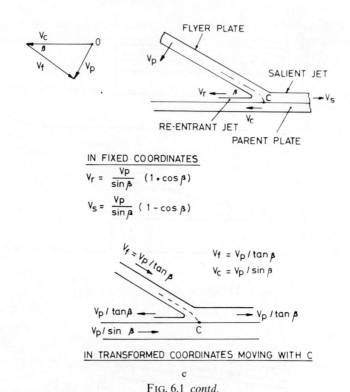

Fig. 6.1 contd.

be accounted for by either a pressure welding or pure melting mechanisms. The interfacial grain deformation and wave formation suggest that the mechanism of welding must be associated with a flow process. Otto[11] established experimentally the fact that the interfacial shear occurs during the welding and attributed the weld to the effect of the heat generated by shearing at the interface. This could cause sufficient heating of the boundary layers to achieve a bond and, could also account for the wave formation.

The proposed shearing mechanism was considered inadequate by Bahrani,[4] since it ignored the existence of a jet produced normally in an oblique collision and consequently did not assign to it any role in the bonding process.

Most of the discussed investigations were carried out using microscopy (SEM and TEM) which is limited in the degree of resolution by the technique of preparation employed. Recently, however, Hammerschmidt and Kreye[22] reported an improved technique in connection with the TEM analysis, which enabled them to examine the transition zone at the interface with a much higher resolution than previously possible.

These tests revealed that a very thin layer of melt is formed at the interface, most of which is ejected from the system by the jet. The estimated rate of cooling of the remainder—due to the contact with the relatively large bulk of the welded components—is of the order of magnitude of 10^5 K/s. Such a high rate of cooling leads to the formation of ultrafine grains of random orientation and about $1-2\,\mu$ thick. The existence of this amorphous layer within the bonding zone has been observed in various combinations of metals and is interpreted by the authors as the basic mechanism of explosive welding.

We must, however, make the distinction between the theories which describe the exact manner in which welding occurs and are based on the well understood physical phenomena, and the assertions which are used to explain the observed facts in terms of such theories. While the philosophical arguments continue, we have to be satisfied with a mechanism, based naturally on experimental facts, which is capable of encompassing all the relevant technological parameters and can therefore enable us to establish a working range of welding parameters applicable to any given material combination. On this basis, it is generally accepted by many investigators[2,3,11-22] that the well known phenomenon of jet formation at the collision point is an essential condition for welding. This is because in fact, it is through the agency of the jet that chemically clean mating surfaces are produced, free of films and contaminants, that make it possible for the atoms of two materials to meet at interatomic distances when subjected to the explosively produced pressure waves. The high collision pressure can thus produce a condition which 'opens' up the electron media of the welded metals and produces a bond across the interface.

Since the interfacial pressure during the collision can reach values of the order of magnitude of 10^2 Mbar, the strength of the metals involved is clearly very small by comparison and the constituent metals can therefore be treated—transiently—as fluids. The problem of interfacial jet formation thus becomes one of fluid mechanics and the condition limiting the occurrence of the jet can be established.

Much of the early work on jet formation was carried out in connection with hollow charges and penetration devices,[5-10] but more recently Cowan and Holtzman,[14] among others, have shown that jetting occurs in oblique collision between plates in the absence of an oblique shock wave attached to the propagating collision point. This criterion has been accepted as the limiting condition for jet formation. Figure 6.2 shows—for a selection of materials—the critical collision angle β, for jetting to occur as a function of the collision velocity. Ishi et al.[23] provided further experimental evidence of the occurrence of the jet

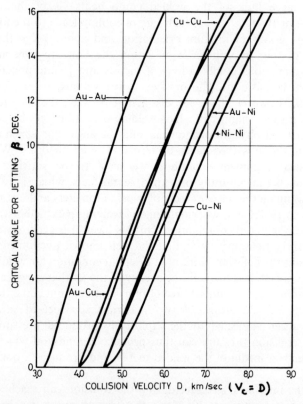

FIG. 6.2. The relationship between the calculated values of β and the collision point velocity for initially parallel plates (Ref. 14).

during explosive welding and studied the effect of asymmetry of collision, in various metal combinations, on the process of wave and jet formation and on the mechanics of the metal flow within the collision region.

The requirements for welding to occur can be summarised as follows:

(i) The occurrence of the jet at the interface.
(ii) An increase in the pressure—associated with the dissipation of

kinetic energy—to a sufficient level and for a sufficient length of time to achieve interatomic bonds, and for these to reach stability. In this case, the pressure is determined by the impact velocity, whereas the time available for bonding is governed by the velocity of the collision point.

A general, theoretical analysis of the phenomena involved is presented in the next section with a view to providing means of definition of the welding parameters and of explanation of the process of wave formation.

6.3 PARAMETERS OF THE EXPLOSIVE WELDING PROCESS

Considering the general case of plate welding arrangement, shown in Fig. 6.1, the welding process can be divided into three basic stages.

(i) The detonation of the explosive charge.
(ii) The deformation and acceleration of the 'flyer' plate or component.
(iii) The collision between the welding components.

After the initial impact between the components has taken place the process is assumed to continue in a steady manner. The contribution of (i) is to provide the energy required for the deformation and acceleration of the flyer element across the stand-off distance.

During stage (ii), the flyer element acquires momentum and a velocity V_p at the point of collision with the stationary component, where the kinetic energy is dissipated and welding occurs (stage (iii)). In the case of multiple collisions, as in the welding of multilayer composites this process is repeated at each interface, Fig. 6.3(a). The kinetic energy is reduced at each interface as interlaminar impact takes place due to progressive increase in the accelerated mass. In some cases, an additional buffer is used to protect the top surface from damage by the detonation products and, if necessary, a driver plate is used to artificially increase the mass and hence the momentum of the accelerated components.

The welding parameters which influence both the final conditions in the collision region and the mode of energy dissipation, are thus contributed by the three stages of the process and can be summarised as follows:

(i) The impact velocity V_p.
(ii) The collision point velocity V_c.
(iii) The dynamic angle of collision β.

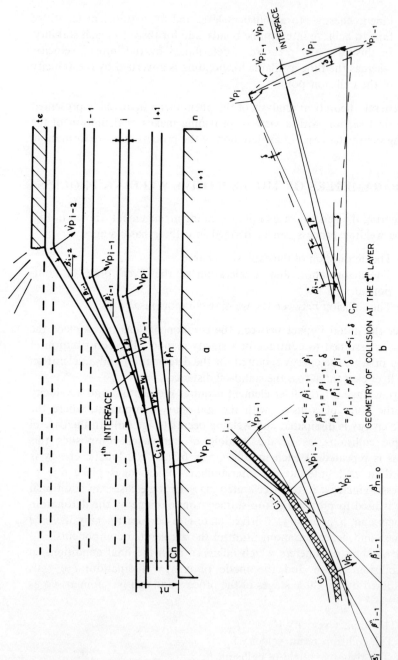

FIG. 6.3 Multiple impacts. (a) Multiple collisions in composite foil welding, (b) velocity diagram at the ith interface.

The magnitudes of these parameters can be varied by manipulating the geometry of the system, the charge type, the detonation velocity, the thickness of the buffer or the driver plate, and the stand-off distance. The geometry is defined as either parallel or oblique (Figs. 6.1(a) and (b)). The above-mentioned parameters can be calculated quite easily by considering in detail the mechanics of momentum transfer from the charge to the flyer plate, the continuity of flow through the collision region in transformed coordinates (Fig. 6.1(c)) and the dynamic deformation of the plate.

6.3.1. The Collision Parameters V_p, V_c, β

Different authors have allocated different directions to the plate velocity V_p and consequently the velocity relationships, obtained from the velocity diagrams, differ from one work to another. Figure 6.4 shows a range of the velocity diagrams relative to the collision point. In Fig. 6.4(a), V_p is assumed to bisect the angle between the portion of the plate already accelerated, behind the detonation front, and the undeformed portion. This assumption was justified by Birkhoff[5] for the case of an axisymmetric conical shell by considering the continuity of mass flow through the collision point. Figure 6.4(b) for instance, implies that V_p is perpendicular to the plate after detonation, which means that the accelerated part rotates about the plastic hinge, formed at the detonation front, while the equilateral triangle in Fig. 6.4(c) gives $V_c = V_f$, i.e. the length of the flyer plate remains unchanged. In Fig. 6.4(d), V_p is perpendicular to a line bisecting the initial angle α, and in Fig. 6.4(e), V_p is perpendicular to the plate in its original position. A comparison between the respective configurations shows that the angle θ, which defines the direction of V_p, differs in the five cases; giving

$$\theta_B > \theta_A > \theta_C > \theta_D \tag{6.1}$$

for the case shown in Fig. 6.4(b) and with reference to Fig. 6.4(e), the following relationships can be deduced:

$$\left.\begin{array}{l} V_p = D \sin(\beta - \alpha) \\ V_f = V_p \cot \beta \\ V_c = V_p \sin \beta \\ \beta = \sin^{-1}(V_p/V_c) = \alpha + \delta \end{array}\right\} \tag{6.2}$$

Similar relationships are easily obtained for other configurations. Several empirical and semi-empirical formulae for the prediction of V_p have been

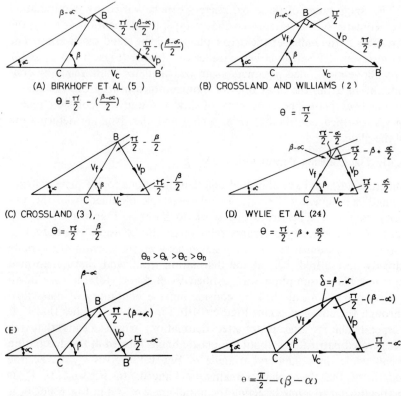

FIG. 6.4. Velocity diagrams, (a–d) Configuration proposed by various authors, (e) diagram based on V_p normal to the original position.

suggested and experimental results indicate good agreement with predicted value.[26]

The well-known Gurney equation, introduced during the Second World War gives the plate velocity as a function of the explosive energy, and of the charge to plate mass ratio.

For one-dimensional detonation:

$$V_p = \sqrt{(2E_K)} \left(\frac{3R^2}{R^2 + 5R + 4} \right)^{1/2} \tag{6.3}$$

The equation was examined later by Stern who suggested a slightly different alternative:

$$V_p = \sqrt{(2E_K)} \left(\frac{5/3 R^2}{R^2 + 5R + 5/4} \right)^{1/2} \tag{6.4}$$

where E_K is the explosive kinetic energy, and

$$R = \frac{\text{charge mass per unit area}}{\text{plate mass per unit area}} = \frac{m_c}{m_p}$$

Deribas et al.[12] obtained experimental data for low detonation velocity, granular explosives and fitted them to the following equation:

$$V_p = 1.2D \left[\frac{(1 + \frac{32}{27}R)^{1/2} - 1}{(1 + \frac{32}{27}R)^{1/2} + 1} \right] \quad (6.5)$$

From the gas dynamics, they obtained the expression:

$$V_p = \frac{3}{4}D \left\{ 1 + 2/R \left(\frac{R+3}{R+6} \right) \left(1 - \left[1 + \frac{3R(R+6)}{(R+3)^2} \right]^{1/2} \right) \right\} \quad (6.6)$$

where D is the detonation velocity.

Equations (6.3) to (6.6) are based on one-dimensional detonation (in a plane perpendicular to that of the flyer component and parallel to D).

An approximate two-dimensional solution was obtained by Duval and Erkman (reviewed in ref. 26) by multiplying the velocity V_p from (6.3) by a factor of ϕ, where

$$\phi = \left(1 - \frac{V_G}{2E_K} \right)^{1/2} = 0.7 \text{ to } 0.83 \quad (6.7)$$

and V_G is the particle velocity of the detonation gases behind the detonation front.

The accuracy of the adjustment factor ϕ decreases as the deviation of the explosive material from ideal behaviour increases. Table 6.1 summarises typical values of E_K, E_0 (total energy released by the explosive per unit mass) and ϕ. A similar correction was carried out by Takizawa et al.[33] This accounts for the deviation of the detonation velocity from the ideal value, and is given by

$$V_p = \frac{2.23}{(t_e)^{0.35}} \frac{D}{D_0} \sqrt{(2E_a)} \sqrt{\frac{0.6R}{1 + 0.2R + 0.8R}} \quad (6.8)$$

where V_p is the plate velocity m/sec, t_e is the explosive thickness in mm, D is the detonation velocity of charge of thickness t_e, D_0 is the ideal detonation velocity of an infinitely thick charge, E_a is the available energy from adiabatic expansion of explosive gases, and R is the charge to plate ratio.

TABLE 6.1
EXPLOSIVE CONSTANTS FOR USE WITH MODIFIED GURNEY EQUATION (Ref. 26)

Explosive	Heat of explosion E_0 (cal/g)	Gurney energy E (cal/g)	Detonation velocity (m/s)	Explosive density (g/cm^3)	$\sqrt{(2E)}$ (m/s)	ϕ
Detasheet 'D'	870	870	7 200	1.40	2 700	0.71
Pentolite (cast)	1 220	1 220	7 470	1.66	3 170	0.73
Pentolite (powder)	1 220	900	5 150	0.95	2 730	0.76
TNT (cast	1 160	1 160	6 700	1.56	3 100	0.74
TNT (powder)	1 160	800	4 800	1.0	2 580	0.77
Nitrostarch dynamite (70%)	1 050	430	3 300	1.0	1 810	0.78
AN dynamite (40%)	800	240	3 200	1.25	1 400	0.81
AN–8% Al	1 100	205	2 300	1.05	1 300	0.80
AN–6% fuel oil	890	310	2 540	0.82	1 600	0.80
Trimonite No. 1	1 260	185	3 000	1.1	1 240	0.81

6.3.2. Limiting Conditions for Welding

While the various welding mechanisms discussed earlier differ from one another in so far as their explanations as to how welding occurs are concerned, they all almost agree that it occurs as a direct result of high velocity collision. Experimental results indicate that there are certain critical values for both the collision and the geometrical parameters which have to be observed, i.e. either exceeded or not exceeded depending on the system, in order that an acceptable weld is obtained. These can be summarised as follows:

(i) Since a jet is required at the collision region, a minimum collision angle β must be exceeded. For a given metal the value of such an angle is a function of the collision velocity. In this connection Walsh et al.,[8] and Harlow et al.,[27] calculated the critical values of β as a function of the collision velocity and of the dynamic equation of state. Similar results were obtained by Cowan and Holtzman.[14] These are shown in Fig. 6.2.

(ii) The collision velocity V_c and the plate velocity V_p should be less than the velocity of sound in either welding component. This condition was established experimentally, and was used by El-Sobky and Blazynski[28] as an explanation of the condition necessary for the reflected stress waves not to interfere with the incident wave at the current collision point. It is also known that at supersonic velocities, the dynamic pressure is not held for a sufficiently long period to create physical conditions conducive to the adjustments of interatomic diffusion and of equilibrium within the collision region. Although Wylie et al.[24] suggested that the velocity of sound may be exceeded by up to 25%, the subsonic collision appears to be more satisfactory. As V_c is related to D and β, it may be adjusted either by reducing the detonation velocity or by introducing an initial angle of obliquity. The velocity of sound, or more precisely, the velocity of stress wave propagation provides an upper limit for V_p and V_c.

(iii) A minimum impact pressure must be exceeded (and hence a minimum plate velocity V_p), in order that the impact energy should be sufficient to produce a weld. It is suggested that the impact energy required is related to the strain energy and to the dynamic yield strength of the flyer material.[24] An upper limit for the energy is also required to avoid excess heating and possibly melting by viscous dissipation and thus the formation of brittle layers. Obviously such an upper limit should be sought in terms of the melting energy of the lower melting point of the weld combination.

(iv) A sufficient stand-off distance, S, has to be provided so that the flyer component can accelerate to the required impact velocity. A satisfactory value lies between 0.5 to 1.0 times the thickness of the flyer component.[24] Ezra[26] suggests different values of S which depend on the specific gravity of the flyer material. These are, in multiples of thickness $(\frac{1}{2}-\frac{2}{3})$, $(\frac{1}{2}-1)$, $(\frac{2}{3}-2)$ for specific gravities of less than 5, between 5–10, and greater than 10 respectively.

From the above summary of the important physical parameters, one can clearly conclude that the interaction between the geometrical and dynamic collision parameters on the one hand and material characteristics on the other, makes it very difficult to separate their respective efforts completely (see next section). Consequently, some authors attempt to estimate the total energy required and, from this, the explosive charge load in preference to a theoretical approach based on the mechanics of the process. Blazynski and Dara[29] for instance, used the dimensional analysis to derive the following equation relevant in the welding of duplex cylinders:

$$E = 0.226 \, \rho D^2 L t^2 \left(\frac{d}{t}\right)^{3/2} \quad \text{(in S.I. units)} \qquad (6.9)$$

where E is the explosive energy, ρ is the density of the flyer material, L is the length of assembly, t is the flyer thickness, d is the flyer cylinder thickness and D is the detonation velocity.

Using the same method Kowalick and Hay[30] obtained, for a flat component:

$$R = K\rho t c^2 \beta^2 \qquad (6.10)$$

where K is a constant, R is the charge load ratio, and c, t are the sound velocity in the flyer material, and flyer thickness respectively. Each of these two relationships is essentially an approximation since accurate correlation between the dimensionless groups could not be established. Carpenter et al.[31] derived another expression on the basis of an energy balance between the explosive energy and the sum of kinetic and elastic–plastic energy of deformation of the flyer plate. This is in the form of

$$L_e = \rho_e t_e \alpha \frac{\beta^2 t Y}{S} \qquad (6.11)$$

where the suffix e denotes explosive material and Y is the yield strength.

The above expressions, though not exact, are reported to yield a reasonable degree of success and can be used together with accumulated experience to produce the necessary information. However, the relatively large number of parameters and their interdependence makes it necessary to devise a more systematic method of determination of the working range of basic parameters. Christensen et al.,[16] Wylie et al.,[24] amongst others, suggested a graphical construction of the diagram giving the range of upper and lower limits of each parameter within which an acceptable weld is to be expected. To construct such a diagram they accepted the formation of waves as an indication of the presence of a good weld. This is not necessarily true, since in some cases the waves appear on poor quality interfaces and separation—due to the effect of stress waves—occurs on the interfaces after the welding. The latter effect was noted by El Sobky and Blazynski.[28] Bearing this in mind, and judging by the experimental evidence, the wave formation can be accepted as a reasonable criterion because it leads to the determination of a range of parameters within which manipulation of conditions is possible and therefore elimination of the secondary effects such as the coincidence of stress waves with the collision region can be made. This method has become known as the 'weldability window'.

The critical parameters used to establish a weldability window are: (i) the critical collision angle for jet formation β_L, (ii) the collision point velocity V_c, and (iii) the kinetic energy and impact pressure in the collision region associated with the impact velocity, V_p. All these quantities can be represented simultaneously in a $\beta - V_c$ coordinate system. Figure 6.5 shows a schematic representation of the

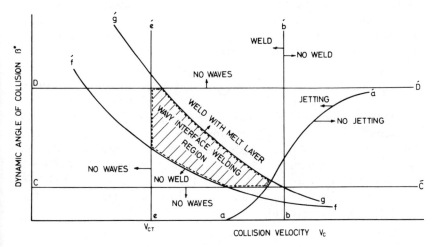

FIG. 6.5. Weldability window.

collision for a monometallic system. Line aa' represents the critical angle β_c necessary for jet formation. A set of such curves is given in Fig. 6.2 and represents the results obtained by Abrahamson,[32] and Cowan and Holtzman.[14] For detailed information on the theoretical calculation of β the reader is referred to the papers by Walsh et al.,[8] Harlow et al.,[27] Abraham,[32] and Cowan and Holtzman.[14] Line bb' represents the upper limit of V_c estimated at 1.2 to 1.5 times the sound velocity.[12,16] It is, however, experimentally evident that approaching the upper limit of V_c restricts the choice of other parameters within the weldability window.

The lower and upper limits of the dynamic angle β have been experimentally determined by Bahrani and Crossland[4] amongst others. A lower limit of $2°-3°$ and an upper limit of $31°$ have been suggested, and the data published are in general agreement with these values. The upper and lower values of the initial angle α in an inclined system have been determined[4] as $18°$ and $3°$ respectively, whereas the limits of the actual dynamic angle—corresponding to these values—can be estimated by using any of the velocity diagrams discussed earlier, e.g.

$$\beta = \tan^{-1} \frac{V_p \cos(\alpha/2)}{V_c - V_p \sin(\alpha/2)} \qquad (6.12)$$

and are represented by lines cc' and dd' in Fig. 6.5. As a lower limit of the collision velocity, the analysis of Cowan et al.[34] gives an expression which defines the minimum transition velocity above which a wavy interface is obtained. In analogy with fluid flow, a value for the equivalent Reynolds number is given by:

$$R_c = \frac{(\rho_F + \rho_B)V_c^2}{2(H_F + H_B)} \qquad (6.13)$$

where the subscripts C, F, B denote the collision region of the flyer and base components respectively, and the diamond pyramid hardness, H, is taken as a measure of the strength. The transition occurs at $R_c = 10.6$ and this defines the minimum value for V_c indicated by line ee' in Fig. 6.5.

The remaining factor to be determined is the impact velocity V_p which is responsible for the impact pressure and for the impact energy. The impact pressure is given by:

$$P_i = \rho V_p C$$

where C is the bulk sound velocity.

As a certain minimum pressure must exist, it follows that a minimum velocity V_p should also exist. This was determined by Wittman[35] and is given by

$$V_p = \frac{5(HEL)}{\rho C} \qquad (6.14)$$

where HEL is the Hugoniot elastic limit,* or alternatively by

$$V_p = \left(\frac{\sigma_{TU}}{C}\right)^{1/2} \qquad (6.15)$$

where σ_{TU} is the ultimate tensile stress.

Both equations give values for V_p and P_i in agreement with Wittman's experimental results.

* $HEL = \frac{1}{2}(K/G + 4/3)Y_0$, where K is bulk modulus, G is the shear modulus and Y_0 is the tensile yield stress, defined as the normal stress at yield for a perfectly plastic solid.

Once the minimum pressure has been provided the quality of the weld will depend on the amount of kinetic energy dissipated in the collision region. Wylie et al.[24] suggest that the impact energy required should be related to the total strain energy up to the level of the dynamic yield, i.e.

$$E_i = \frac{\lambda \sigma_D^2}{2E} = \frac{1}{2}\rho t V_p^2 \qquad (6.16)$$

where E_i is the impact energy, E is the Young's modulus, σ_D is the dynamic yield strength, λ is a constant, and t is the flyer thickness.

This approach appears to be reasonable but it depends on the availability of precisely defined values for the dynamic yield stress, σ_D.

The line ff' representing the lower limit of V_p in Fig. 6.5 can thus be calculated.

The condition for melting to occur can be used as a basis for calculating the upper limit of the impact energy and subsequently V_p. Adiabatic heating—due to the entrapment of a portion of the jet-energy— causes transient temperature rises which in turn, may cause melting of either or both welded components followed by a very rapid cooling. The thermophysical properties of the unalloyed mixture at the interface as well as the thermal conductivity of the flyer and base will govern this process. An expression for the maximum value of V_p in terms of the thermal properties is given by Wittman as

$$V_p = \frac{1}{N} \frac{(T_m C)^{1/2}}{V_c} \left(\frac{KCC_h}{\rho t} \right)^{1/4} \qquad (6.17)$$

where N is a constant, T_m is the melting point, K is the thermal conductivity, C_h is the specific heat, V_c is the collision point velocity, and t is the flyer thickness. Equation (6.17) is represented by the line gg'.

The shaded area of Fig. 6.5 represents the range of parameters within which an acceptable weld can be obtained. Experimental results provide in general, evidence of the veracity of the arguments used in establishing the above criteria. In the case of bimetallic welding, the weldability window should be established for each material and then superimposed on one another. This will result in an area of overlap which will represent the working range of parameters applicable to the combination in question.

Table 6.2 lists some welding data for mono- bi- and trimetallic combinations.

TABLE 6.2
WELDING PARAMETERS TO OBTAIN A CONTINUOUS WAVY INTERFACE (VARIOUS AUTHORS)

System	Materials	Energy (kJ/cm^2)		Velocity V_p (m/s)	
		Min.	Max.	Min.	Max.
Mono-	Mild steel (MS)	20	74	400	820
	Stainless steel (SS)	44–66	86.5	270	330
	Copper (Cu)	3.5–4.8	13.5	255	585
	Hard copper	>10	57.5	330	760
	Aluminium (Al)	10	40	500	960
Bi-	Cu–Brass	12.1			
	MS–SS	11.5			
	MS–Cu	10.7		Not available	
	MS–Al	6.2			
	MS–EN 30B	10.5			
	Brass–MS	4.3			
Tri-	Br–Cu–MS	16.1		Not available	
	MS–Br–Cu	35.5			

6.4. INTERFACIAL WAVES

6.4.1. Introduction

The interfacial periodic deformation, usually referred to as the 'interfacial waves', is perhaps the most discussed aspect of explosive welding.

The appearance of such periodic surface deformation is not confined to interfaces in explosive welding, but remarkably similar effects can be found on solid surfaces subjected to erosive action of high speed liquid film, jet drags, e.g. in pump, steam turbine blades and, of course, at the interface of air and sea. In fact the wave formation in explosive welding can be regarded as a special case of a general phenomenon of interfacial wave formation under certain flow circumstances.

The presence of the jet in the collision region, and the transient fluid-like behaviour under high pressure in the collision region[8,14,27] have led many investigators to seek an explanation and a characterisation of these waves in terms of a flow mechanism of one kind or another.

6.4.2. Mechanisms of Wave Formation

Various mechanisms have been suggested for wave formation, but on close examination we find that they are mostly of qualitative or semi-qualitative nature and as such do not provide the basis for an analytical treatment. An important point emerged recently[46,47] that is, that there exists a variety of wave types and therefore a possibility that each of the proposed mechanisms is capable of describing one type only but not the other. Hence the need for a closer look at the fluid mechanics of the collision zone in order to be able to establish a mechanism with a built in means of explaining the various type of waves.

Figure 6.6 shows three types of waves. Figure 6.6(a) shows a high degree of rotation with no sign of any appreciable melting. In Fig. 6.6(b) we note a large wave with a front but no back vortex in a multilayer Cu–brass–stainless steel wire reinforced composite. Obvious phase changes at the front vortex and at the crest of the wave are present. In Fig. 6.6(c), a multiple interface is shown which comprises waves with front and back vortices, plane interfaces and various degrees of asymmetry.

A brief review of the existing mechanisms is presented here followed by an analysis of the mechanics of flow in the collision region. According to the analogies used in each case, the existing mechanisms can be classified into four groups:

(i) *Jet Indentation Mechanisms.*
Abrahamson[32] postulated that waves are formed because of the indentation action of the salient jet on the base plate and the periodic release of the hump formed ahead of the collision point by the material removed by the indentation. This allows the salient jet to overtake the hump and the process continues. Bahrani *et al.*[36] gave a more detailed description of how the waves were formed but attributed the indentation to the re-entrant jet and the formation of vortices to the trapping of the re-entrant jet material between the hump and the salient jets. Fig. 6.7(a) illustrates this feature.

(ii) *Flow Instability Mechanism*
While the indentation mechanisms indicate that waves are formed essentially within the collision zone, the flow instability mechanisms suggest that they are created ahead of the collision point, e.g. Hunt,[37] or behind the collision point, e.g. Robinson,[38–40] as a result of a velocity discontinuity across the interface which involves the re-entrant jet and salient jet respectively. The process of wave formation is treated as a Helmholz instability. In Hunt's work it is postulated that the discontinuity is a sharp one and that the re-entrant jet is necessary for

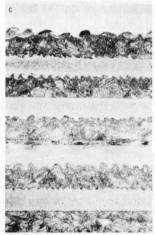

FIG. 6.6. Types of welding waves, (a) Rotation without appreciable melting, (b) large wave with front vortex only, (c) multiple interface waves.

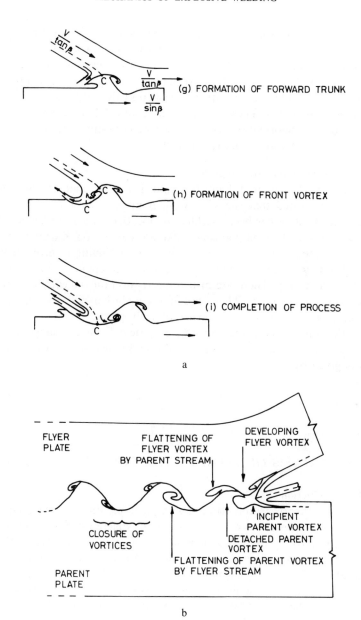

FIG. 6.7. The process of wave formation. (a) the Bahrani mechanism (Ref. 36), (b) the Kowalick and Hay mechanism (Ref. 30).

wave formation. It is also assumed that the re-entrant jet must remain in contact with the parent plate and that the fluid behaviour does not extent to the salient jet. In Robinson's work, calculations show that the strain rates within the salient jet are sufficiently high to justify fluid like behaviour and that the velocity profile across the interface has one or two inflection points. These assumptions enabled him to compute the surface of the deformation as a surface of constant vorticity, and he obtained a wave with a rolling up crest.

(iii) Vortex Shedding Mechanisms
Cowan and Holtzman,[14] Kowalick and Hay,[41,42] and Onzawa and Ishi,[43,44] suggested that the stagnation point acts as a solid obstacle, once a steady state has been established, and that waves are formed due to a vortex shedding mechanism—analogous to a von Kármán vortex street—initiated at the stagnation point and continuing to grow behind the collision zone. Fig. 6.7(b) shows the vortex mechanism suggested by Kowalick and Hay. There are also some reports on the effects of stress waves on the interface. These are not conclusive and the reader is referred to the literature for a review of such effects.[47]

In addition to these mechanisms, empirical formulae suggested by Deribas *et al.*[47] give satisfactory results in predicting the wavelength. This is given by

$$\lambda = 32\left(t_1 \sin^2 \frac{\gamma_1}{2} + t_2 \sin^2 \frac{\gamma_2}{2}\right) \qquad (6.18)$$

6.5. ANALYSIS OF FLOW IN THE COLLISION REGION

6.5.1. Introduction
The various theories summarised above postulate different explanations of the process of wave formation and of the difference in the predicted position, relative to the collision point, where this process takes place. The very nature of explosive welding process renders the direct observation of the wave formation, which takes place on a microscopic scale in a matter of a few microseconds practically impossible. However, since it is understood that wave formation is essentially a fluid flow phenomenon, it is possible to model the actual welding system by a kinematically similar real fluid system that allows slowing down of the process in time and thus makes visual observation possible. Such a

'liquid analogue' utilises layers of a suitable fluid and a mechanism to achieve oblique collision was constructed.[45,48] It proved to be successful in reproducing the collision geometry, and the phenomena of jet and wave formation. Modestly high speed photography (100–180 m/s) was quite adequate to observe the details of the process with an equivalent speed reduction of about 2000 times.[45,46] One fundamental difference between the real and the analogous systems, is that in the latter real liquids are used and therefore the waves formed continue to grow. While in the real system, the metals cease to behave as fluids and the waves freeze permanently.

The liquid analogue shows that waves are due to a combination of the two components of flow deformation; one occurring ahead of the stagnation point and the other behind it. Figure 6.8(c) shows one example of the fluid analogue results.

The experimentation with the fluid analogue also throws some light on the sequence of events which occur within the collosion region and was the basis of the following general analysis, which can be extended to include metals.

6.5.2. The Interfacial Pressure Profile

Since the fluid behaviour of metals depends on the magnitude of the pressure in the collision region, it is essential to examine pressure distribution on the interface around the stagnation point. The pressure profile in oblique collision of a jet on a solid surface was calculated by Taylor,[49] and can be expressed by an equation of the following form, with the x-axis taken along the interface and the origin coinciding with the stagnation point:

$$P(x) = \frac{1}{2}(a - q^2)V_f^2 \qquad (6.19)$$

where q is an auxiliary variable related to x by:

$$x = \frac{t}{\pi} \left\{ \log\left(\frac{1-q}{1+q}\right) - \cos\beta \log\left(\frac{1+q^2 - 2q\cos\beta}{1-q^2}\right) - 2\sin\beta \tan^{-1}\left(\frac{q\sin\beta}{1-q\cos\beta}\right) \right\} \qquad (6.20)$$

where t is the flyer plate thickness, β is collision angle and V_f the flyer plate velocity.

The behaviour of the materials can be assumed to vary between rigid

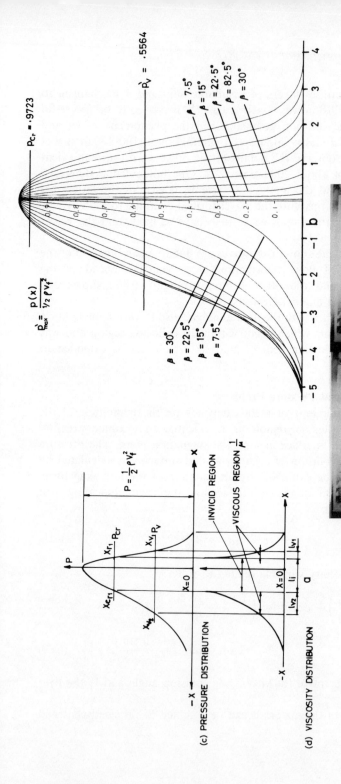

FIG. 6.8 Collision region flow, (a) Pressure and velocity distributions, (b) pressure and the dynamic angle β, (c) an example of liquid analogue (Refs. 45 and 46).

and inviscid. Such a variation is continuous and is a function of the pressure only. As a representative property, the viscosity μ can be expressed as:

$$\mu = \mu(P) \quad (6.21)$$

which takes the value of zero for inviscid behaviour and infinity for rigid behaviour. The viscosity can thus be expressed as a function of x in the form of

$$\frac{1}{\mu} = f(x) \quad (6.22)$$

Figure 6.8(a), shows the pressure distribution and a possible distribution of the inverse of the viscosity. From this graph, it is seen that the entire collision region lies between the points of intersection of the function $f(x)$ with the x-axis and that there exist two viscous regions of length l_{v1} and l_{v2} ahead and behind the stagnation point respectively, as well as an inviscid region l_i. These are defined by pressures P_v and P_{cr} which are the pressures at which viscous and inviscid behaviour can be significant. P_v and P_{cr} are thus material properties.

Figure 6.8(b) shows the dependence of the pressure profile on the collision angle β (in dimensionless coordinates). It is apparent that the asymmetry of the pressure distribution increases as β decreases with the steeper portion of the curve ahead of the collision point.

It can therefore be postulated that the flow at any typical point on the interface is viscous at the beginning and lasts for a time of the order of magnitude of l_{v1}/V_c over a length l_{v1} (where V_c is the collision point velocity) and it is followed by an inviscid flow of length l_i for a time equal to l_i/V_c, and a final viscous region of length l_{v2} for a time l_{v2}/V_c. It must be noted that the inviscid region can be of zero length if $P_0 \leqslant P_{cr}$, and that pure viscous flow may occur. If $P \leqslant P_v$ no significant flow will take place at all.

The flow region can be defined as $l_{vi} + l_i + l_{v2}$, the components of which are plotted against β in Fig. 6.9. A similar trend is observed to that of wavelength as a function of β, as reported in refs 43, 39.

6.5.3. The Flow Pattern in the Collision Region

The wave theories reviewed in the previous section are all based on a 'steady state' instability of one type or another. They cannot therefore account for the occurrence of the front and back vortices, which

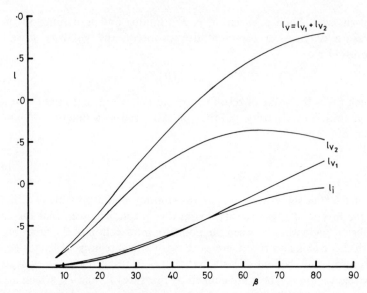

FIG. 6.9. The variation of length L with β in the collision region (Ref. 21).

demands an explanation in terms of a flow process that is reversed in direction on either side of the collision point.

It is also realised that there are various types of waves, see Figs. 6.6 and 6.7, which cannot all be explained by one of these mechanisms.

According to the description of the flow zone presented here, these problems can be solved since a clear source of periodicity and reversibility can be clearly defined. This becomes clearly apparent on noting that, while the magnitude of the pressure can be responsible for the transient nature of the material, it is the magnitude and direction (sign) of the pressure gradient that determine the direction of the flow. Figure 6.10 shows the distribution of the pressure profile at various collision angles. The pressure gradient is negative ahead of the stagnation point, decreasing from zero at infinity to a minimum value and reaching zero again at $x=0$; to rise to a maximum positive value behind the collision region $(x - ve)$. This variation is the source of reversibility, as the flow will change direction at any part on the interface once it is overtaken by ths stagnation point. The frequency of this change, at the same V_c, increases as β decreases (see Fig. 6.10). Here the distance between the highest negative and positive values decreases with β. This is

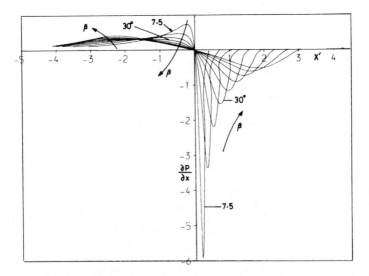

FIG. 6.10. The variation of pressure gradient with β (Ref. 21).

consistent with the observation that the wavelength is proportional to the collision angle.

It is also noted that the ratio of the absolute value of the pressure gradient (ahead of the stagnation point) to that behind it, is greater than one and that it increases as β decreases. This is also consistent with the observation that waves are asymmetrical and tend to elongate in the direction of V_c, i.e. in the detonation direction (see Fig. 6.6(a)).

While this theory is still being developed further, initial computer predictions of wave shapes appear to be promising, but a large volume of data is required—particularly about the shape of the waves and dimensions of the vortices. The advantage of this approach is that it provides the basis for quantitative analysis, as well as simplifying the complexity of the problem.

The above analysis can be summarised as follows; the collision region is defined by the pressure profile and the characteristic values of P_v, P_{cr}. The characteristic lengths l_v, l_i are related to wavelength, the periodicity is due to the sign reversal of the pressure gradient, the asymmetry of waves is due to the asymmetry of the profile of the pressure gradient and the sizes of the front and back vortices are proportional to the minimum and maximum values of the pressure gradient.

REFERENCES

1. CARL, L. R. *Metal Progr.*, **46** (1944), 102–103.
2. CROSSLAND, B. and WILLIAMS, J. D. *Metallurgical Reviews* (1970), 79–100.
3. CROSSLAND, B. *Metals and Materials*, Dec. (1971), 401–413.
4. BAHRANI, A. S. and CROSSLAND, B. *Proc. Inst. Mech. Eng.*, **179** (1964), 264.
5. BIRKHOFF, G., MCDOUGALL, D. P., PUGH, E. M. and TAYLOR, G. J. *Appl. phys.*, **9** (1948), 563.
6. PUGH, E. M., EICHELBERGER, R. J. and ROSTOKER, N. *J. Appl. phys.*, **23** (1952), 532.
7. EICHELBERGER, R. J., and PUGH, E. M. *J. Appl. phys.*, **23** (1952), 537.
8. WALSH, J. M., SHREFLER, R. G. and WILLIG, F. J. *J. Appl. phys.*, **24** (1953), 349.
9. EICHELBERGER, R. J. *J. Appl. phys.*, **26** (1955), 398.
10. EICHELBERGER, R. J. *J. Appl. phys.*, **27** (1956), 63.
11. MCLELLAND, H. T. and OTTO, H. *Proc. 4th Int. Conf. of the Center for High Energy Rate Forming*, University of Denver, Colorado, (1973).
12. DERIBAS, A. A. *Explosive Welding*, 1967, Siberian Academy of Science, USSR.
13. CARPENTER, S. H., WITTMAN, R. H. and OTTO, H. E. *Proc. 11th MTDR. Conf.*, 1970, Pergamon Press, Oxford.
14. COWAN, G. R. and HOLTZMAN, A. H. *J. Appl. phys.*, **34** (1963), 928.
15. BERGMAN, O. R., COWAN, G. R. and HOLTZMAN, A. H. *Trans. Met. Soc. AIME*, **236** (1966), 646.
16. CHRISTENSEN, K. T., EGLY, N. S. and ALTING, L. *Proc. 4th Int. Conf. of the Center for High Energy Rate Forming*, University of Colorado (1973).
17. EL-SOBKY, H. and BLAZYNSKI, T. Z. *Proc. 5th Int. Conf. of the Center for High Energy Rate Forming*, University of Denver, Colorado (1975).
18. BEDROUD, B., EL-SOBKY, H. and BLAZYNSKI, T. Z., *J. Metals Technology*, **1** (1976), 21.
19. JOHNSON, W. *Impact Strength of Materials*, Edward Arnold Co., 1972, London.
20. WYLIE, H. K., WILLIAMS, P. E. G. and CROSSLAND, B. *Proc. 3rd Int. Conf. of the Center for High Energy Rate Forming*, University of Colorado (1971).
21. EL-SOBKY, H. and BLAZYNSKI, T. Z. *Proc. 7th Int. Conf. of the Center for High Energy Rate Forming*, University of Leeds (1981).
22. HAMMERSCHMIDT, M., and KREYE, H. *ibid.*
23. ISHI, Y., ONZAWA, T. and SEKI, N. *J. Japan Welding Soc.*, **40** (1971), 523–534.
24. WYLIE, H. K., WILLIAMS, P. E. G. and CROSSLAND, B. *Proc. 3rd Int. Conf. of the Center for High Energy Rate Forming*, University of Denver, Colorado (1971).
25. SMITH, E. G., LABER, D. and LINSE, V. D. *Ibid.*
26. EZRA, A. A. *Principle and Practice of Explosive Metal Forming, 1973*, Industrial Newspaper Ltd., London.
27. HARLOW, F. H. and PRACHT, W. E. *The Physics of Fluids*, **9** (1966), 1951–59.
28. EL-SOBKY, H. A. and BLAZYNSKI, T. Z. *Proc. 15th Int. MTDR. Conf.* (1974), Macmillan Press, London.

29. BLAZYNSKI, T. Z. and DARA, A. *Proc. 11th Int. MTDR. Conf.* (*1970*), Pergamon Press, Oxford.
30. KOWALICK, J. F. and HAY, R. *Proc. 2nd Int. Con. of the Center for High Energy Rate Forming*, University of Denver, Colorado (1969).
31. CARPENTER, S., WITTMAN, R. H. and CARLSON, R. J. *Proc. 1st Int. Conf. of the Center for High Energy Rate Forming*, University of Denver, Colorado (1967).
32. ABRAHAMSON, G. R. *J. Appl. Mech.* (1961), 519–28.
33. TAKIZAWA, Y., IZUMA, T., ONZAWA, T. and FUJITA, M. Unpublished Research Report, Tokyo Institute of Technology, (1975), private communication.
34. COWAN, G. R., BERGMAN, O. R. and HOLTZMAN, A. H. *Met. Trans.*, **2** (1971), 3145–55.
35. WITTMAN, R. H. *Proc. 4th Int. Conf. of the Center for High Energy Rate Forming*, University of Denver, Colorado (1973).
36. BAHRANI, A. S., BLACK, T. J. and CROSSLAND, B. *Proc. Roy. Soc.*, **A296** (1967) 123–36.
37. HUNT, J. N. *Phil. Mag.*, **17** (1968) 669–80.
38. ROBINSON, J. L., *Phil. Mag.*, **31** (1975), 587–97.
39. ROBINSON, J. L. *Proc. 5th Australian Conf. on Hydraulics and Fluid Mechanics*, 1974.
40. ROBINSON, J. L. *Proc. 5th Int. Con. of the Center for High Energy Rate Forming*, University of Denver, Colorado, (1975).
41. KOWALICK, J. F. and HAY, D. R. *Met. Trans.*, **2** (1971), 1953–58.
42. KOWALICK, J. F. and HAY, D. R. *Proc. 3rd Int. Conf. of the Center for High Energy Rate Forming*, University of Denver, Colorado (1971).
43. ONZAWA, T. and ISHI, Y. *Trans. Japan Welding Soc.*, **4** (1973), 101–108.
44. ONZAWA, T. and ISHI, Y. *Proc. 5th Int. Conf. of the Center for High Energy Rate Forming*, University of Denver, Colorado (1975).
45. EL-SOBKY, H. A. *Experimental Investigation of the Mechanics of Explosive Welding by means of Liquid Analogue. Ibid.*
46. EL-SOBKY, H. A. and BLAZYNSKI, T. Z. *Proc. 6th Int. Conf. of the Center for High Energy Rate Forming*, Essen, W. Germany (1977).
47. DERIBAS, A. A., KUDINOV, V. M., MATVEENKOV, F. I. and SIMONOV, V. A. *Fizika Goreniya*, **4** (1968), 100–107.
48. EL-SOBKY, H. A. Ph.D. Thesis, 1975, University of Leeds.
49. TAYLOR, G. I. *Phil. Trans. Roy Soc.* **260** (1966), 96–100.

Chapter 7

EXPLOSIVE WELDING IN PLANAR GEOMETRIES

M. D. Chadwick and P. W. Jackson

*International Research and Development Co. Ltd,
Fossway, Newcastle upon Tyne, UK*

7.1. INTRODUCTION

Of the various explosively welded components now available, clad plate, with a current world-wide production rate of about 25 000 m^2 per annum, is in greatest demand by far. Although the clad plate has extensive direct application in the simplest planar forms in which it is produced, that is to say as rectangles or discs (for tube-plates), other forms may be obtained readily, using suitable fabrication techniques, from the flat clad plate. Notable examples of these are flat-or spherical-ended cylindrical heat-exchangers and pressure vessels made of structural steel in which the clad inner surface is a relatively thin layer of corrosion-resistant material such as stainless steel or titanium. For some applications, a single large plane clad plate can be machined to provide numerous transition joints which may be either planar themselves or tubular.

Commercial cladding involves, for reasons of economy, plates and buffer layers of large area and constant thickness and subsonically detonating explosives which will allow welding to occur with a constant stand-off gap between the cladding and backer plate. The last condition, however, imposes a restriction in that the cladding rate, or the collision point velocity, is in this case, equal to the detonation velocity of the explosive which may not be optimum for the combination of metals involved. For small-scale welding, the planar components to be bonded may be mutually inclined and/or tapered and optimum bonding conditions may then be obtained by one or more of a variety of welding techniques. Examples of such minor applications include seam and spot welding of thin sheets, the butt and scarf welding of plates and the welding of foils in a stack by the use of a driver plate.

Finally, in research and development work into the welding of new materials, the use of planar geometries can be a convenient way to establish parameters which can then be applied to other geometries with the aid of appropriate equations.

7.2. MATERIAL COMBINATIONS AND FLYER THICKNESSES

The many combinations of metals and metallic alloys successfully bonded by explosive techniques up to 1964, as reported by Pocalyko and Williams of Du Pont,[1] are shown in Fig. 7.1. At this time stainless steel flyer plates of up to 15 m^2 in area and up to 25 mm in thickness had been used successfully. According to Linse et al.,[2] by 1967 the number of combinations had risen to over 260. Table 7.1 gives current data on cladder materials and thicknesses for commercial cladding (under license to Du Pont) by the Nobel's Explosives Company (NEC), Scotland. Recently Otto and Carpenter[3] demonstrated the successful cladding of large areas of steel with lead, which may become a commercial process.

Some additional combinations welded successfully in small scale experiments by the authors include:

Aluminium	to Al–Zn–Mg alloys
Aluminium	to Maraging steel
Cobalt	to Mild steel
Copper	to Duralumin
Copper	to Nickel
Copper	to Niobium
Copper	to Vanadium
Platinum	to Titanium
Hadfield's steel	to Medium carbon steel
Armour plate	to Medium carbon steel

The factors determining weldability include the ductility, melting-point

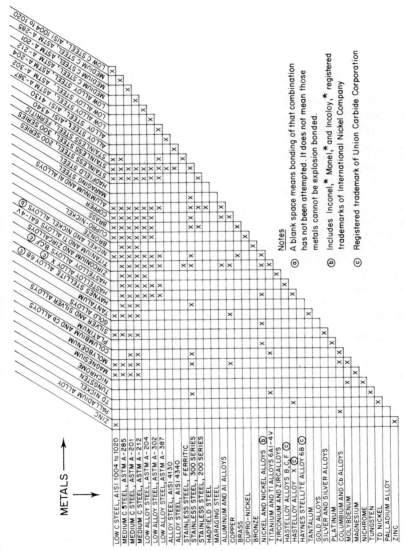

FIG. 7.1 Combinations of metals joined by explosive welding up to 1964 (ref.1).

TABLE 7.1
VARIOUS CLADDING MATERIALS AND THICKNESSES*

Cladding metal	Thickness for routine cladding (mm)	
	min.	max.
Aluminium	6	50
Aluminium–bronze (CA106)	1.5	20
Brasses	1.5	20
Copper (OF, DHP and DLP)	1.5	22
Cupro-nickel (90/10)	1.5	22
Cupro-nickel (70/30)	1.5	22
Hastelloy (B, B2, C, C–276, C4 and G)	1.5	13
Incoloy (800 and 825) Inconel (600 and 625) Monel (400, 404 and K–500)	1.5	20
Nickel (Gd. 200, 201)	1.5	20
Nickel–silver (Du Pont)	1.5	20
Platinum	0.4	ND
Silicon bronze (Du Pont) (Everdur-1015 and high silicon 655)	1.5	16
Stainless steel (austenitic and ferritic)	1.5	25
Tantalum	0.5† or 1.5	>6
Titanium	1.5	20
Zirconium	ND	12

ND = Not determined.
* Recent NEC data (Kelomet) except where indicated. The metals given, and others, are also available from Du Pont (Detaclad).
† If copper interlayer is used.

and/or freezing range, density and thickness of one or both components. Du Pont[4] and others have suggested that an elongation of at least 5% in a standard tensile test or a minimum Charpy V-notch impact value of 14 J is required of the flyer or cladding material to avoid cracking during cladding and, when one of the components of a desired composite is relatively brittle, it is normal practice to make this component the target or backer plate. Thus Wright and Bayce[5] and Williams[6] have clad spheroidal graphite cast iron and antimony respectively with soft steel. Richter[7] found, however, that sometimes the flyer behaved in a more ductile fashion than the target and also exhibited less stretching. Spheroidal graphite cast iron, for example, could only be welded without cracking with the explosive parameters he used if it were made the flyer. This apparent conflict may be explained by differences in the level and/or duration of the hydrostatic loading of the flyer in different investigations. TZM molybdenum, which is too brittle to be resistance-welded, can be successfully welded by explosive techniques.[5]

Materials of very low melting-point, such as bismuth–cadmium fusible alloys, melt so readily under impact that it is difficult to avoid the presence of substantial quantities of melt at the critical time when rarefaction waves reach the temporarily weakened interface. Alloys with a higher melting-point but with a wider freezing range can produce similar difficulties especially if there is also a large difference in density between the components, as explained below.

Welding becomes more difficult (the 'weldability window' becomes smaller) the greater the difference between the densities of the two components and a difference of 9 g cm^{-3} has been suggested as a limiting value.[4] The importance of density stems from the response of the materials in the vicinity of the collision point where, if the pressure is very high relative to the yield strength, they behave as fluids whose movements around (wave formation), and removal from (jetting) the collision point are governed by the laws of hydrodynamics. Given the presence of sufficient pressure these movements therefore depend on the densities but not on the strengths of the materials. When the components have similar densities, the welded interface shows a symmetrical wave pattern which is very nearly sinusoidal for an appreciable range of collision point velocities. As the difference in density between the components increases the loss of symmetry becomes more marked[8,9] and, although the interface may retain a regular pattern with repeating features, it becomes flatter and more jetted material tends to be trapped, possibly because the less dense and therefore faster component of the jet tends to turn towards the denser component. This type of interface is much less desirable than an interface with well-developed ripples, on the crests and troughs of which the melt tends to be isolated by appreciable lengths of direct solid/solid bond (Section 7.4.4). Thus it is difficult to weld, for example, aluminium–zinc–magnesium alloys directly to steel because of the wide freezing range and relatively low density of the alloy. By using an interlayer of pure aluminium between the aluminium alloy and the steel, a satisfactory bond can be achieved, because the higher melting-point, zero freezing range and higher thermal conductivity of the aluminium reduce both the amount of melt produced and the time in the liquid state to levels with which the rather flat Al/steel interface can cope. The Al/Al–Zn–Mg interface is sinusoidal, because of the similarity of densities, and melt is therefore isolated at the crests and troughs of the well developed waves.

As indicated in Table 7.1, flyer thicknesses between 0.4 mm and 50 mm (aluminium) are used in commercial practice. In small scale tests, flyers as thin as 0.025 mm[5] and steel flyers as thick as 50 mm[10] have been welded successfully. Difficulties with foils may arise for several reasons (Section 7.7.1), for example their thickness may be comparable with the amplitude of the waves which may penetrate the foil. Although the minimum flyer velocity required for welding reduces as the flyer thickness increases (Section 7.4.2), an increase in the mass and/or thickness of the flyer can cause problems. Thus the thicker explosive layers employed require a greater 'run-up' distance,[11] to establish the steady detonation pressure and collision front necessary for welding, while the stiffness of a thick plate can also have an adverse effect by delaying the achievement of a turning angle and/or flyer velocity sufficient to produce jetting. The first of these problems is aggravated by the fact that with thick layers of explosive the increases in flyer velocity obtained by further increases in explosive thickness progressively diminish.

7.3. BASIC WELDING GEOMETRIES

7.3.1. Parallel Geometries

In the case of the parallel (constant stand-off distance) geometries (Fig. 7.2), the collision point velocity, V_c, is equal to the detonation velocity, V_d, so the choice of explosive is important to ensure that V_c is in the 'working range' for the materials (Section 7.4.4). Provided the detonation velocity is constant during cladding, the relationship $V_d = V_c$ holds whether the flyer has a constant velocity or is accelerating at the instant before impact (i.e. V_c is independent of stand-off distance). This is because, in the parallel case, the turning angle θ of the flyer at the instant before impact is also equal to the collision angle β the sine of which increases approximately linearly with flyer velocity V_p. Thus:

$$V_c = \frac{V_p}{2\sin\theta/2} = \frac{V_p}{2\sin\beta/2} \simeq \frac{V_p}{\sin\beta} \simeq V_d = \text{constant} \qquad (7.1)$$

The geometries are employed in commercial large scale cladding practice,

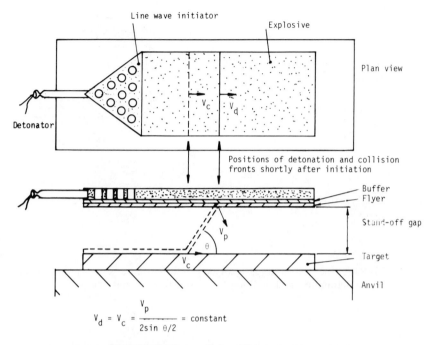

FIG. 7.2 Basic parallel welding geometry.

where the length of the cladding is too great to permit the use of a static angle (Fig. 7.3) between flyer and target because of the excessive stand-off gaps incurred even at distances of say, only 0.25 m from the initiation point. Although used successfully in special cases, large stand-off gaps often lead to excessive distortion of the flyer, debonding due to excessive loss of impelling pressure at the time of impact and an increased incidence of lack of bonding around the periphery of the plate due to exaggerated edge effects[12,13]

Generally, without special (and often costly) precautions, there will be unbonded areas adjacent to the initiating/booster charges because steady detonation/collision conditions are not developed instantly. Thus corner initiation (Fig. 7.4(a)) is often an economical technique for producing maximum bond area with least mass of initiating explosive.[14] If the clad plate is to end up as an annular disc, simple central initiation, which normally produces a small unbonded region immediately under the detonator/initiating charge, is preferable (Fig. 7.4(b)). Even this small unbonded area can be avoided by suitable dimpling of the centre of the flyer plate.[15]

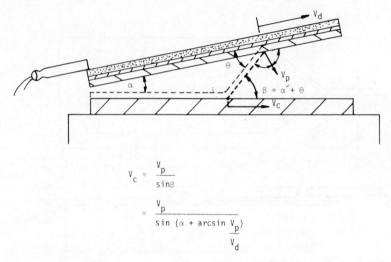

FIG. 7.3 Basic inclined welding geometry. Dotted lines show positions of detonation and collision fronts shortly after initiation.

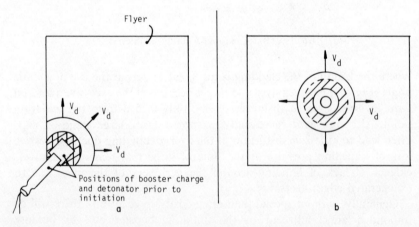

FIG. 7.4 Examples of unbonded areas (shaded) associated with initiating charge. (a) Corner initiation. (b) Central initiation.

The 'balanced' parallel sandwich technique (Fig. 7.5), which obviates the use of an anvil, is attractive in practice if the major surface of each plate can be held in the vertical plane, to avoid the composite being thrown into the air from reaction with the ground. Holding the plates in

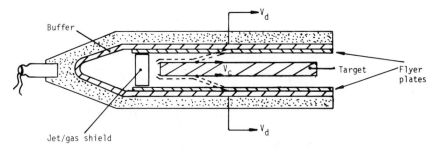

FIG. 7.5 Balanced parallel geometry (schematic). Dotted lines show positions of collapsed flyers when the detonation fronts have reached position shown.

such a position, however, increases handling difficulties and in addition the explosive must be positively fixed to each flyer plate. Obviously sidethrust may create problems if the momenta of the two flyer plates differ appreciably. A further disadvantage of the balanced sandwich technique is that the subsonically detonating explosives required are usually in powder or granular form and thus are difficult to pack uniformly when the plane of the plates is vertical. In the case of Trimonite No. 1 powder it was found[16] that at a constant packing density of $1.12\,\mathrm{g\,cm^{-3}}$, a change in the particle size distribution could account for a change in detonation velocity of 27%.

7.3.2. Inclined Geometries

The basic inclined geometries are shown in Figs. 7.3 and 7.6. In the first, initiation of the charge is made simultaneously along the length of the 'hinge' end of the flyer so that a plane detonation front travels in a direction at right angles to that edge. The directions of movement of the detonation front, the flyer plate and the collision front are thus all contained within the same plane. In the second inclined geometry, Fig. 7.6(a), the charge is initiated so that a plane detonation front travels in a direction parallel with the 'hinge'. In this case the stand-off distance along the detonation front at any given time is not constant, unlike the first case, but increases linearly with distance from the hinge. The collision front therefore lags behind the detonation front progressively as the distance from the hinge increases. The collision front vector therefore makes an angle ϕ to the hinge in the plane of the target given by

$$\tan\phi = \frac{\sin\alpha}{\tan\theta} \tag{7.2}$$

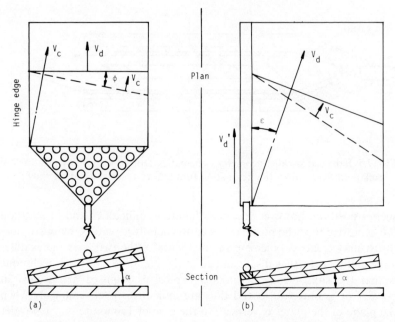

FIG. 7.6 Inclined geometries producing oblique collision fronts.

where α is the static or preset dihedral angle between flyer and target.

In further variations of the method, the charge may be initiated progressively along the entire length of the hinge, using a strip of explosive of higher detonation velocity (V'_d) than the main charge (Fig. 7.6(b)) or it may be initiated at one end of the hinge. In the former variation, the detonation front vector in the main charge makes a constant angle ε with the hinge given by

$$\cos \varepsilon = \frac{V_d}{V'_d} \tag{7.3}$$

In this case the detonation front in the main charge is inclined to the hinge so the collision front vector makes an even more oblique angle with the hinge than in the case of the second inclined geometry.

In any of the inclined geometries the main charge may have a supersonic or subsonic detonation velocity, since the collision point velocity is now less than the detonation velocity. For the first case, where the detonation front remains parallel with the 'hinge' end, the collision point velocity is given approximately by:

$$V_c = V_d \frac{\sin \theta}{\sin(\alpha + \theta)} \simeq \frac{V_p}{\sin\left(\alpha + \arcsin \frac{V_p}{V_d}\right)} \tag{7.4}$$

For the second case, where the detonation front is normal to the 'hinge' (Fig. 7.6(a)) the collision point velocity is given by:

$$V_c = V_d \frac{\tan \theta}{(\tan^2 \theta + \sin^2 \alpha)^{1/2}} \simeq \frac{V_p}{(\sin^2 \theta + \cos^2 \theta \sin^2 \alpha)^{1/2}} \tag{7.5}$$

For the general case when the detonation front is inclined to the hinge (Fig. 7.6(b)) the equation for the collision point velocity is rather complex.[17]

Equations (7.4) and (7.5) show the collision point velocity to be very sensitive to changes in the static angle α, an increase of a few degrees sometimes reducing V_c by a factor of two. Nevertheless, even with α equal to only a few degrees, the stand-off gap can become large at modest distances from the hinge which limits the applicability of the inclined geometries. Thus the techniques are unsuitable for welding sheets which are long in the direction at right angles to the hinge edge. If the sheet is short but wide, i.e. it has a relatively large hinge length, of the two basic inclined techniques the oblique technique is preferable because of its simpler initiation systems.

The advantages of the inclined techniques are:

(a) The collision point velocity, collision angle and impact velocity can be varied over a wider range than in the parallel case so that optimum bonding conditions can be achieved more often.
(b) Supersonically detonating explosives, which are generally more consistent and more easily handled (usually as 'plastic' explosives) than subsonically detonating explosives, can be used. This is especially important when the materials weld only within a very limited range of conditions.
(c) Bonding conditions are usually established at smaller distances from the initiation point than in the parallel case because a finite collision angle is required for welding and the use of a preset angle assists in the achievement of this, especially if the plate is stiff by virtue of its thickness.

With regard to (c), although the strain energy of the flyer plate is usually negligible compared with its kinetic energy under steady welding conditions, this strain energy is not always negligible in the region where the charge is initiated. Here the detonation pressure has not achieved its maximum level and the flyer velocity is therefore low. In addition, over the 'run-up' region in the explosive this reduced detonation pressure will be acting on a 'beam' of small length so the bending resistance of the flyer (which is proportional to the square of its thickness) will tend to reduce its velocity and turning angle. Typically, at a flyer velocity of a few hundred metres per second, the strain energy will amount to less than 1% of the kinetic energy. To achieve this velocity more quickly a booster of supersonically detonating explosive is often employed to initiate the main charge. Alternatively, or in addition, the plate may be bent suitably over a small distance to ensure that the collision angle is sufficiently large in the critical early stages of impact.

A difficulty with supersonically detonating explosives is the very high shock pressure generated in the flyer (since the detonation pressure is proportional to the square of the detonation velocity) which can thus spall unless it is separated from the explosive by a suitable buffer or attenuator. This buffer must be in close contact with the flyer to give uniform shock transmission and to prevent surface damage occurring by secondary jetting in areas of separation, i.e. jetting caused by oblique collision of the buffer with the surface of the flyer. The role of a buffer in commercial cladding, where spalling is not normally a problem with the subsonically detonating explosives used, is less critical, being mainly to prevent burning and embossment, from the explosive constituents, of the upper surface of the flyer. Often two or three layers of a suitable paint provide adequate protection.

7.3.3. Parallel/Inclined Geometries
When two or more flyer plates are superimposed for welding to a single target plate with the simple parallel geometry of Fig. 7.2, the collision point velocity between each pair of adjacent plates is equal to the detonation velocity of the explosive. Therefore it is unlikely that this velocity will be the optimum value for each combination of materials in the composite. If, however, two or more parallel plates are impacted by an inclined plate,[18] which is referred to as a driver or an impactor (Fig. 7.7), the collision point velocity (given by eqn. (7.4)), although again equal for each pair of metals, is no longer equal to the detonation velocity and can be varied over a wide range to obtain a value which is a

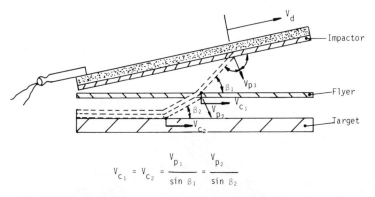

FIG. 7.7 Welding geometry and mode of collapse using an impactor.

best compromise. Moreover, the impact velocities also can be varied, by varying the thicknesses of driver, buffer (if used) and explosive as well as the detonation velocity.[19] If the driver plate is not itself to become welded, however, it does constitute an additional cost, although it may obviate the use of a buffer. The technique has attractions only in special cases (Section 7.6.6) because of the limitation on weld length incurred by the preset angle between the driver and upper plate.

7.3.4. Double Inclined Geometries

When an interlayer of metal is required between two layers of different metals, for example in the welding of aluminium alloys to titanium or steels,[20,21] the interlayer can be incorporated in three ways. The interlayer-to-be can be first bonded, to the target and/or the flyer plate, using the simple parallel or inclined geometries described above (electroplating has been used for applying some interlayers),[22,23] and the three-component composite then obtained with a second firing. Secondly, as in the commercial bonding of tantalum to steel using an interlayer of copper,[14,24] the flyer, interlayer and target can be disposed parallel to but spatially separate from each other, and bonded in a single firing. As indicated earlier, the latter technique, although practically attractive because of its simplicity, depends on the flyer/interlayer and interlayer/target interfaces being formed satisfactorily with the same collision point velocity which is equal to the detonation velocity of the explosive. Thirdly, as Hayes and Pearson[25] demonstrated in 1962, three (or more) metals can be bonded in a single firing using the double inclined geometry of Fig. 7.8. Here the collision point velocities for the

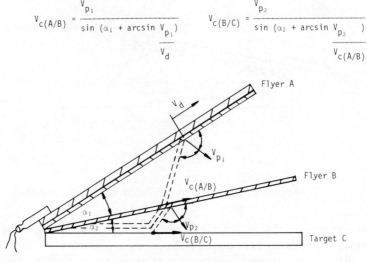

FIG. 7.8 Double inclined geometry.

two combinations are no longer equal and can be varied over a wide range, by varying for example, the two static angles α_1 and α_2, to achieve optimum conditions at each interface of the composite. Thus if good bonding between say, the interlayer and the lower plate requires a comparatively high collision point velocity, α_2 can be made smaller than α_1 (or even negative). Variations on the double inclined geometry have been discussed in a recent patent.[26] Equations expressing the collision point velocities in terms of the welding parameters (Fig. 7.8) naturally include the thickness of the interlayer but, if this is very small relative to the thickness of the primary flyer, the equations reduce to:

$$(1) \quad V_c(A/B) \simeq \frac{V_p}{\sin\left(\alpha_1 + \arcsin\dfrac{V_p}{V_d}\right)} \tag{7.6}$$

$$(2) \quad V_c(B/C) \simeq \frac{V_p}{\sin\left(\alpha_1 \pm \alpha_2 + \arcsin\dfrac{V_p}{V_d}\right)} \tag{7.7}$$

where α_2 is reckoned negative if the stand-off gap between B and C decreases in the direction of detonation.

7.3.5. Geometries Producing Welding Conditions Transiently

When the explosive loading is confined to an area which is small relative to the total area of the flyer (Fig. 7.9), welding conditions can be obtained over narrow regions even when supersonically detonating explosives are used in conjunction with components disposed parallel.

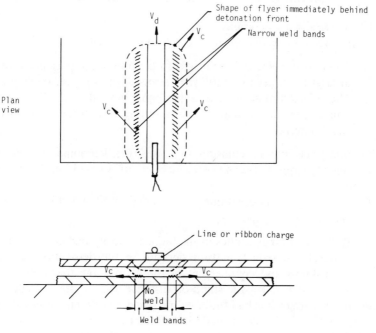

FIG. 7.9 Subsonic collision fronts produced transiently by a supersonically detonating strip charge.

Immediately under the supersonically detonating explosive and throughout its length the collision point velocity is supersonic, but due to the inertial restraint of unloaded flyer material around the charge, large collision angles are developed laterally as the flyer plate decelerates—a phenomenon sometimes referred to as 'flap' or 'lag'. Thus, over a small distance, the transient impact velocities and collision angles are often suitable for welding, even though immediately under the charge they may not be suitable for welding. (In contrast, when cladding large areas with uniform layers of explosive, flap may prevent bonding in peripheral zones.)

This is the basis of some methods of explosive seam and spot welding of sheets (Sections 7.6.2 and 7.6.5) which depend additionally on the provision of sufficient stand-off gap between the sheets to allow the dynamic collision angles to develop.

7.4. SELECTION OF BONDING PARAMETERS

7.4.1. General Considerations

In explosive welding two primary conditions are met:

(1) The impact velocity is sufficient to generate an impact pressure of at least ten times the static yield stress of the stronger component.
(2) The collision angle exceeds some minimum value which, with the impact pressure available, allows jetting of a surface layer from each material.

The impact velocity and collision angle together determine the collision point velocity (eqns. (7.1) and (7.4)) which, to satisfy condition (2), must be subsonic.

Given these primary conditions, the strength of the bond between two given metals then depends on:

(i) the actual magnitudes of the impact velocity and collision angle developed (which in turn depend on the detonation velocity, thickness and confinement of the charge, the thickness of the flyer and buffer, the stand-off distance and the static angle)

and (ii) the surface finish of the colliding plates as this influences the ease with which jetted materials (metal, oxides and other contaminating films) escape from between the plates.

Many of the variables in explosive welding are interdependent or have important secondary effects. For example, if the surfaces are not smooth the impact velocity used will need to be higher than normal (Section 7.4.7) to ensure that the wave amplitude and/or jet thickness is large compared with the height of the surface asperities.[27] In turn the higher impact velocity will only be achieved if, for example, the stand-off gap is sufficiently large (Section 7.4.5). Also the thickness and/or material of the flyer, which acts as a dynamic tamper,[28] can in some cases influence the detonation velocity of the explosive.[29,30] More well known is the increase in detonation velocity with the thickness of the explosive.[29,31,11] Both of these effects can be appreciable with Trimonite No. 1[16] (Fig. 7.10). The displacement/time curve of a flyer[32,33,34] and the duration of

EXPLOSIVE WELDING IN PLANAR GEOMETRIES 235

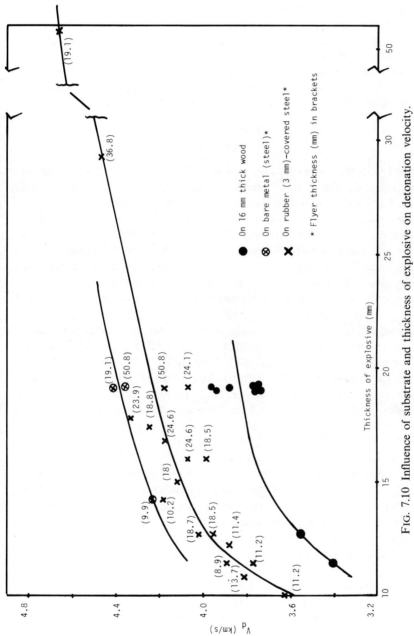

FIG. 7.10 Influence of substrate and thickness of explosive on detonation velocity.

the impact pressure both depend on the thickness of the flyer also. The greater duration of the impact pressure obtained with thick flyers may partly explain why they do not require such a high impact velocity for welding as thinner flyers. Obviously the thinner the flyer the sooner will shock waves, reflected by the upper surface of the flyer, reach the welded interface which may be disrupted if, for example, cooling has not occurred sufficiently in the meantime. The higher explosive loading

$$\frac{C \text{ (mass of explosive)}}{M \text{ (mass of flyer)}}$$

used with thin flyers in most cases implies a greater residual impelling pressure at the time of impact which presumably counteracts any tendency to rebound. Apart from its potential disruptive action, however, the reflected shock wave has the effect of decompressing the highly stressed fluid-like material behind the collision point so 'freezing' it. Thus for the same pressure (impact velocity), the fluid-like behaviour will continue for a longer period with thicker flyers. This is compatible with the observed increase in wave size with flyer thickness at a given collision angle.[35,36,37] Therefore if a minimum wave amplitude must be exceeded to give an adequate weld strength (Du Pont[4] suggest a minimum amplitude of 5 µm), then obviously as the impact velocity is reduced the thinner flyers will cease to weld before the thicker flyers. An alternative but related interpretation is that jetting should also continue for a longer period in the case of thicker flyers. Assuming that a minimum thickness of material from target and flyer must be removed by jetting to allow welding (this minimum thickness would obviously depend on surface condition—Section 7.4.7), then if this thickness is just achieved with a thick flyer it will not be achieved with a thin flyer.

The influence of flyer thickness and surface finish on the minimum impact velocity required for welding two steels together[16] are given in Fig. 7.11. Note that the impact velocity reaches a limiting value of about 120 m/s for the thickest flyers which corresponds to an impact pressure of about ten times the yield stress of the steel (Section 7.4.2). Below this velocity welding is not possible regardless of the thickness, kinetic energy or surface finish of the flyer.

7.4.2. Impact Velocity
(a) *Equations for Minimum Impact Velocity*
The pressure P developed in each metal is given by[38]

$$P = \rho u U \tag{7.8}$$

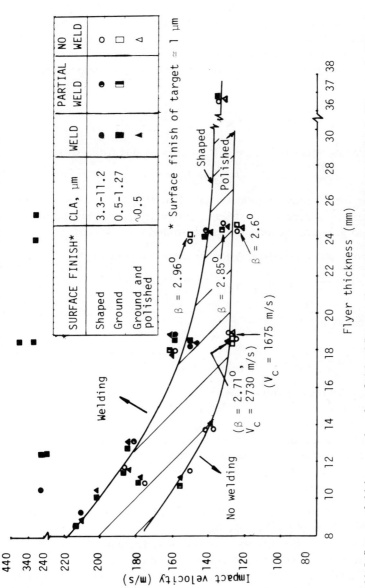

FIG. 7.11 Influence of thickness and surface finish of flyer on the minimum impact velocity for welding steels. *Experimental details*. For shaded area, $V_c = 2500$–2970 m/s (with the one exception shown)*; $\beta = 2.6$–$4.9°$ (those $<3°$ indicated)*. Explosive: Trimonite no. 1 powder (density 1.12 g/cm³). Hardness (HV 30): Flyer = 160 ± 30, Target = 120 ± 30. *Estimated accuracy $\pm 12\tfrac{1}{2}\%$.

where ρ is the initial density of the metal, u is the rate at which its impacted surface is displaced and U is the velocity of the shock wave through the metal which is approximately equal to the bulk longitudinal sound wave velocity.

From conservation of momentum it follows that for two dissimilar metals A and B impacting at a relative velocity of V_p the impact pressure is given by:

$$P = P_A = P_B = \frac{\rho_A V_p U_A}{1 + \rho_A U_A / \rho_B U_B} \qquad (7.9)$$

For similar materials ($\rho_A = \rho_B = \rho$, $U_A = U_B = U$) this reduces to

$$P = \tfrac{1}{2} \rho V_p U \qquad (7.10)$$

Assuming from limited empirical data[39] that $P_{min} \simeq 10\sigma$, where σ is the quasi-static yield stress of the stronger component, the minimum impact velocity to generate this pressure can be found from:

(i) dissimilar metals (with A the stronger component)

$$V_{p(min)} \simeq 10\,\sigma_A \left[\frac{1 + \rho_A U_A / \rho_B U_B}{\rho_A U_A} \right] \qquad (7.11)$$

(ii) similar metals

$$V_{p(min)} \simeq \frac{20\,\sigma_A}{\rho_A U_A} \simeq \frac{20\,\sigma_A}{\sqrt{(\rho_A E_A)}}, \quad \left(\text{since } U_A \simeq \sqrt{\frac{E_A}{\rho_A}} \right) \qquad (7.12)$$

where E_A = Young's modulus of A.

Hence the value of $V_{p(min)}$ is influenced primarily by the yield stress of A and to a lesser extent by the density and elastic modulus. Calculated and experimentally determined minimum impact velocities for welding various combinations of metals are given in Table 7.2.

The calculated minimum values refer to ideal explosive welding conditions, namely:

(i) The impacting surfaces of flyer and target are smooth and free of thick films of foreign material.
(ii) The flyer/buffer thickness is sufficient to maintain the impact pressure for more than a critical period (Section 7.4.1).
(iii) The efficiency of the support system (section 7.4.6) is sufficiently high to minimise the movement of the target plate and hence of the welded surface.

TABLE 7.2
MINIMUM IMPACT VELOCITIES FOR WELDING

Metal combination	Density ($kg\,m^{-3}$)	Bulk sound velocity (m/s)	Assumed yield stress (MN/m^2)	$V_{p(min)}$ (m/s) Estimated*	$V_{p(min)}$ (m/s) Measured	Comments
Aluminium/Aluminium	2 700	6 400	35	41		
6061–T651 Al alloy/ 6061–T651 Al alloy	2 700	6 400	276	319	270 (ref. 40)	Flyer thickness 6.35 mm
Copper/Copper (half-hard)	8 960	4 900	150	68	~200 (ref. 41) 130 (ref. 42) 240 (ref. 43)	Flyer thickness 1.1 mm
Steel/Steel	7 870	6 000	200	85	90 (ref. 44) 120 (ref. 43) ~125 (ref. 16) ~165 (ref. 16) 130 (ref. 45)	Threshold for jetting (mild steel to stainless steel) Flyer thickness ⩾ 25 mm } Fig. 7.11. Flyer thickness 10 mm Flyer thickness 10 mm (welded *in vacuo* with oxide-free surfaces).
Titanium 115/Titanium 115	4 500	6 100	250	182	220 (ref. 43)	
Molybdenum/Molybdenum	10 200	6 400	400	123		
Aluminium/Titanium	2 700/4 500	6 400/6 100	35/250	236		
Aluminium/Steel	2 700/7 870	6 400/6 000	35/200 35/470‡	158 372	~460 (ref. 16)	Flyer thickness 3 mm
Titanium/Steel	4 500/7 870	6 100/6 000	250/200	144	~200†	Flyer thickness 3 mm
Nickel/Steel	8 900/7 870	5 800/6 000	150/200	81	~200†	Flyer thickness 3 mm

* From eqns. (7.11) and (7.12).
† Obtained by extrapolating data taken from ref. 4.
‡ Fortiweld.

(iv) The collision angle is not so high as to reduce significantly the normal component of flyer velocity.

(b) *Maximum impact velocity.*
Attempts have been made to predict a maximum impact velocity in terms of such quantities as the flyer thickness, the collision point velocity and the thermal conductivity of the metals.[40,46] However, although some of the experimental observations agreed qualitatively with the theories (e.g. the limiting maximum impact velocity for welding increased with conductivity and decreased with increasing collision point velocity and flyer thickness), the predicted weldability ranges for different combinations clearly conflict with experiment.

(c) *Impact velocities used in practice*
In practice, while it is obviously desirable to minimise the explosive loading to reduce material costs, blast and noise levels, loadings higher than the minimum appropriate to the thickness of flyer used are required, especially in commercial cladding operations, to guarantee consistency of welding with the variations in surface finish, flatness and support likely to be met with large plates.

Examples of the velocities employed in commercial cladding for various cladding materials (generally applied to a steel backer) in thickness of 2 mm and above are:

Copper and cupro-nickel	400 ms^{-1}
Stainless steel, nickel and Monel	440 ms^{-1}
Titanium	600 ms^{-1}

These values are higher than the minimum levels given in Table 7.2. by a factor of two or more. In fact flyer velocities in excess of 1000 m/s, with a 100-fold or greater increase in kinetic energy, can be employed in some cases (for example steels to steels[27]) with satisfactory results, provided the collision point velocity is kept within certain limits (Section 7.4.3). Nevertheless values only fractionally above the minimum impact velocities may be required in specialised small scale welding operations where it may be necessary to minimise the amount of distortion or melt produced. For example, Chadwick et al.[47] found this to be true for welding zirconium alloys to stainless steel.

7.4.3. Explosive Loading
Once a flyer velocity appropriate to the various conditions pertaining

has been selected, the explosive loading to produce this velocity can be calculated from equations derived by Gurney[48] and Duvall and Erkman.[49] These equations give the terminal velocity V'_p of the flyer and therefore the use of an adequate stand-off gap is implicit. The Gurney relationship for the 'open sandwich' (explosive not covered) planar geometry (see also Chapter 6) is:

$$V'_p = \sqrt{(2E)}\left(\frac{\sqrt{(3)}}{1+2M/C}\right) \quad (7.13)$$

where M is mass of flyer including buffer, per unit area, C is mass of explosive, per unit area, E is kinetic energy of explosive products, per unit mass, and $\sqrt{(2E)}$ is the 'characteristic Gurney velocity' for explosive and geometry used. The kinetic energy is typically 75% of the measured heat of detonation, ΔH_d, which is related to the detonation velocity by the equation:[28]

$$V_d = \sqrt{(2\Delta H_d(\gamma^2 - 1))} \quad (7.14)$$

where γ is the adiabatic exponent for the detonation products.

Combining eqns. (7.13) and (7.14) gives:

$$\frac{V'_p}{V_d} = \frac{1.5}{\sqrt{(\gamma^2-1)}}\left(\frac{1}{1+2M/C}\right) \quad (7.15)$$

As an example, for Composition B explosive, $\gamma = 2.7$ giving:

$$\frac{V'_p}{V_d} = \frac{0.60}{1+2M/C} \quad (7.16)$$

The numerical constant in eqn. (7.16) generally increases as V_d decreases because γ tends to decrease with V_d.

If the charge is covered with a layer of non-fragmenting inert material (tamper) of mass N (per unit area), the explosive loading can be reduced according to a modified Gurney equation derived by Kennedy[28] which shows the efficiency of a tamper to be more marked the lower the value of C/M.

$$V'_p \simeq \sqrt{(2E)}\left(\frac{1+A^3}{3(1+A)}+\frac{N}{C}A^2+\frac{M}{C}\right)^{-1/2} \quad (7.17)$$

where $A = 1+2M/C/(1+2N/C)$ and V_d is again measured in the test. The latter equation is obviously more appropriate in those cases where, by virtue of the set-up, the presence of a tamper is unavoidable. For example with the vertical double 'closed sandwich' arrangement (Fig. 7.5), the container itself may be a significant tamper.

7.4.4. Collision Angle/Collision Point Velocity

Regardless of the magnitude of the impact velocity, welding is not possible unless plastic flow can occur ahead of the collision point. In practice this means that the collision point velocity must be less than the sonic velocity in the metals. For optimum welding conditions, however, the collision point velocity must be kept well below the sonic velocity and within a range which is appropriate to the materials and flyer velocity being used. It also appears that some minimum collision angle must be exceeded to produce welding although this angle is rather small, e.g. $\sim 2\frac{1}{2}°$ for steel/steel,[16] 4° for nickel/steel and 7° for titanium/steel.[4] The finite angle will produce an additional forward impetus to the jet as it moves ahead of the collision point. It is difficult, in view of such small angles, to accept the simple descriptive indentation theory of wave formation of Bahrani et al.[50] according to which the typical wave slope of $\sim 45°$ would require the flyer to collide with the target at a similar angle. In fact waves become more peaked (wave slope increasing to say 60°) by *decreasing* the collision angle to a few degreees.

The collision point velocity ($\simeq V_p/\sin \beta$) is important because it is an imposed velocity which also influences the behaviour of the jet and the final form of the interface.

If the collision point velocity falls below a certain value, which is material-dependent, the weld interface becomes flat. This is associated with the unimpeded escape of the jet (free jetting). With increasing collision point velocity the volume of trapped jet increases,[4,5] possibly because the collision point overtakes and traps a large part of the jet whose effective velocity is reduced (from a maximum value of $2V_d$) by turbulent oscillation and interaction with the colliding surfaces. There is evidence[16,27] to suggest that jet trapped ahead of the collision point can suppress the formation of waves (See also Fig. 7.12). In the limit the weld interface may contain a continuous layer of melt which will only rarely give a satisfactory joint because it will contain most of the oxide originally present as a film on the surfaces of the metals and may contain shrinkage pores. Moreover, for some combinations of metals, the high temperature and intimate mixing of the two metals in the melt gives brittle intermetallic compounds.

At intermediate collision point velocities the jet behaviour is regular, a steady wave pattern at the interface is established and maximum bond

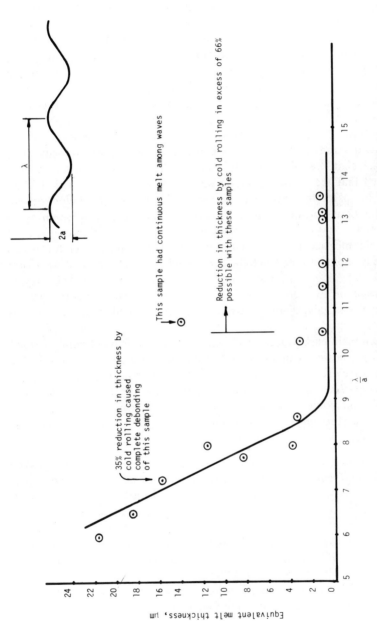

Fig. 7.12 Effect of wave shape on the amount of trapped melt. Flyer: Titanium 3.2 mm thick, target: 0.08% C Steel. (Data taken from ref. 4).

strength is obtained. Effective deformation and scouring of the surfaces of the target and flyer alternately by the regular oscillation of the jet,[44] just ahead of the collision point, prior to it escaping as a free jet,[51,41] may contribute to this strength. Or the fluid-like oscillation of the interface within the high pressure region immediately behind the collision point[52,53] may cause strengthening by extending the area of the interface and isolating any trapped jet/melt as islands at the crests and/or troughs of the waves. The amplitude of the waves can be increased appreciably by increasing the collision angle and/or the impact velocity (via an increased stand-off distance) at a constant collision point velocity. (For parallel cladding of steel with titanium and nickel, Cowan et al.[54] showed that the wave amplitude was proportional to β^2.) This finding may be used to maximise the strength of the bond for certain combinations forming brittle compounds since the increase in wave amplitude, a (at constant wavelength λ) increases both the total area of the weld interface and the separation of pockets of melt. Even for metals which do not form intermetallic compounds, the more peaked waveform (small λ/a) can give stronger joints in the as-clad condition. This may be the result of an increased weld area discussed above and the greater depth of shock-hardening observed with the peaked waves.[55] There is some evidence to suggest[16,56] that with $\lambda/a \leq 7$ the joints obtained between plain carbon steels will be satisfactory. For other combinations, e.g. Ti/Fe, however, or in general when the composite is to be rolled, a flatter waveform (large λ/a), which gives a lower proportion of trapped melt (Fig. 7.12), may be required. In this case the joint strength may increase on rolling whereas interfaces with a peaked waveform may crack, even though they may be much stronger in the as-clad condition. This inferior behaviour during rolling has several causes:

(i) A low ductility shocked zone often accompanies the peaked waveform in steels because the peaked wave implies a high impact velocity and/or a high collision point velocity.
(ii) The accentuated waveform itself constitutes a stess-raiser.
(iii) The volume of melt is higher with the peaked waveform (although from the results of static tests[57] this melt can in some cases cover up to 70% of the interface without ill-effects) which results from a high collision point velocity.
(iv) Adiabatic shear bands, melted zones, and the direct metal/metal interfaces of the waves are unfavourably oriented with respect to the rolling plane. The adiabatic shear bands in particular are often sites of incipient cracking.[58]

In some cases, an even lower collision point velocity may be preferred to achieve a flat interface and so avoid melt completely. Chadwick and Graham[59] and Czajkowski[60] found that flat interfaces between aluminium and maraging steel and between aluminium and mild steel respectively could be raised to surprisingly high temperatures (ca 480 °C) without nucleation and growth of intermetallic compounds. In contrast the commercially clad rippled aluminium/steel interface[10] shows deterioration above 260°C.

A flat interface between commercially pure aluminium and D20 aluminium alloy was found by Kruskov et al.[20] to be at least as strong as the D20 alloy in short time tests (~15 min) at temperatures up to 300 °C. Bahrani[55] found that in the case of stainless steel/mild steel clads annealed at 850 °C for 30 min the flat interface was stronger than a wavy interface. This was evidently due to the greater depths of decarburisation exhibited by the wavy interface especially around the wave peaks where, according to Trueb[61] the shock-induced strain and diffusion rates are maximum.

In general, however, maximum static bond strength is usually achieved with a well developed rippled interface, that is, when the collision point velocity is appreciably in excess of that value where the interface just becomes flat.[62] The latter, known as the transition velocity V_T (the transition being from turbulent to laminar flow),[63] has been shown by Cowan et al.[54] to depend on certain properties of the metals being welded, in accordance with hydrodynamic theory, viz:

$$V_T = \left(\frac{2Re \, (H_f + H_t)}{\rho_f + \rho_t} \right)^{1/2} \tag{7.18}$$

where Re is the Reynolds number appropriate to the flow process, H is hardness (Nm^{-2}), ρ is density (kgm^{-3}), and the subscripts f and t refer to flyer and target.

Published values[54] for V_T fall in the range 1600 ms^{-1} (nickel to nickel and copper to aluminium) to 2500 ms^{-1} (magnesium to magnesium). The Reynolds number corresponding to these values has an average value of 10.6 and a standard deviation of 18%.

From the calculated or experimentally determined value of V_T a suitable collision point velocity to use can be determined. Thus Christensen et al.[64] suggest for steels

$$V_c = V_T + 400 \text{ ms}^{-1} \tag{7.19}$$

while Stivers and Wittman[65] suggest that the following equations may have general applicability:

(i) $V_c = V_T + 50 \text{ ms}^{-1}$ for $V_T \geqslant 2500 \text{ ms}^{-1}$ (7.20)
(ii) $V_c = V_T + 100 \text{ ms}^{-1}$ for $2000 \text{ ms}^{-1} \leqslant V_T < 2500 \text{ ms}^{-1}$ (7.21)
(iii) $V_c = V_T + 200 \text{ ms}^{-1}$ for $V_T < 2000 \text{ ms}^{-1}$ (7.22)

Recent results of Prümmer[66] for welding molybdenum to Inconel and to Kanthal ($V_T = 2900 \text{ ms}^{-1}$ and 2000 ms^{-1} respectively) indicate optimum collision velocities of $V_T + 600 \pm 500 \text{ ms}^{-1}$ and $V_T + 1200 \pm 600 \text{ ms}^{-1}$ respectively which are at variance with the Stivers and Wittman equations. Du Pont[4] have given results which suggest that with V_c between 2000 and 2500 ms^{-1} satisfactory bonding is achieved for many combinations.

All of these suggestions are based on the fact that waveform and melt distribution are governed to a large extent by the value of V_c, which can be varied. For cladding with the parallel geometry, the detonation velocity of the explosive used should be close to the optimum collision point velocity. If, at the thickness required to give this velocity, the explosive would give too high a flyer velocity, the buffer thickness can be increased to compensate. If on the other hand, the flyer velocity would be too low, the buffer thickness can be decreased or the charge tamped. Other factors such as the thickness and density of each component are fixed by the materials being welded.

7.4.5 Stand-off Distance

The flyer is accelerated initially by a shock wave, resulting from the detonation pressure, and then by the expanding gaseous products of detonation. If the stand-off distance between flyer and target is sufficiently large, the flyer will eventually attain a terminal velocity given by the appropriate Gurney equation. In normal cladding, when the detonation front travels parallel with the plane of the flyer, the terminal velocity is achieved in the time it takes the gases to expand to seven times the volume of the explosive.[67] In special cases, e.g. in explosive spot welding, when the detonation front travels normal to the flyer surface, the terminal velocity is achieved in the time it takes the gases to expand to twice the volume of the explosive. It follows that, in either case, the higher the detonation velocity and density of the explosive the smaller is the stand-off distance required. Various workers[33,34,68,69,80] have shown that, in cladding, large acceleration distances (e.g. 25 mm or more) are needed before the theoretical flyer velocity (e.g. eqn (7.13)) is approached, although an appreciable velocity may be attained after a displacement of say only 0.5 mm. Takizawa et al.[33] have commented that the values of V_p and β measured by different investigators differed because most were obtained in the acceleration stage or in the apparent steady-state zone rather than the real steady-state zone. The stand-off distances required for the latter in their experiments were 10–20 mm depending on the thickness of flyer and explosive. This finding is particularly important when minimum values for V_p and β are being investigated or where a constant value of V_p is needed in an analysis. Thus the small stand-off distances used by Wylie et al.[70] might well invalidate their conclusions on minimum impact velocities, kinetic energies and collision angles for welding and at the same time account for some of the conflicting results of their work. Similarly the increase in wave amplitude in multiple foil cladding with increase in distance from the top foil observed by El-Sobky[71] was found by Al-Hassani and Salem[72] to reverse when a larger stand-off distance was used.

The stand-off distance S required to achieve a specified fraction of the terminal velocity V_p' may be obtained by combining the theoretical results of Kury et al.[67] and Aziz et al.[73] Whence for planar geometries:

$$S = 3 k x_e C/M \qquad (7.23)$$

where $k \simeq 0.4$ and $\simeq 0.7$ for impact velocities of 70% V_p' and 100% V_p' respectively and x_e = thickness of explosive. In normal cladding practice 70% V_p' would be more appropriate since it is neither necessary nor desirable to achieve terminal velocity because: (a) a useful fraction of the terminal velocity is reached early in the acceleration phase, and (b) at terminal velocity there is virtually no impelling pressure to counteract rebound forces which may tend to separate flyer and target while the weld interface contains some material which is still molten.

Thus according to Stivers and Wittman,[65] a practical minimum value of stand-off distance for cladding can be obtained from the following empirical equation:

$$S = 0.2(x_e + x_f) \qquad (7.24)$$

where x_f is the thickness of flyer.

In some cases of parallel cladding where it is desired to improve the wave form by increasing the collision angle, without increasing the collision point velocity, a stand-off distance larger than given by eqn. (7.24) (but less than given by eqn. (7.23), with $k = 0.7$) can be used.

When such abnormally large stand-off distances are required it becomes important to ensure that potentially disruptive forces and edge effects are minimised. Thus the thickness of the explosive and the efficiency of the anvil system (Section 7.4.6) as well as the area of the flyer and explosive relative to the area of the target may also have to be increased.

In the case of a flyer being accelerated by impact with a solid body (an impactor), the stand-off distance required to achieve terminal velocity is smaller than those given by eqns. (7.23) and (7.24) by an order of magnitude.[19] This is because the impactor delivers its energy in the time it takes the impact shock wave to travel twice the thickness of the impactor.

7.4.6. Anvil

As indicated in Section 7.4.5, if the stand-off gap is increased substantially without an appropriate increase in the thickness of the explosive, the impelling pressure may drop to such a level that it cannot counteract the disruptive force arising when the compression wave produced on impact is reflected as a tension wave at the upper surface of the flyer plate. Disruptive forces of similar magnitude can arise from the underside of a thin or poorly supported target plate. The thickness of the target plate which can be used is determined in some cases by the support system. If the anvil is not in intimate contact with the underside of the target, or if the anvil has an acoustic impedance which does not match that of the target, a tensile wave will arise at the target/anvil interface and this will tend to separate target and flyer. With appropriate overcharging, debonding is not normally a problem with ill-fitting anvils provided the target thickness is at least twice, and preferably at least three times the flyer thickness.[69] In this case the role of the anvil is to provide a convenient working surface, to limit the movement of the welded composite (which might otherwise be buried to a depth of several feet) and to limit the distortion of the composite. Under these circumstances an anvil of explosively compressed sand[3] may be preferred to a steel or concrete anvil. Thus the sand anvil is less expensive and can be levelled easily after each welding operation. Also, since the sand 'gives', there is no tendency for elastic recoil. In contrast, with a steel anvil the composite may bounce from and back on to the anvil resulting in damage to either or both. Moreover, since the initial rebound does not occur simultaneously across the composite the latter will invariably end up curved.

If a target is very thin relative to the flyer plate it is necessary to have the target in intimate contact with an appropriate metal anvil to ensure that the shock wave travelling from the flyer/target interface is transmitted into the anvil from which it cannot return. Adhesive bonding may be employed in such a case.

The target/anvil thickness is an important welding parameter when the assembly moves appreciably as shown by Trutnev et al.[74] From conservation of energy and momentum they showed that:

$$E_w = \eta E_f \qquad (7.25)$$

where E_w is energy consumed at the welded interface, E_f is kinetic energy of the flyer before impact and

$$\eta = \frac{M_t}{M_f + M_t} = \text{collision energy utilisation factor} \qquad (7.26)$$

where M_t and M_f are the masses of target and flyer. Thus a thin target produced less melt than a thick target under otherwise identical conditions.

This finding is compatible with the decreasing wave amplitude with depth in a stack of foils, and with the abrupt increase in amplitude when the last foil collides with the anvil as observed by Al-Hassani and Salem.[72]

In many cases of explosive welding, including non-planar geometries, the anvil can be an expensive item. Thus the use of a balanced or symmetrical welding system (Fig. 7.5) whereby a second explosive charge replaces the anvil—apparently first suggested by Davenport[75]—is economically attractive.

7.4.7 Surface Finish

Surface finish and surface condition are important for several reasons. First, the size of the surface asperities determines the degree of jet trapping and therefore the degree of additional melting at the weld interface as the kinetic energy of the jet converts to heat. Secondly, if the wave size is small compared with the surface asperities, it will be impossible for a steady waveform to develop and the distribution of even normal quantities of melt will be unpredictable. Alternatively, if, in the latter case, a flat bonded interface is required, then it must be ensured that the thickness of the material stripped from target and flyer by jetting exceeds the height of the asperities. Therefore in all of these cases the effect of the rough finish may be at least partly overwhelmed by increasing the explosive loading (see for example Fig. 7.11) to increase the dimensions of the wave and jet. To the authors' knowledge, however, the quality of the weld obtained will always be inferior to that obtained with a smooth finish, this difference being particularly marked with such combinations as zirconium to stainless steel[47] and aluminium to maraging steel.[59] Increased explosive loadings are also undesirable on the grounds of increased costs and blast levels.

Thirdly the method of preparing a surface can determine the degree to which it is work-hardened. Highly reflecting surfaces can be produced by burnishing but the resulting surface hardening can impair welding when the preferred low explosive loading is used. Removal of burnished layers by emerying can leave the surfaces in a condition more suitable for welding.

A secondary benefit of employing smooth finishes is that they are more tolerant of contaminating films such as oil, grease and water, evidently because these can escape with the jet more readily when the surface is smooth. Chladek[76] made the interesting observation that water, which was the most harmful agent, was more easily removed from remote locations on the intended clad area than from areas close to the initiation point. This was attributed to the high thermal capacity of the water which could only be adequately vaporised if jetting and air shocks preceding the collision point were well developed. Rust, being a solid, tended to be retained in the weld interface. Chadwick and Lowes[77] found that a film of condensed water on smooth surfaces of thick steel plates delayed the onset of welding slightly and also influenced the collision process in the later stages of welding when the water was probably present as a cloud of steam at high pressure.

Ideally, therefore, if welds which are consistently of the highest quality are required, the intended welding surfaces should be as smooth as possible. In commercial cladding, however, some sacrificing of weld quality may be necessary because high degrees of polish are not feasible economically. Nevertheless welds of an acceptable quality are generally obtained with a finish of better than 2–3 μm (100–120 μin) CLA which is readily achieved.

In some situations, where access to one or both surfaces is restricted, much rougher finishes may have to be tolerated. Chladek[76] found in the case of roughly planed surfaces that the use of a lower collision point (detonation) velocity or welding in the direction parallel to the planed grooves had a mitigating influence. Chadwick and Lowes[77] showed that an acceptable weld between two thick plates of carbon steel of which the flyer had a roughness of 12 μm (460 μin) CLA (measured in the direction of welding) was obtainable by increasing the impact velocity sufficiently. In similar studies[78] longitudinal grooves 0.4–2.5 mm wide and 0–1.6 mm deep in the flyer plate were readily filled by plastic flow and trapped jet.

Hampel[27] from studies of tube to tube-plate welding, found that if the sum of the roughness values of target and flyer was less than one-tenth of the normal wave amplitude, a good quality weld with regular symmetrical waves was obtained. If the sum was about equal to the normal wave amplitude, a wavy interface with periodic layers of melt was obtained. If the sum was equal to twice the normal wave amplitude, the wave amplitude continuously diminished and the melted layer was continuous. For rougher finishes wave formation was suppressed completely although occasional layers of melt gave some bonding but if the sum of the roughness values exceeded three times the normal wave amplitude welding was prevented completely.

Hampel[27] also determined that flat welds obtained with low explosive loadings had adequate shear strength if smooth surfaces had been employed.

7.5. DIRECT MEASUREMENT OF BONDING PARAMETERS

7.5.1. Introduction

As discussed in Section 7.4, a knowledge of the welding parameters V_d, V_c, V_p and β is desirable for any explosive welding operation. These parameters can be estimated with sufficient accuracy, for many practical purposes, from appropriate theoretical or empirical data, but where conditions are critical, and when the flyer is still accelerating significantly at the instant before impact, more exact data can prove valuable. Hence direct measurements of the parameters have been made by a number of methods, as outlined below.

7.5.2. The Dautriche Method

Basically, the Dautriche test (Fig. 7.13) gives the ratio of the distances of detonation, travelled in exactly the same time, of two explosives. Thus if the detonation velocity of one of the explosives is known the other can be calculated from

$$V_d = V_d^* \frac{L}{2D} \qquad (7.27)$$

The method depends on the 'pressure spike' arising where two

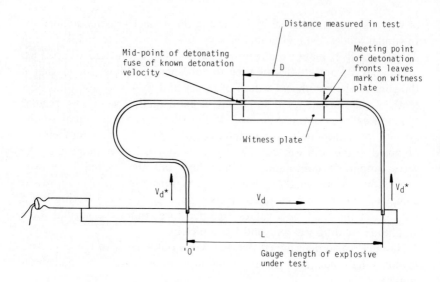

FIG. 7.13 Dautriche test for measuring detonation velocity.

detonation fronts meet head-on. This meeting point is shown by a witness mark on a soft metal plate. The method is extremely simple and inexpensive and the accuracy of measurement can be made to match that of the 'standard' explosive. The detonation velocity of the standard explosive is measured by one of several absolute methods described below.

7.5.3. Wire and Pin Contactor Methods

The detonation velocity of an explosive layer or cylinder can be determined by triggering electronic signals from current-carrying wires or probes placed in the explosive a known distance apart in the detonation direction and recording these using an oscilloscope.

Pin contactor methods for measuring plate velocity were used originally (by, for example, Minshall[79]) to check the efficacy of plane shock wave generators and to measure the subsequent free surface movement of the plate receiving the detonation shock wave at normal incidence. More recently Kury et al.,[67] Hoskins et al.,[80] Crossland et al.[81] and Christensen et al.[64] have employed contactor pins to determine the axial detonation velocity of an explosive contained within a metal tube and/or the accompanying radial velocity of the tube. Signals from pins held at a fixed radial distance from, but at different distances along the tube give the detonation velocity while pins held at varying radial distances at a fixed station along the tube give the change in tube velocity with radial displacement.

Deribas et al.[82] determined the collision point velocity during the explosive welding of inclined plates by measuring the time between the closing of special contacts laid along the target plate. Hoskins et al.[80] and Shribman and Crossland[83] also applied the pin contactor technique to flat plates accelerated by a detonation front travelling along the plate but they used it to measure the flyer velocity. Most other workers, however, appear to favour high speed photographic techniques for this purpose (Section 7.5.4) probably because the flyer acceleration process generally involves large displacements and therefore numerous pins would be required.

7.5.4 High Speed Photography

Although explosive welding techniques were not developed until around 1959, high speed photography had been in use for at least the preceding decade to study the motion of colliding metal plates and the accompanying jetting process, for example in lined shaped charge cutting devices. In most of these early applications of high speed photography, using high speed framing or streak camera techniques, measurements were again made of the plate motion under the influence of a detonation wave entering the flyer plate at normal incidence. For example Shreffler and Deal[32] and Walsh and Christian[84] used a Bowen type framing camera to take 25-frame shadowgraph records of the flyer undersurface at rates of up to 1.1 frames/μs with effective exposure times as small as 0.2 μs/frame. In most explosive welding applications the detonation shock wave enters the plate at grazing incidence and Bergmann et al.,[51] Cowan et al.[54] and Deribas et al.[82] have followed the flyer movement in this case also, mainly by the use of a framing camera technique. The jet and collision angle developed during explosive welding have also been photographed by Deffet and Fosse[85] and Bergmann et al.[51] In such experiments luminescence from jet/air interaction in the stand-off gap and smoke from the detonation process are avoided by evacuating the stand-off gap and by gas-shielding. Bergmann et al. showed that the jet travelled progressively ahead of the collision point at a velocity which was appreciably lower in the presence of air.

7.5.5. Flash Radiography

The much shorter exposure times (of the order of 10^{-9} sec) associated with flash X-ray techniques and the higher penetrating ability of the X-ray beam compared with light sources give high resolution images regardless of luminescence and smokey detonation products. In early techniques each exposure required a separate X-ray source so they were not suitable for relatively long-time events (e.g. flyer acceleration) requiring multiple exposures. The techniques were ideal, however, for providing high definition pictures of jets produced by the detonation of shaped/lined charges or by the oblique collision of two plates in explosive welding. More recently the technique has been developed to determine bonding parameters.

A single radiograph (e.g. ref. 86) can show the positions of the detonation and collision fronts at the same instant and thus enables the instantaneous turning angle θ and instantaneous collision angle β to be measured directly, from which the instantaneous value of the ratio V_p/V_d may be obtained (eqns. (7.1) and (7.4)). Thus if V_d is known from other measurements (Sections 7.5.2 and 7.5.3), V_p and V_c can be estimated.

Smith et al.[30] used a two channel flash X-ray technique to obtain images of the detonation and collision fronts at two positions along the assembly and with an accurately known time interval. Two such images can give the collision angle directly at each position while their displacement in distance and time can give the average velocities, between the two positions, of the detonation front, flyer and collision front.

In neither of the above cases is the rate of increase of V_d, V_p or V_c determined. By using a pulsed X-ray system, however, Takizawa et al.[33] were able to follow the movement of the flyer over large displacements.

7.5.6. Velocity Probe

Wittman[40] described the use of pressure-activated continuous writing velocity probes[87] to determine the position/time plots of the detonation and collision fronts simultaneously, from which V_d and V_c could be obtained directly. Each probe consisted of a closed-ended aluminium tube containing a skip-insulated resistance wire which made electrical contact with the closed end of the tube. One probe was laid on the upper surface of the flyer plate and aligned in the direction of detonation and the other was laid on the upper surface of the target plate and aligned in the direction of welding. Under the influence of the detonation and collision pressure, respectively, progressive collapse of each tube from its closed end on to the resistance wire continuously reduced the effective length of the wire causing corresponding changes in voltage which were monitored by an oscilloscope.

The method also allows V_p and β to be calculated from the same plots but these values refer to the instant of impact; changes in V_p and θ across the stand-off gap are not detected. Hence acceleration data would only be obtainable by repeating the experiment several times with different values of the initial stand-off distance.

7.5.7. Slanting Wire Methods

A simple electronic method[88] by which the collision parameters could be measured in a single experiment was developed by Prümmer.[89] A roof-shaped double slanted resistance wire is positioned as in Fig. 7.14, the

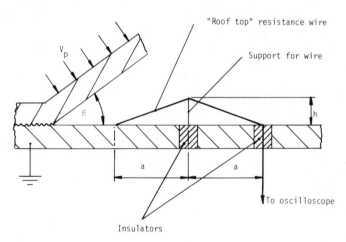

FIG. 7.14 Double slanted wire technique (ref. 89).

angle between wire and target plate being made somewhat less than the anticipated collision angle. As the wire is progressively impacted by the approaching flyer a progressive shortening of the circuit occurs which can be monitored by an oscilloscope.

It is assumed for the analysis that the velocity and turning angle of the flyer are practically constant (but not necessarily at terminal values) between the roof top and the target plate. Because of the differing collision point velocities between the flyer plate and each half of the wire, the oscilloscope registers the movement as two distinct sloping sections on the trace. If the time of duration of collision on the ascending and descending parts of the wire are t_1 and t_2 respectively,

$$\beta = \tan^{-1}\left[\frac{h(t_1+t_2)}{a(t_2-t_1)}\right] \tag{7.28}$$

$$V_c = \frac{2a}{t_1+t_2} \tag{7.29}$$

$$V_p = \frac{2ah}{\sqrt{(h^2(t_1+t_2)^2 + a^2(t_2-t_1)^2)}} \tag{7.30}$$

which, if h/a is small, reduces to

$$V_p = \frac{2h}{(t_2-t_1)} \tag{7.31}$$

Recent improvements in the electronic circuitry and the increased accuracy achieved by the technique were reported by Held.[90] Despite this the technique provides no information on the acceleration of the flyer over the total stand-off gap.

Smith and Linse,[34] however, used a variation of the method to measure the change in turning angle and therefore in velocity of the flyer plate over almost the full stand-off distance. Their set-up is shown schematically in Fig. 7.15. In this case a single-slanted resistance wire was angled right up from target plate to flyer plate, with the slant angle somewhat greater than the anticipated collision angle. A coaxial resistance probe (Section 7.5.6) was also incorporated into the buffer to obtain an independent estimate of the detonation velocity.

Let the measured detonation velocity be V_d and let the measured average rate of change of the effective length of the resistance wire over a known time interval Δt, be V_w. From Fig. 7.15:

$$V_w = \frac{D_w}{\Delta t} \quad \text{and} \quad Z = V_w \Delta t \cos\psi$$

The instantaneous value of turning angle θ is given by:

$$\cot\theta = \frac{Z + V_d \Delta t}{D_F} = \frac{Z + V_d \Delta t}{D_w \sin\psi}$$

$$= \frac{V_w \Delta t \cos\psi + V_d \Delta t}{V_w \Delta t \sin\psi}$$

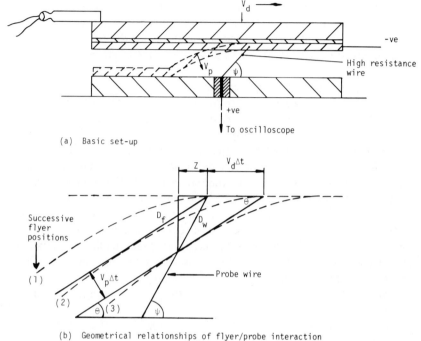

(a) Basic set-up

(b) Geometrical relationships of flyer/probe interaction

FIG. 7.15 Single slanted wire technique (ref. 34).

or

$$\theta = \cot^{-1}\left[\frac{\cos\psi + V_d/V_w}{\sin\psi}\right] \quad (7.32)$$

The instantaneous value of $V_p (= 2V_d \sin\theta/2)$ is then obtained. V_w can then be redetermined for different values of Δt (or different vertical displacements of the flyer) and the corresponding values of V_p computed. For all of the explosive loadings used by Smith and Linse, V_p was still increasing appreciably after displacements of 5 mm which is consistent with eqn. (7.23).

7.6. MISCELLANEOUS WELDING GEOMETRIES FOR SHEETS AND PLATES

Various aspects of the cladding techniques for large plates, which account for about 80% of the market for all explosively welded products, have been described in Sections 7.3 and 7.4 as well as in Chapter 5. This section deals with various other techniques which have been developed and used for special purposes. The applications for the products of these techniques as well as those for clad plate are described in Section 7.8.

7.6.1. Lap Welding of Narrow Plates

In any lap joint, a weld of width at least equal to the thickness of the plate or sheet is desirable. Therefore, in the case of a plate, welding conditions produced transiently by a line charge (Fig. 7.9) may not persist laterally over a sufficient distance to produce the required width of weld. A charge design producing steady subsonic conditions in a direction normal to the longitudinal axes of the components is therefore required. Thus if the components are disposed parallel in the overlap region a subsonically detonating explosive is required. Alternatively, since both the width of the components and the distance of overlap are small, the components may be mutually inclined locally and a supersonically detonating explosive used to give an oblique collision front (Fig. 7.6). A single or double-lap geometry may be employed using a single or balanced charge arrangement (Fig. 7.16) the latter obviating the use of an anvil.

Willis[91] has reported the welding of aluminium busbars, with cross-sections of 50 mm × 4 mm (5 g charges), 50 mm × 6 mm and 50 mm × 10 mm (10 g charges), using the geometry shown in Fig. 7.16(d). Similar charge configurations were used to produce cross-over, T- and L-shaped joints.

7.6.2. Seam/Line Welding of Sheets

For seam welding of sheets a long narrow weld is required. Welding conditions produced transiently by line or ribbon charges of supersonically detonating explosive can persist over lateral distances of several mm so the width of weld obtained is usually adequate for sheets up to a thickness of 2–3 mm. The welding mechanism for this case was described in Section 7.3.5. While the lateral extent of welding is thus limited, the length of weld possible along the joint is unlimited provided the stand-off gap is shielded (for example with tape) to prevent entry of detonation products as welding progresses along the seam.

Seam welding was first used by Addison et al.[92] on aluminium alloys. Later Wright and Bayce[5] mentioned the seam welding of sheets of TZM molybdenum alloy as well as aluminium. Kameishi et al.[93] welded a large number of metals in combination. Viewed in section some of the

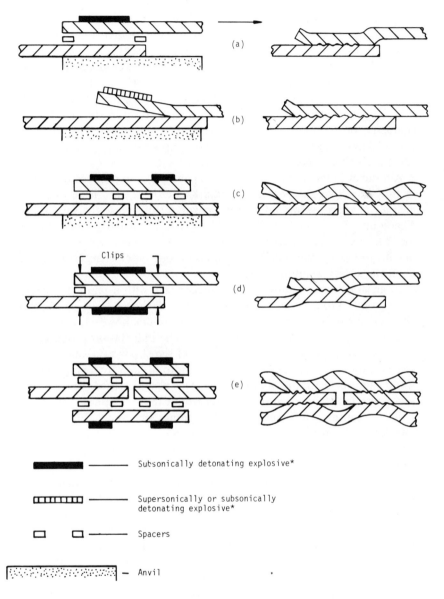

FIG. 7.16 Examples of basic geometries used to produce lap joints between rectangular bars.

arrangements for seam welding sheets are similar to those for lap welding plates although the types of charge used and the welded zones obtained differ for the two cases. Whereas Addison et al.,[92] Polhemus[94] and Bement[95] spaced, and sometimes even inclined the components in the region of overlap (Fig. 7.17(a)–(d)), Kameishi et al.[93] used some sheets simply in contact and achieved welding by shaping the charge instead (Fig. 7.17(e)). Regardless of the technique used, however, the line charges characteristically produce two narrow parallel bonded zones as shown in Fig. 7.17 (see also Fig. 7.9), whereas in the lap welding of plates (Section 7.6.1) each charge produces a single weld band.

A number of balanced line charges in parallel[95] can produce a radiator section from two sheets (Fig. 7.17(d)). Kameishi et al.[93] have used line charges of zigzag form and also in networks to increase the weld areas. The joint shown in Fig. 7.17(c) has given particularly good results for various sheet materials between 0.4 mm and 2.2 mm thick as demonstrated by Bement.[95] The opposing but offset charges weld and align the sheets simultaneously.

In general Bement found a parallel geometry with a minimum stand-off gap of 0.25 to 0.5 mm, depending on sheet thickness, to be satisfactory although a gap of 3 mm still worked well. Titanium, however, gave better results if a preset angle of 10° was used. The indentation produced by welding was accompanied by a thinning of the sheets of about 5%. The indent radius was increased from 0.125 mm to 3.2 mm by taping suitable metal shims beneath the ribbon charge. This also increased the weld area and protected the surface of the sheet from explosive attack.

In evaluating the Bement seam welding technique (Fig. 7.17(c)), Otto and Wittman[96] compared the properties of the joints with those obtained by automatic gas tungsten arc welds. The explosive welds gave superior tensile strength, yield strength and flexural fatigue strength at high and low strains. Usually the joint strength exceeded that of the parent metal whereas the fusion welded joints tended to fail in the heat-affected zones at lower stresses.

These techniques are suggested as being complementary to, rather than replacements for, conventional seam welding techniques. Explosive techniques are preferable, however, under the following conditions: (i) When thermal degradation of a high strength, heat-treatable alloy cannot be tolerated, (ii) when very long lengths or very large areas are involved, (iii) when conventional welding equipment may not be readily available for welding on location, and (iv) when the metals, dissimilar or otherwise, cannot be welded by conventional techniques.

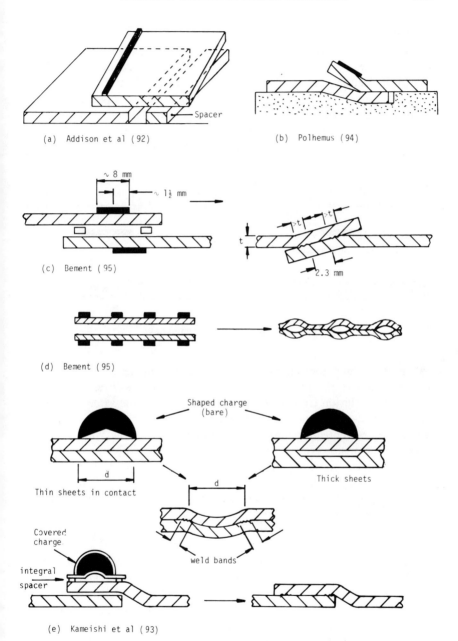

FIG. 7.17 Examples of line welding geometries.

7.6.3. Scarf Welding

Scarf welding of plates can be carried out using the geometries shown in Fig. 7.18. These geometries are expensive to achieve but avoid the abrupt change in section associated with lap joints. In the first geometry (Fig. 7.18(a)) welding with a subsonically detonating explosive is preferable but a supersonically detonating explosive could be used if detonation is made

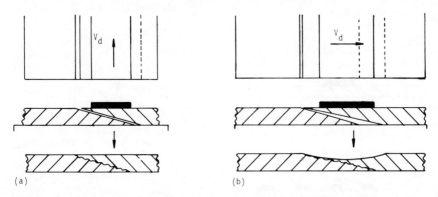

FIG. 7.18 Scarf joints (schematic).

to proceed along the length of the intended joint, i.e. in a direction perpendicular to the Figure, since the varying thickness flyer will cause an oblique collision front to be established in much the same way as a varying stand-off distance will with a constant thickness flyer (Fig. 7.6). If detonation were required to proceed directly across the width of the prepared area, however, a subsonically detonating explosive would be necessary unless the components were inclined as shown in Fig. 7.18(b), but this would result in an unavoidable reduction in joint thickness. Addison[97] has demonstrated the production of weld lengths of 11 m in a single firing using the geometry shown in Fig. 7.18(a).

7.6.4. Butt Welding

Wright and Bayce[39] and Du Pont[98] have demonstrated true butt welding of plates using the basic geometry shown in Fig. 7.19(a). Welding of the abutting surfaces occurs by the progressive lateral deformation which results when a shock wave of sufficient strength enters the plane of the plates at normal incidence. Conditions for producing these welds are rather critical and only certain materials such as aluminium have the required ductility to be joined in this way.

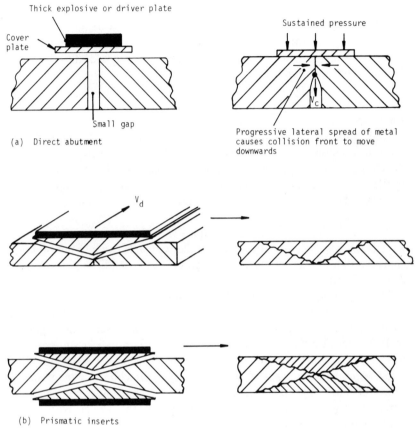

FIG. 7.19 Butt joints (schematic).

Denyachenko[99] has joined 8 mm and 12 mm thick aluminium plates using the geometries shown in Fig. 7.19(b), the detonation vector being parallel to the line of abutment. The joints produced can be regarded as butt joints or as double scarf joints. The areas which actually butt together do not bond but the prismatic inserts used to bridge the two plates can be likened to the fusion weld deposits in the V-preparations of standard fusion weld butt joints. The geometry of Fig. 7.19(b) is similar to that used by Velten[100] to join aluminium alloy reaction rails in a linear motor transport system. In this case flyers of constant thickness were used but these were sufficiently thick to completely fill the V-section. Excess flyer was ground off.

Both Velten and Denyachenko found the joints to be stronger than the parent plates. The latter found the tensile strength to be 33% higher in the as-welded condition. Ductility was lower, however, although this was restored by heating to 210 °C for 2 h. The joints could then be bent through 180° without cracking.

7.6.5. Spot Welding

Fusion spot welding is often used to fix a loose lining of corrosion-resistant material into a mild steel vessel but the local melting can dilute the cladding and impair its corrosion resistance. Brittle compounds and brittle heat-affected zones may also be produced for some combinations of metals. These problems do not arise under optimum explosive welding conditions.

In spot welding large areas by explosive techniques the sheets to be joined are conveniently disposed parallel to each other. Spot welding charges, like those for line welding, produce conditions transiently (Section 7.3.5).

The Asahi Kasei Kogyo Kabushiki Kaisha Corporation, Japan,[101] patented techniques for spot welding two or more sheets together using small charges (~1.2 g) which were contained in a closed-ended cylinder of PVC placed either in direct contact with the top sheet or separated from it by a buffer consisting of a thin disc of PVC, wood or double-sided adhesive tape. Apart from the charge the techniques are basically those of line welding (Fig. 7.17(e)). For a total flyer thickness of 2 mm–4 mm, a 0.1 to 1 mm deep hollow in the target plate, most conveniently produced by grinding, was found to be necessary for good joints whereas for thinner flyers the target and lower flyer could be placed in contact, provided the surfaces were clean. The minimum diameter of the hollow required was equal to the diameter of the explosive charge. A container having the inner surface of its base of convex shape was also more beneficial than a base of constant thickness. Shaping the base is of course equivalent to shaping the charge which is particularly efficacious in line welding (Section 7.6.2). The indentation produced by the spot welding process could be smoothly blended with the sheet by bevelling the perimeter of the PVC container. Several spot welds were achieved in a single firing. Applications mentioned were: lining of chemical vessels, building construction, shipbuilding and vehicle manufacture. Persson[102] has developed an automatic explosive spot welding machine.

7.6.6. Patch Welding

An impactor method (Fig. 7.7) may be used for the formation of either small patches of weld or long narrow seams.[103] A small parallel gap between the flyer and the target may be used, even when the flyer is relatively thick. Such a gap may be achieved in practice by, for example, dimpling the flyer. Test patch welds have been achieved between stainless steel, copper and brass linings from 1.5 to 4.8 mm thick and a mild steel substrate with initial parallel stand-off gaps of 0.25 to 0.4 mm. The welds were all ductile, as demonstrated by their resistance to peeling. With this impactor method, the charge package does not have to make intimate contact with the flyer surface and it is therefore less critical to set up than conventional explosive welding processes. Welds may be produced in any orientation or position with equal facility.

7.6.7. Channel Welding

Channelled structures may be obtained by standard cladding techniques provided that suitable temporary support can be given to the walls of the intended channels. A low melting point metal such as Wood's metal can be used for this purpose. Holtzman and Rudershausen[86] demonstrated the production of a small bronze heat exchanger of channelled construction by using temporarily supported U-sections sandwiched between parallel plates.

Linse[104] described the fabrication of more complex channelled structures for turbine applications. Ribbed panels of Type 304 stainless steel or nickel-base alloy 718 were clad with Hastelloy-X cover skins. Plain carbon steel support tooling used between the ribs was subsequently removed by acid leaching.

A simple channel structure in stainless steel[16] was produced by machining channels 10.6 mm deep and 12.6 mm wide into a stainless steel plate, filling these with aluminium bars of similar section to support the ~3.2 mm thick ribs, welding on a 1.6 mm cover plate of stainless steel and then dissolving out the aluminium with caustic soda.

7.7 WELDING OF FOILS

7.7.1. Theoretical Considerations

A single foil may be welded to a thick substrate, or numerous foils of similar or dissimilar thickness and composition may be welded together with or without the interposition of reinforcing wires.

A foil differs in several respects from a normal flyer. First the foil does not readily keep its shape and, even when the foil rests freely on the flat surface of a target, stand-off gaps comparable with the foil thickness will normally be present. Secondly the small mass of the foil enables it to reach a high velocity very quickly; in fact the response to the detonation pressure may be so rapid that the pressure at the gas/foil interface may be momentarily relieved by an appreciable amount. The acceleration of the foil is therefore even less uniform than the acceleration of a flyer of normal thickness and the foil therefore exhibits oscillations or flutter in flight.[105,30] Such a phenomenon may be observed even for flyers as thick as 1 mm. The foil also responds rapidly to decelerating forces which may arise in two ways. First, as discussed in Section 7.4.5, the duration of the pressure generated when a freely moving flyer impacts a target is proportional to its thickness, so a foil, in the absence of a sufficient impelling pressure on its upper surface, will tend to separate from the target. Secondly, when a foil (or any other flyer) collides with a target, a shock wave, accompanied by a high temperature and pressure, is generated in the air gap ahead of the collision point[106-108], which can influence the behaviour of the flyer here. In particular the shock pressure will oppose and may overwhelm the impelling pressure resulting in a decrease in the velocity of the foil and an increase in the collision angle and stand-off gap ahead of the collision point. The subsequent collision process will be unstable since any momentary stoppage of the foil will remove the driving force behind the shock wave and so relieve the opposing air pressure. Buckling of the foil may therefore be expected.

The shock pressure, P_s, in the air gap is given by[108]

$$P_s \simeq P_0 + 1.1\rho_0(V_c)^2 \qquad (7.33)$$

where P_0 is initial pressure of air in stand-off gap and ρ_0 is initial density of air in stand-off gap.

The temperature T_2 (°K) behind the shock front is related to P_s and the initial temperature T_1 by the equation

$$T_2 = \frac{0.75R}{C_v} \frac{P_s}{P_0} T_1 = 0.3 \frac{P_s}{P_0} T_1 \qquad (7.34)$$

where R is gas constant and C_v is molar heat capacity (measured at constant volume) of the air (C_v is $\frac{5}{2}R$).

For high collision point velocities, P_s and T_2 may attain high values; T_2 for example may reach 18 000 °C.[107] Hence, as well as the effect of the high pressure on the foil, the foil may also be preheated by the shock wave with possible adverse effects. For an impelling pressure P_i (from the explosion) the acceleration of the foil of thickness x and density ρ is given by

$$\frac{dV}{dt} = \frac{P_i - P_s}{x\rho} \qquad (7.35)$$

Obviously, because x is small, dV/dt is sensitive to pressure changes and may even become negative for large P_s.

In view of these considerations the following precautions, where possible, are advisable when welding foils:

(i) Use a former on which to roll the foil flat and adhesive bond the foil to the former; also stretch the foils in a multiple foil assembly. These precautions avoid small or negative collision angles which give high values of V_c,
(ii) Use a thick driver plate, rather than apply the explosive charge directly to the foil or former, and so accelerate the foils in a uniform manner,
(iii) Evacuate the assembly to reduce ρ_0 (eqn. (7.33)),
(iv) Use progressive rather than planar compaction to reduce V_c (eqn. (7.33)).

7.7.2. Welding of Single Foils and Simple Multi-Foil Laminates

A foil may be desirable as a cladding if for instance the foil material is extremely expensive, e.g. platinum. Platinum foil has been used to clad titanium (Section 7.2 and ref. 21) but since platinum is inert it will have an indefinite life in a corrosive environment despite its small thickness. For other purposes even common cladding metals may be adequate in foil-thicknesses. Thus Willis[109] reported the explosive line welding of 1600 m² of 0.2 mm thick aluminium foil to a single substrate. In contrast, high alumina ceramic substrates of area only 4.83 mm × 0.76 mm have been clad with foils of aluminium and other metals between 2.5 µm and 17.8 µm thick using less than 1 mg of explosive.[110] Even smaller quantities of explosive have been used in electronic circuits.[111]

The possibilities of producing laminates consisting of layers of the same or different materials by the explosive compaction of foils in a stack have been recognised and investigated by a number of workers. Holtzman and Rudershausen[86] mentioned the fabrication of two types of laminates, one consisting of sixteen 0.5 mm thick alternate layers of stainless steel and carbon steel, the other of eleven 0.125 mm thick alternate layers of aluminium and Hadfield's manganese steel. The latter was demonstrated to have much greater bullet resistance than either constituent metal of the same mass per unit area. Wright and Bayce[5] referred to the bonding of over a hundred sheets of foil in a single firing and to unusual combinations of properties obtainable in laminated dissimilar metal structures (e.g. strength and lateral heat conduction, hardness and crack-arresting ability). The toughness of laminates has been demonstrated for example by Embury et al.[112] and even simple clads have good crack-arresting properties if the cladding is ductile.[113]

A quite different application of multi-foil welding was reported more recently by Willis.[109] Each end of a twelve layer stack of 1 mm thick aluminium foils was welded to the ends of two busbars to provide a flexible connection of similar current-carrying capacity to the busbars.

Attempts at analysing the collision process in a stack of uniform thickness, uniformly spaced foils have been made by El Sobky[71] and Al-Hassani and Salem.[72] The latter authors correlated the changes in wavelength and amplitude of the waves between layers of foils of different thicknesses and densities with the kinetic energy lost in each successive collision process. By adjusting the mass per unit area of successive foils to ensure the same loss of kinetic energy in successive collisions, interfacial welds of near constant ripple size could be achieved.

Apart from the small-scale speciality applications mentioned above, duplex-clad tantalum–copper–steel and honeycomb (both described in Section 7.8) have been produced commercially. Otherwise, multiple explosive welding of sheets or foils for the production of laminates in planar form has not as yet been widely exploited industrially. Tubular laminates may have important applications (see Blazynski, Chapter 8).

7.7.3. Wire-Reinforced Composites

Introduction
Fibre-reinforcement of metal matrices offers the prospect of significantly better structural materials.[114,115] In principle, most benefits are derived from using low density fibres which possess high modulus and strength over a wide temperature range. Fibre materials which possess this combination of properties are non-metals or ceramics such as carbon, boron or alumina, but the problems involved in the fabrication of high quality composites using such brittle materials are formidable. Thus the use of explosive techniques to produce fibre-reinforced metals has been

restricted to those employing instead relatively ductile metal wires of only moderately high specific strength and modulus.

Consolidation methods
Several workers[116-125] have used explosive techniques to produce composites from stacks consisting of alternate layers of wires and foils. Many combinations of materials have been used, with the wires laid unidirectionally or in planar mesh form, as detailed in Table 7.3. To consolidate fibre/foil assemblies the high pressure produced by the detonation of a high explosive can be applied in two ways: (i) directly—to the upper surface of the assembly (Fig. 7.20(a)), or (ii) indirectly—to a sufficiently massive driver plate, held well above the assembly, which then transfers its momentum to the stack (Fig. 7.20(b)).

In the first method, the upper plate is subjected to a very high peak pressure (the detonation pressure of the explosive) followed by a rapidly decreasing pressure. In the second method, the massive driver plate receives the impulse from the explosion and imparts a comparatively constant pressure to the stack. In either method, the pressure can be applied progressively along the stack as in the normal explosive welding situation or to all points on the upper surface simultaneously as in plane shock wave generation systems (Fig. 20(b)). Progressive application of pressure is advisable on theoretical grounds for the reasons given in Section 7.7.1. In addition the escape of jetted material is facilitated.

Placing the stack of foils and wires in a plastic bag and evacuating prior to compaction avoids the risk of delamination or blister formation from trapped air pockets. Under optimum conditions, densities approaching theoretical rule-of-mixtures values have been achieved, with the residual porosity confined to the interstices between the mating foils and wires.

With evenly spaced unidirectional wires in an aluminium matrix, no wire damage was noted by Slate and Jarvis,[117] even with $\sim 40\%$ volume fraction of relatively brittle tungsten and beryllium wires. However, Wylie et al.,[121] working with nominally unidirectionally aligned stainless steel wires in aluminium ($v_w \sim 8\%$ max.) noted a tendency for fracture to occur at the cross-over points of misaligned wires. Much earlier a similar effect was detected by Jackson et al.[126] in 50 volume % silica fibre-reinforced aluminium composites consolidated by hot pressing, and the marked benefit of more careful fibre alignment was subsequently demonstrated. Bhalla and Williams[123] also noted that fracture of high strength stainless steel wires in an aluminium matrix occurred frequently

TABLE 7.3
EXPLOSIVELY FABRICATED WIRE REINFORCED COMPOSITES

Wire	Matrix	Maximum v_w(%)	Unidirectional or mesh	Plate dimensions (mm)	Maximum UTS (MN/m^2)	References
0.1 mm dia. tungsten	Aluminium	10	Unidirectional		370	
0.15 mm dia. tungsten	Copper	17	Unidirectional	100 × 13 × 6 approximately	754	116,117
0.11 mm dia. beryllium	Aluminium	38	Unidirectional		369	
AFC–77 steel	6061 aluminium alloy	Not stated	Unidirectional	Not stated	669	118
AM–355 stainless steel	1100 aluminium	Not stated	Unidirectional	Not stated	462	119
0.25 and 0.5 mm dia.	C129Y columbium (niobium) alloy	14.7	Unidirectional	Not stated	938	120
TZM molybdenum alloy						
0.25 mm dia. high tensile steel	Aluminium	8	Unidirectional	600 × 200 × 5 approximately	283	121
0.13–0.27 mm dia. high tensile stainless steel	Aluminium	24	50/10 steel/Al mesh	500 × 300 × 6 approximately	453	123
0.125 mm dia. tungsten	Mild steel	5.1	Unidirectional	90 × 22 × 10 approximately	268	122
Various steel meshes	Aluminium and 7005 aluminium alloy	20	Mesh	300 × 300 × 8 approximately	290	124
Kantal and chrome–nickel	Various alum. alloys	10	Mesh			
Tungsten and molybdenum	Various alum. alloys	32	Unidirectional	250 × 100 × 3 approximately	445 (as fabricated to 960 (aged)	125
Maraging steel	Various alum. alloys	35	Sheet + uni-directional wire			

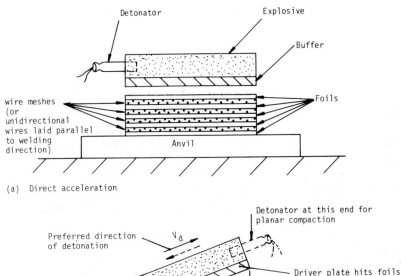

FIG. 7.20 Geometries for welding foils to unidirectional wires or to wire meshes.

at the cross-over points of the mesh, a problem which was alleviated by using a special mesh with a stainless steel wire warp and an aluminium wire weft. (The stainless steel wire volume fraction in this latter case was only 24%, but nevertheless the special mesh had the practical advantage of maintaining the wire spacing during preparation for consolidation.)

Bonding mechanisms

The mechanism by which bonding is achieved in wire-reinforced composites is controversial. The direct compaction method (Fig. 7.20(a)) has some similarity with the well-known parallel arrangement used for explosive cladding (Fig. 7.2) or for the welding of multi-foil laminates[127,72] and the frequent evidence of local jetting and ripple weld formation at foil/foil interfaces supports this view. However, the bonding mechanism in many wire reinforced composites is more complex. Thus satisfactorily bonded composites have been produced[123] using planar compaction (Fig. 7.20(b)), a technique which does not lead to welding between stacks of foils without the presence of wires, either in unidirectional or mesh form. The physical presence of the wires is thus important for bonding in this case. Possibly foils are bonded in areas between adjacent rows of wires because oblique subsonically travelling collision fronts (Section 7.3.2, Fig. 7.6) are produced here. Such fronts should arise since the foils will have minimum velocity at the tops of the wires and maximum velocity at the mid-points of the lines joining the centres of adjacent wires. Nevertheless it would still seem advisable to use progressive rather than planar impact to facilitate the escape of jetted material and air (if present). In addition explosive welding on a small scale may occur between wire and foil as the collapsing foil follows the circumference of the wire to give a range of collision angles. This possibility, and similarly that of adjacent wires collapsing plastically to give suitable collision angles and collapse velocities for welding across their common interface, is also suggested by the evidence of small-scale jetting and wave formation between dynamically compressed metal powders.[128] There is little evidence, however, of an extensive wire/foil bond since in most cases the wires separate from the matrix under stress. There are conflicting results in this respect even in the same system. Thus Wylie et al.[121] reported bonding between steel wires and aluminium while Bhalla and Williams[123] found clean separation of the wires from the matrix.

Mechanical Properties

Mechanical properties data on explosively consolidated wire-reinforced composites are sparse. In most cases, only the tensile strength of composites has been determined as shown in Table 7.3. Except in cases where extensive wire damage occurred, the tensile strength measured in a direction parallel to the wire axes was fairly accurately represented by a rule of mixtures law, i.e.

$$\sigma_c = \sigma_m v_m + \sigma_w v_w \qquad (7.36)$$

where σ_c is tensile strength of composite, σ_m is stress in the matrix at the breaking strain of the wires, v_m is volume fraction of matrix, σ_w is tensile

strength of wires, and v_w = volume fraction of the axially aligned wires.
The corresponding tensile (Young's) modulus of composites also approximated to the weighted mean of the moduli of the constituents. However, more scatter has been observed in the results when wire meshes have been used, probably due to random straightening of the kinked wires on loading.

Very few fatigue data have been obtained from explosively fabricated composites. Test results were reported[117] on two tungsten-copper composites which indicated lives of 5.6 and 8.9×10^5 cycles at a fatigue stress (rotating-bending) of 309 MN/m² (corresponding to fatigue stress/UTS ratios of 0.39 and 0.43 respectively). Bhalla and Williams[123] using fluctuating tension tests carried out sufficient tests on 24% volume fraction stainless steel wire–aluminium composites to produce an S/N curve. The fatigue stress/UTS ratio at 10^5 cycles was approximately 0.4 and at 10^7 cycles approximately 0.24; it was noted that these data compared favourably with published data on aluminium matrix composites fabricated by other means.

In summary, explosive techniques offer an interesting alternative means of consolidating 'model' composites consisting of alternate layers of metal wires and foils. The main advantages over other fabrication methods would appear to be the absence of wire/matrix thermal degradation effects and the potential for producing large areas of composites. However, the achievable wire volume fraction appears to be somewhat limited since the highest reported was 40% and several workers did not achieve 20%.

7.8. APPLICATIONS

7.8.1. Introduction

Whilst the previous sections have indicated the considerable scope offered by explosive welding techniques for planar geometries, only in relatively few instances have the techniques been exploited commercially. This can be attributed in part to a resistance within industry to change from an established technique to a relatively new one. Also while explosive welding will generally produce a superior product, it is not always the most economical technique. Certainly in the case of simple routine fusion welding, brazing and riveting of structures *in situ*, explosive welding will rarely compete as an economical or even practical alternative. Nevertheless the process has now become well established for a number of applications including the production of clad plate for heat exchangers and pressure vessels, tube to tube-plate welding, transition and other joints between dissimilar metals, honeycombs and a number of specialised applications. The properties of clad plate and tube-to-tube-plate joints have been reviewed by Chadwick and Jackson[129] and discussed by Cleland and Blazynski respectively in Chapters 5 and 8 of this book.

7.8.2. Clad Plate

Clad plate has been a valuable construction material for chemical plant vessels and heat exchanger tube-plates for many decades. Its use in vessel manufacture is primarily to combine the cheapness of common structural materials, such as carbon steels, with the high corrosion resistance of expensive liners, such as stainless steel, at a cost nearer that of the structural materials. The product was thus well established before the commercial availability of explosively welded clad plate in the early 1960s. Prior to that time clad plate was produced primarily by fusion strip weld cladding and/or hot roll-bonding. These high temperature processes limited both the range of sizes and the combinations of materials that could be produced and also, in some cases, the quality and consistency of the product. The advent of explosive cladding[86] greatly increased both the range and the quality of clad plates available. Although stainless steel remains the most widely used cladding material, more exotic materials such as titanium, zirconium or tantalum are finding increasing application. There appears to be no upper limit on the area of plate that can be clad in a single firing (the largest to date is approximately 27 m^2), whilst individual tube-plates up to 450 mm thick and weighing over 50 tonnes have been clad.[10]

Explosively clad plate may be further processed by hot and/or cold rolling. Reductions of 90% or more are possible even with titanium–steel clads,[130] although in this case rolling temperatures must be carefully controlled to minimise the growth of intermetallic compounds while maintaining ductility. The explosive cladding process can also be combined with roll-bonding to produce a high quality and competitively priced product. For example, in the British Steel Corporation's 'Colclad' process, a slab of stainless steel (or a nickel alloy) is explosively clad directly with a thin layer of mild steel, which is then hot roll-bonded to a slab of structural steel.[131] All stainless steel/structural steel clad in the UK is now produced exclusively by this process.

Clad plates can be cut and formed using processes similar to those used for unclad structural materials.[106] However, the joining of clad plates by fusion welding requires some care to maintain the integrity of the cladding across the joint. In the case of clads involving compatible metals such as stainless steel–carbon steel, a V-preparation in both backer and cladder is made; the backer materials are then welded first, and finally the groove in the stainless steel is filled in with stainless steel. For plates involving reactive or incompatible metals, such as titanium–steel, special procedures are essential to avoid the formation of brittle intermetallic compounds in the fusion weld zone. The batten-strap technique,[132] provides a widely-used solution to the problem enabling vessels several metres in diameter and over ten metres long, with either flat or domed ends, to be fabricated. Duplex tantalum–copper–steel clads have also been welded using a modified batten-strap technique.[24] In this instance the copper acts as a heat sink, preventing melting of the steel backer despite the high melting point ($\sim 3000\,°C$) of the tantalum layer. The development of this duplex clad product and the associated joining technique led to their application in the construction of chlorine plant vessels 18.3 m long × 2.4 m diameter.

Perhaps the most commonplace use of clad plate has been in the production of strip for United States coinage.[10] In 1965 the world-wide shortage of silver forced the US government to reduce or eliminate the silver content of 10 cent, 25 cent and 50 cent pieces. In the first two coins, the original homogeneous 90% silver alloy was replaced by a clad coin consisting of cupro-nickel on both sides of a copper core which occupied 66% of the total thickness. The resultant coin had a weight and overall electrical conductivity similar to the original silver alloy and thus could be used interchangeably with existing coins in automatic vending machines. For the 50 cent coin, the original homogeneous silver alloy was replaced by a coin of 40% overall silver content consisting of 80% silver alloy surfaces clad on a 20% silver alloy core.

In both cases, the thin strip supplied to the mint for coin production was produced by explosively bonding the two cladding plates to a relatively thick core. In the case of the cupro-nickel/copper clads the plate was reduced from the original 125 mm composite thickness to 12.5 mm by hot rolling and then to the final gauge by cold rolling, whereas the silver clads were processed by cold rolling with intermediate annealing. Strip produced thus was of comparable quality to that produced using more traditional processes such as hot-roll bonding. Figures released by the US government in January 1968 indicated that approximately 17% of the cupro-nickel/copper and 49% of the silver coinage strip purchased had involved explosive cladding.

Apart from pressure vessels, heat exchangers and tubeplates, clad plate is also used to produce:

(i) Chemical retorts for use at high temperature, e.g. copper-clad stainless steel vaporiser vessels for handling adipic acid in nylon manufacture,
(ii) Containers for water and nuclear waste (copper-clad stainless steel and cupro-nickel clad mild steel),
(iii) Vessels for treatment of municipal waste water containing appreciable quantities of chloride ions (titanium clad mild steel),
(iv) Cooking utensils having good thermal properties, toughness and good appearance (copper and/or mild steel clad with stainless steel),
(v) Dual-hardness armour plate (armour plate bonded with a soft-steel or with aluminium),
(vi) Hard or corrosion-resistant edges on tools, earth-moving vehicles and factory equipment (e.g Hastelloy-B on mild steel),
(vii) Bimetallic strip for thermostats, (α-brass/Invar),
(viii) Jewellery foil (gold alloy-clad nickel rolled down after explosive welding from 38 mm to 0.5 mm).

7.8.3. Dissimilar Metal Joints

Perhaps the greatest advantage of explosive welding is its ability to form strong metallurgical bonds between almost any combination of metals many of which would be difficult or impossible to weld by other means.

Electrical contacts for specialised uses require a combination of features, including high electrical conductivity, high wear resistance and low cost, which are not usually found in a single material. Holtzman and Rudershausen[86] described wear-resistant electrical contacts consisting of silver, dispersion-hardened with cadmium oxide, bonded to a carbon steel backing. The bonded strip was cold rolled to give a 75% reduction in thickness prior to stamping out contact buttons approximately 1.2 mm thick. Turner and Dawson[133] reported high current-carrying copper contacts made much more erosion-resistant by cladding with a 0.5 mm thick layer of molybdenum and low inertial electrical components made using silver strip (copper may also be used) explosively welded on to aluminium bars. In the latter case a good electrical joint was produced which was unaffected by corrosion in damp atmospheres (in contrast to experience with brazed joints between the same materials) and several thousand of these joints have been satisfactory in service over a period of several years.

Because of the portability of small explosive charges, many joints can also be made conveniently on site rather than under factory conditions. According to Willis,[109] electrical connections involving copper and/or aluminium have been made in this way, including:

(i) Electrical supply: earthing strip connections between aluminium and aluminium or between copper and aluminium (Fig. 7.16(d)).
(ii) Transport: rail-to-rail copper–steel current carrying joints, as used on the Isle of Man electric railway (Fig. 7.16(c)).
(iii) Chemical plant: copper–aluminium busbar connections.

In the USA copper earthing is mandatory in mines for electrical conductors such as rail tracks and water pipes. According to Velten,[100] attaching copper to steel underground by arc welding results in welds of variable quality and introduces a significant risk of gas explosion. By using detonating fuse as a connecting line and with the mine cleared of all personnel, over a mile of earthing devices can be fitted in a single explosive firing. By 1973 over 4000 of these devices had been fitted in Colorado mines and tunnels at a unit cost approximately half of that for conventional welding.

The adaptability of explosive welding processes to site use is further illustrated by the joining of the 6061-T6 aluminium alloy reaction rails which formed part of a high speed ground transportation system involving a linear induction motor.[100] Fusion welding the rails resulted in weak joints which failed in service. An explosive process was developed in which the rail ends were bevelled on both sides, alignment inserts were fitted and then an aluminium alloy collar was placed over the aligned sections and explosively welded to both sides simultaneously. After grinding excess material away the rail was within the required straightness tolerance. The strength of these explosively welded joints exceeded that of the parent rails. In Japan an aluminium/steel reaction rail for a linear motor automatic system is used in marshalling yards.[134]

7.8.4. Transition Joints

Transition joints are used in all-welded structures to connect dissimilar metals which cannot normally be welded directly together satisfactorily. Each component of a transition joint must be sufficiently thick to enable fusion welding of each component to be carried out without degrading the explosively welded interface. Hence transition joints, in contrast to the clad plate discussed in the previous section, normally consist of a relatively thick flyer bonded to a thick target and are of somewhat more modest area. One consequence of the latter feature is that an inclined welding geometry can also be considered as well as a parallel one (Figs. 7.2 and 7.3) and thus there is more flexibility in the choice of welding parameters. If a particularly thick transition joint is required, the thickness of the cladding can be increased by explosively welding a second flyer over the first, although this has the obvious disadvantage of increasing the cost of the joint considerably.

Examples of applications for planar transition joints include:

(a) Aluminium–copper busbar connections which enable the maximum benefit to be gained from the low contact resistance of copper and the low cost of aluminium. Two variations exist: the aluminium and copper may be permanently bonded to produce a continuous conductor or the aluminium may be bonded to a thin pad of copper which may be bolted to a copper busbar and dismantled when required.

(b) Aluminium–steel, used in aluminium electrolysis plants. Steel supports are used to hold the graphite electrodes and current is fed to these via aluminium busbars and an aluminium–steel transition joint. The temperature of the molten electrolyte is such that the aluminium–steel interface is subjected to temperatures where interdiffusion can occur with the formation of brittle intermetallic compounds. Although this thermal degradation cannot be eliminated completely, there is some evidence of degradation proceeding more slowly if the proportion of melt in the original weld interface is minimised, possibly because there are fewer nuclei for the formation of intermetallic compounds.[59] Hence a weld with a flat melt-free interface may be preferred. An alternative means of improving the resistance to thermal degradation is to incorporate a diffusion barrier. Thus transition joints containing a 2 mm thick titanium interlayer[21] can withstand higher temperatures than direct aluminium–steel joints, both during fabrication (480 °C vs. 315 °C) and service (425 °C vs. 260 °C).

(c) Aluminium–steel, for the connection of aluminium superstructures, deckhouses, masts and antennae to the steel hulls of ships. This lowers the centre of gravity and increases the stability of the ships. Although aluminium superstructures had been used previously with bolted construction (sometimes via insulating gaskets), crevice corrosion had resulted in severe maintenance problems. The use of welded joints has reduced the maintenance requirements and enabled the benefits of aluminium superstructures to be more fully realised.[135]

In addition to the examples given above of planar transition joints, tubular transition joints are sometimes produced by machining rings from clad plate. An example of such a product is an aluminium–stainless steel tubular transition joint for use as a cryogenic coupling, both for pipework and for connections to liquefied gas storage vessels. Attempts have been made to improve the mechanical properties of such joints by the incorporation of one or more interlayers, such as silver, nickel or titanium. An aluminium alloy–stainless steel joint incorporating a 0.75 mm thick silver interlayer is now available commercially with a guaranteed minimum tensile strength at $-196\,°C$ of $280\,MN/m^2$.[21] The impact resistance of such joints is, however, relatively poor (approximately $0.04\,J/mm^2$). Better all-round mechanical properties were obtained from a five-layer composite joint consisting of A5083 aluminium alloy–pure aluminium–titanium–nickel–304 stainless steel,[136] with the impact values increased to approximately $0.3\,J/mm^2$ which is similar to that obtained from the aluminium alloy. However, the inevitably much increased fabrication cost of such multi-layered joints may preclude their widespread commercial use.

7.8.5. Honeycomb

One of the most sophisticated applications to date for explosive welding is the production of explosively bonded honeycomb as described by Velten.[100] Metal foil is gravure printed with a stop weld, cut and stacked. The stop weld is alternated between the stacked sheets such that the only area where metal to metal contact can occur is along the intended node lines of the honeycomb. The explosive shock impulse accelerates the top sheet to impinge on the second sheet which in turn impinges upon the third, and so on. Up to 600 sheets may be bonded in one shot depending on the foil gauge. After welding the stack can be cut and expanded to form the honeycomb structure.

The explosively bonded honeycomb exhibits superior mechanical properties to that produced by alternative methods, such as diffusion bonding. The explosively welded honeycomb has been produced and sold primarily in 5 mm cell size, 0.025–0.05 mm foil gauge, in titanium or stainless steel, although the process has also been demonstrated with a wide variety of other materials.

A major advantage of explosively welded honeycomb is cost savings in comparison with all alternative manufacturing processes.

Most metal honeycomb is purchased as panels by aircraft manufacturers. To fully utilise the superior mechanical properties of the explosively welded honeycomb, processes are under development for the attachment of facing skins using a combination of laser welding and diffusion bonding. Honeycombs are also used for filter holders, radiation collimators and vacuum tube grids as reported by Wright and Bayce[5] who have demonstrated the production of honeycombs in a non-planar geometry. An all-copper honeycomb in the form of a disc with a solid rim was produced by electroplating aluminium wires with copper and packing these into a copper tube which was implosively welded to the wires. The aluminium was subsequently removed by dissolving in caustic soda.

7.9. CONCLUSIONS

Following a few years of development work in the late 1950s, the production of clad plate was exploited commercially very quickly and has continued to increase over the last twenty years. Over much of this period, the important parameters for producing high integrity welds between most combinations of metals and alloys appear to have been well established although the welding mechanism itself has yet to be elucidated. Various methods of measuring and controlling the parameters for welding in planar and tubular geometries are now available. The market for tubular components, however, remains a minor one compared with that for clad plate.

In this chapter a wide range of planar geometries and possible products have been considered. It is notable that many applications were proposed back in the 1960s, but, despite the often superior properties of explosive welds compared with other joints, relatively few have been widely applied industrially. The reasons for this are many and varied—sometimes cost, sometimes a reluctance to change from established techniques and products and sometimes unwarranted suspicions within industry of the effects of explosives, which are renowned more for their destructive than their constructive properties.

In looking into the future, for perhaps the next twenty years, it appears likely that there will be further consolidation of existing uses, such as clad plate, aided by further developments in fabrication techniques for forming and joining plates. It also appears probable that further applications will be developed industrially as the technique becomes better known and costs become more competitive.

ACKNOWLEDGEMENTS

Much of the information contained in this chapter is derived from experience gained at IRD over a period of nearly twenty years. In addition to the referenced items, our thanks are therefore due to our many colleagues at IRD who have unwittingly contributed to this chapter. Thanks are also due to Mr. D. B. Cleland of Nobel's Explosives Co. Ltd., Scotland for supplying information on clad plate.

REFERENCES

1. POCALYKO, A. and WILLIAMS, C. P. Clad plate products by explosion bonding, *Welding J.*, **43** (1964), 854–61.
2. LINSE, V. D., WITTMAN, R. H. and CARLSON, R. J. Defence Metals Information Center, Memo No. 225 (1967).
3. OTTO, H. E. and CARPENTER, S. H. Explosive cladding of large steel plates with lead, *Welding J.*, **51** (1972), 467–73.
4. DU PONT. UK Patent No. 1,168,264.
5. WRIGHT, E. S. and BAYCE, A. E. Current methods and results in explosive welding, Central Institute for Industrial Research, Oslo, Sandefjord/Lillehammer (1964), Vol. 2, 448–72.
6. WILLIAMS, J. D., Ph.D. Thesis, Queen's University of Belfast, (1969).
7. RICHTER, U. Influence of explosion cladding on the properties of the base material, *Proc. 5th High Energy Rate Fabrication Conf.*, Denver (1975), 4.14. 1–15.
8. HUNT, J. N. Wave formation in explosive welding, *Phil. Mag.*, **18**, (1968), 669–80.
9. LUCAS, W. Ph.D. Thesis, Queen's University of Belfast (1970).
10. STONE, J. M. The properties and applications of explosion-bonded clads, *Select Conference on Explosive Welding* (1968), Hove (The Welding Institute), Paper No. 10, 55–62.
11. COOK, M. A. *Science of High Explosives*, Reinhold, New York, (1958).
12. POPOFF, A. A. US Patent No. 3,258,841.
13. RUPPIN, D. The explosion welding of metals—investigation into the unstable processes associated with movement of cover plates, *Colloquium on Welding by Thermochemical or Mechanical Energy*, I. I. W. Meeting, Warsaw, 1968.
14. DU PONT. UK Patent No. 923,746.
15. HOLTZMAN, A. H. Canadian Patent No. 784,458.
16. CHADWICK, M. D. Unpublished work.
17. CHUDZIK, B. UK Patent No. 1,042,952.
18. CHADWICK, M. D. Some aspects of explosive welding in different geometries, *ibid.* ref. 10, Paper No. 2, 21–7.
19. CHADWICK, M. D. Explosive welding using an impactor, *Proc. 7th Int. Conf. High Energy Rate Fabrication*, Leeds (1981), 152–63.

20. KRUSKOV, Yu. N. et al. Mechanical properties of explosively welded titanium–aluminium composites at elevated temperatures, *Svar. Proiz.* (Welding Production), **22** (1975), April, 34–6.
21. ANDERSON, D. K. C. Explosive cladding—available products, properties and applications, *Explosive Welding* (The Welding Institute), London, (1975), Chapter 3, 8–11.
22. BOES, P. J. M. Some aspects of explosive welding, Publication No. *103* (1962), Tech. Centre for Metalworking, TNO, Delft, Holland.
23. VERBRAAK, C. A. Explosive forming can cause problems, *Met. Prog.*, **83** (1963), 109–12.
24. BOUCKAERT, G. P., HIX, H. B. and CHELSUS, J. Explosive-bonded tantalum–steel vessels, *ibid.* ref. 7, 4.4. 1–25.
25. HAYES, G. A., and PEARSON, J. Metallurgical properties of some explosively welded metals, *NAVWEPS Report 7925*, NOTS TP 2950 (ASTIA AD-278354), June 1962.
26. UK Patent No. 1,369,879.
27. HAMPEL, H. Some aspects of explosive tube to tubeplate welding in heat exchangers, *ibid.* ref. 19,173–85.
28. KENNEDY, J. E. Gurney energy of explosives: estimation of the velocity and impulse imparted to driven metal, Sandia Laboratories (New Mexico), *Report No. SC-RR-70790* (1970).
29. JONES, H. A theory of the dependence of rate of detonation of solid explosives on the diameter of the charge, *Proc. Roy. Soc.*, **189 A** (1946), 415–26.
30. SMITH, E. G. Jr., LABER, D. and LINSE, V. D. Explosive metal acceleration studies using flash X-Ray techniques, *Proc. 3rd Int. Conf. of the Center for High Energy Forming*, Vail, Colorado (1971), 1.4.1–26.
31. EYRING, H., POWELL, R. E., DUFFEY, G. H. and PARLIN, R. B. The stability of detonation, *Chem. Rev.*, **45** (1949), 69.
32. SHREFFLER, R. G. and DEAL, W. E. Free surface properties of explosively driven plates, *J. Appl. Phys.*, **24** (1953), 44–8.
33. TAKIZAWA, Y., IZUMA, T., ONZAWA, T. and FUJITA, M. An experimental study of the acceleration zone and the terminal velocity of flyer plate driven by explosive, *ibid.* ref. 7, 4.18, 1–42.
34. SMITH, E. G. Jr., and LINSE, V. D. The acceleration characteristics of explosively driven flyer plates, *Proc. 6th Int. Conf. High Energy Rate Fabrication*, Essen (1977), 1.1.1–15.
35. WATANABE, M., MURAKAMI, Z., FUKUYAMA, I., MUKAI, Y., MAKIHATA, T. and MATSUSHITA, M. The effect of bonding conditions on the wave mode formed at explosive bonded interfaces, *International Conference on Advances in Welding Processes* (1970), London.
36. ONZAWA, T. and ISHII, Y. Wave formation in explosive welding of metals, *ibid.* ref. 7, 4.8.1–27.
37. HAMPEL, H. and RICHTER, U. Formation of interface waves and dependence on the explosive welding parameters, *ibid.* ref. 19, 89–99.
38. RICE, M. H., McQUEEN, R. G., and WALSH, J. M. Compression of solids by strong shock waves, in Seitz and Turnbull (editors) *Solid State Physics*, Vol. 6, Academic Press, New York (1958).

39. WRIGHT, E. S. and BAYCE, A. E. US Patent No. 3,313,021.
40. WITTMAN, R. H. The influence of collision parameters on the strength and microstructure of an explosion welded aluminium alloy, *2nd Int. Symp. on Use of Explosive Energy for Manufacturing Metallic Materials of New Properties*, Marianske Lazne (1973) Paper 10, 153–68.
41. MEYER, M. D. Impact Welding using Magnetically Driven Flyer Plates, *Proc. of 4th Int. Conf. of the Center for High Energy Forming*, Vail, Colorado (1973), 5.3.1–23.
42. ANON., *Explosive Welding*, Pacific Factory, March 1962, 6.
43. SHRIBMAN, V. Ph.D. Thesis, Queen's University of Belfast. (1968).
44. COWAN, G. R. and HOLTZMAN, A. H. Flow configurations in colliding plates: explosive bonding, *J. Appl. Phys.*, **34** (1963), 928–39.
45. ZAKHARENKO, I. D. The determining processes for explosive welding, *ibid.* ref. 34, 1.9.1–7.
46. EFREMOV, V. V. and ZAKHARENKO, I. D. Determination of the upper limit to explosive welding, *Fizika Goreniyai Vzryva*, **12**, (1976), 255–60.
47. CHADWICK, M. D. GRAHAM, B. L. and LOWES, J. M. (IRD), Unpublished work on explosive welding of zirconium alloys to type 304 stainless steel.
48. GURNEY, R. W. The initial velocities of fragments from bombs, shells, grenades, Ballistic Research Laboratories (Aberdeen Proving Ground, Maryland), *Report No. 405* (1943).
49. DUVALL, G. E. and ERKMAN, J. O. Acceleration of a plate by high explosive, *Tech. Report No. 1.*, Stanford Research Institute, Project No. GU–2426 (1958).
50. BAHRANI, A. S., BLACK, T. J. and CROSSLAND, B. The mechanics of wave formation in explosive Welding, *Proc. R. Soc.*, **A296** (1967), 123–36.
51. BERGMANN, O. R., COWAN, G. R. and HOLTZMAN, A. H. Experimental evidence of jet formation during explosive cladding, *Trans. TMS–AIME*, **236** (1966), 646–53.
52. GODUNOV, S. K., DERIBAS, A. A., ZABRODIN, A. V. and KOZIN, N. S. Hydrodynamic effects in colliding solids, *J. Comput. Phys.*, **5** (1970), 517–39.
53. EL-SOBKY, H. and BLAZYNSKI, T. Z. Experimental investigation of the mechanics of explosive welding by means of a liquid analogue, *ibid.* ref. 7, 4.5.1–21.
54. COWAN, G. R., BERGMANN, O. R. and HOLTZMAN, A. H. Mechanism of bond zone wave formation in explosion-clad metals, *Metall. Trans.*, **2** (1971), 3145–55.
55. BAHRANI, A. S. Ph.D. Thesis, Queen's University of Belfast (1965).
56. BUCHWALD, J. and FLEISHMAN, S. L. Manufacture and testing of hollow forgings with explosion bonded bores, *ASME Petroleum Division and Pressure Vessels and Piping Division Joint Conference*, Dallas (1968), 68–PET–18.
57. SAKHNOVSKAYA, E. B., SEDYKH, V. S., and TRYKOV, Yu. P. Properties of explosion welded joints between austenitic steel and aluminium alloys, *Svar. Proiz.* (Welding Production), **18** (1971), No. 7, 34–6.
58. HAMMERSCHMIDT, M. and KREYE, H. Microstructural features determining the properties of explosive welds, *ibid.* ref. 19, 60–70.

59. CHADWICK, M. D. and GRAHAM, B. L. (IRD), Unpublished work on explosive welding of an Al–Zn–Mg alloy to maraging steel, (1972–77).
60. CZAJKOWSKI, H. Explosive welding of mild steel–aluminium prefabricates, *Int. Cof. on the Use of High Energy Rate Methods for Forming, Welding and Compaction*, University of Leeds (1973), 14.1–12.
61. TRUEB, L. F. Microstructural effects of heat treatment on the bond interface of explosively welded metals, *Metall. Trans.*, **2** (1971), 145–53.
62. KELLER, K. Investigations of explosive cladding, *Z. Metallkunde*, **59** (1968), No. 6, 503–13.
63. BURKHARDT, A., HORNBOGEN, E. and KELLER, K. Transition to turbulent flow in crystals, *Z. Metallkunde*, **58** (1967), 410–5.
64. CHRISTENSEN, K. T., EGLY, N. S. and ALTING, L. Explosive welding of tubes to tubeplates, *Metall. Constr. Br. Weld. J.*, **5** (1973), 412–9.
65. STIVERS, S. W. and WITTMAN, R. H. Computer selection of the optimum explosive loading and weld geometry, *ibid.* ref. 7, 4.2.1–16.
66. PRÜMMER, R. Explosive welding of a molybdenum–high temperature resistant alloy compound, *ibid.* ref. 19, 186–91.
67. KURY, J. W., HORNIG, H. C., LEE, E. L., MCDONNEL, J. L., ORNELLAS, D. L., FINGER, M., STRANGE, F. M. and WILKINS, M. L. Metal acceleration by chemical explosives, *4th Symp. on Detonation*, (1965), ONR ACR–126, 1–13.
68. SCHMIDTMANN, E. and PAUL, H. U. The elastic-plastic deformation of metal material under extreme dynamic loading, *Arch. Eisenhuttenwesen*, **36** (1965), No. 10, 699–707.
69. CLELAND, D. B. (Nobel's Explosives Co. Ltd.), Private communication.
70. WYLIE, H. K., WILLIAMS, P. E. G. and CROSSLAND, B. An experimental investigation of explosive welding parameters', Queen's University of Belfast, *Dept. of Mech. Eng. Report No. 514* (1970).
71. EL-SOBKY, H. A. Ph.D. Thesis, The University of Leeds (1979).
72. AL-HASSANI, S. T. S. and SALEM, S. A. L. Explosive bonding of multilayer composites (theory and experiments), *ibid.* ref. 19, 208–17.
73. AZIZ, A. K., HURWITZ, H. and STERNBERG, H. M., Energy transfer to a rigid piston under detonation loading, *Phys. Fluids*, **4** (1961), 380–4.
74. TRUTNEV, V. V. et al. Comparative assessment of the quality of the explosive joining of aluminium to titanium, steel and nickel, *Svar. Proiz.* (Welding Production), **20** (1973), No. 7, 19–21.
75. DAVENPORT, D. E. Explosive welding, ASTME Creative Manufacturing Seminars (1961–2), Paper SP 62–77.
76. CHLADEK, L. Effects of microgeometry and physico-chemical state of surfaces on the quality of joints in explosive cladding of metals *ibid.* ref. 40, Paper No. 14, 199–206.
77. CHADWICK, M. D. and LOWES, J. M. (IRD), Unpublished work.
78. LOWES, J. M. (IRD), Unpublished work.
79. MINSHALL, S. Properties of elastic and plastic waves determined by pin contactors and crystals, *J. Appl. Phys.*, **26** (1955), 463–9.
80. HOSKINS, N. E., ALLAN, J. W. S., BAILEY, W. A., LETHABY, J. W. and SKIDMORE, I. C. The motion of plates and cylinders driven by detonation waves at tangential incidence, *ibid.* ref. 67. 14–26.

81. CROSSLAND, B., WYLIE, H. K., WILLIAMS, P. E. G. and BAHRANI, A. S. Explosive welding of cylindrical surfaces, *ibid.* ref. 40, Paper No. 7, 97–133.
82. DERIBAS, A. A., KUDINOV, V. M., MATVEENKOV, F. I. and SIMONOV, V. A. Determination of the impact parameters of flat plates in explosive welding, *Fizika Goreniya i Vzryva* (Combustion, Explosion and Shock Waves), **3**, (1967), No. 2, 291–8.
83. SHRIBMAN, V. and CROSSLAND, B. An Experimental Investigation of the velocity of the flyer plate in explosive welding, *Proc. of the 2nd Int. Conf. of the Center for High Energy Forming*, Denver (1969), 7.3.1.
84. WALSH, J. M. and CHRISTIAN, R. H. Equations of state of metals from shock wave measurements, *Phys. Review*, **97** (1955), 1544.
85. DEFFET, L. and FOSSE, C. Les Bases des Methodes de Placage des Metaux par l'Action des Explosifs, IFCE Conference (1966).
86. HOLTZMAN, A. H. and RUDERSHAUSEN, C. G. Recent advances in metal working with explosives, *Sheet Metal Industries*, **39**, (1962), 399–414.
87. RIBOVICH, J., WATSON, R. W. and GIBSON, F. C. Instrumented card-gap test, *AIAA Journal*, **6** (1968), 1260–3.
88. BARKER, L. M. and HOLLENBACH, R. E. System of Measuring the Dynamic Properties of Materials, *Rev. Sci. Instr.*, **35** (1964), 742.
89. PRÜMMER, R. A. A new and simple method of determination of the parameters of explosive welding and latest results, *J. of the Industrial Explosives Society of Japan*, **35** (1974), No. 3, 121–26.
90. HELD, M. Theoretical and practical aspects of explosive welding *ibid.* ref. 19, 113–31.
91. WILLIS, J. Explosive welding for jointing conductors, *Electrical Times*, 23 July 1970, 43–44.
92. ADDISON, H. J. Jr., FOGG, W. E., BETZ, G. and HUSSEY, F. W. Explosive welding of aluminium alloys, *Welding J., Res. Supplement*, **42** (1963), 359s–64s.
93. KAMEISHI, M., HIGUCHI, R. and NIWATSUKINO, T. Canadian Patent No. 794,093.
94. POLHEMUS, F. C. Explosive welding development at Pratt and Whitney aircraft, *Proc. 1st Int. Conf. of the Center for High Energy Forming*, Denver (1967), 1.3.1–
95. BEMENT, L. J. Small-scale explosion seam welding, *Welding J.*, **52** (1973), 147–54.
96. OTTO, H. E. and WITTMAN, R. H. Evaluation of NASA–Langley Research Center explosion seam welding, *NASA CR-2874* (1977).
97. ADDISON, H. J. Jr. Explosive Welding, *ASME Paper No. 64–MD–47* (1964).
98. DU PONT, UK Patent No. 1,085, 683.
99. DENYACHENKO, O. A. The physical properties of explosion-welded butt joints in aluminium, *Avt. Svarka.* (Automatic Welding), **28** (1975), 56–7.
100. VELTEN, R. Practical Applications of Explosive Welding, *ibid.* ref. 41, 8.4.1–28.
101. ASAHI KASEI KOGYO KABUSHIKI KAISHA CORPORATION. UK Patent No. 1,010,859.
102. PERSSON, I. Explosive welding indoors in serial production, *ibid.* ref. 40, 261–70.

103. JACKSON. P. W. (IRD). UK Patent Application No. 24299/78.
104. LINSE, V. D. The Application of Explosive Welding to Turbine Components, *ASME Paper No. 74–GT–85* (1974).
105. ORAVA, R. N. and WITTMAN, R. H. Techniques for the control and application of explosive shock waves, *ibid*. ref. 7, 1.1.1–27.
106. HOLTZMAN, A. H. Explosive clads, *ibid*. ref. 5, 489–516.
107. DYNAMIT NOBEL AKTIENGESELLSCHAFT. UK Patent No. 1,192,517.
108. CHADWICK, M. D. An assessment of variable angle techniques used to determine minimum collision angles and impact velocities for explosive welding, *ibid*. ref. 34, 1.6.1–15.
109. WILLIS, J. Applications of Explosive Welding, *ibid*. ref. 21, Chapter 10, 40–4.
110. SHAFFER, J. W., CRANSTON, B. H. and KRAUSS, G. Explosive bonding of metal foils to high alumina ceramic substrates, *ibid*. ref. 7, 4.12.1–28.
111. CRANSTON, B. H. UK Patent No. 1,353,242.
112. EMBURY, J. D., PETCH, N. J., WRAITH, A. E. and WRIGHT, E. S. The fracture of mild steel laminates. *TMS-AIME*, **239** (1967), 114–8.
113. PODGORNYI, A. N., GUZ, I. S. and MILESHKIN, M. B. Failure of laminated composites formed by explosive welding, *Avt. Svarka* (Automatic Welding), **28** (1975), 23–5.
114. KELLY, A. and DAVIES, G. J. Principles of fibre reinforcement, *Met. Reviews*, **10** (1965), No. 37, 1–77.
115. CRATCHLEY, D. Experimental aspects of fibre-reinforced metals, *ibid*. ref. 114, 79–144.
116. JARVIS, C. V. and SLATE, P. M. B. Explosive fabrication of composite materials, *Nature*, **220** (1968), 782–3.
117. SLATE, P. M. B. and JARVIS, C. V. Strengthening of metals by explosive incorporation of strong wires, *J. Inst. Metals*, **100** (1972), 217–24.
118. FLECK, J., LABER, D. and LEONARD, L. Explosive welding of composite materials, *J. Composite Materials*, **3** (1969), 699–701.
119. REECE, O. Y. Reported in *Iron Age*, **205** (1970), 60.
120. REECE, O. Y. Molybdenum wire reinforced columbium composites, *ibid*. ref. 30, 2.1.1–11.
121. WYLIE, H. K., WILLIAMS, J. D. and CROSSLAND, B. Explosive fabrication of fibre reinforced aluminium, *ibid*. ref. 30, 2.2.1–26.
122. MCCLELLAND, H. T. and OTTO, H. E. Explosive compaction of composites, *ibid*. ref. 41, 9.1.1–26.
123. BHALLA, A. K. and WILLIAMS, J. D. Production of stainless steel wire–reinforced aluminium composite sheet by explosive compaction, *J. Materials Science*, **12** (1977), 522–30.
124. GONZALES, A., CUYAS, J. C. and CUSMINSKY, G. Explosive welding of aluminium and aluminium alloy sheet steel mesh reinforced composites, *ibid*. ref. 19, 199–207.
125. DABROWSKI, W. The influence of the technological parameters on the mechanical properties of high strength aluminium matrix explosively manufactured composites, *ibid*. ref. 19, 218–23.
126. JACKSON, P. W., BAKER, A. A., CRATCHLEY, D. and WALKER, P. J. The fabrication of components from aluminium reinforced with silica fibres, *Powder Met.*, **11** (1968). No. 21, 1–22.

127. EL-SOBKY, H. and BLAZYNSKI, T. Z. Analysis of the mechanism of collision in multilayered composites, *ibid.* ref. 19, 100–12.
128. RAYBOULD, D. On the properties of material fabricated by dynamic powder compaction (D.P.C.), *ibid.* ref. 19, 261–73.
129. CHADWICK, M. D. and JACKSON, P. W. Explosive welding in pressure vessels and heat exchangers, *Developments in Pressure Vessel Technology—3*, ed. Nichols, R. W., Applied Science Publishers Ltd. Barking, Essex (1980), Chapter 7, 217–65.
130. DU PONT. UK Patent No. 1,062,320.
131. CLELAND, D. B. Industrial application of 'Kelomet' explosively clad metal, *ibid.* ref. 40, Paper No. 18, 231–41.
132. British Standard Code of Practice CP3003: Lining of vessels and equipment for chemical processes, Part 9; Titanium (1970).
133. TURNER, J. C. and DAWSON, P. H. Explosive welding as a manufacturing technique, *Proc. Int. Conf. on Welding and Fabrication in the Nuclear Industry* (BNES), London (1979), Paper No. 42.
134. TAKIZAWA, Y. Explosive cladding industry and researches on its technology in Japan, *ibid.* ref. 40, Paper No. 11, 171–5.
135. HIX, H. B. Commercial explosive bonding, *ibid.* ref. 34, 2.1.1–18.
136. IZUMA, T. and BABA, N. Development of transition joint for cryogenic temperature, *ibid.* ref. 34, 2.14, 1–19.

Chapter 8

WELDING OF TUBULAR, ROD AND SPECIAL ASSEMBLIES

T. Z. BLAZYNSKI

*Department of Mechanical Engineering,
University of Leeds, Leeds, UK*

8.1. INTRODUCTION

Technological utilisation of the more sophisticated materials or integrally bonded material combinations, prevalent in the last decade or so, has made heavy demands on the ingenuity of engineers, first of all, to devise suitable industrial techniques and then to manufacture a variety of components which may well range from the relatively simple mono- and bimetallic systems to complex multi-strand composites.

Although problems arising in the application of explosive welding and cladding processes to axisymmetric geometries are often more subtle than in the plate cladding operation and, on occasions, verge on almost intractable, very considerable progress has been made in this area and an impressive range of semi-fabricates is in existence. Further, the initially purely experimental 'know-how' is being constantly supplemented by reasonably accurate theoretical appreciation of the obtaining physical conditions.

Explosively welded axisymmetric components, generally in the tubular cylindrical form, vary from the mono- and multi-metallic duplex and triplex cylinders, through tube-to-tube plate systems, to materially and geometrically sophisticated composites. They thus comprise bi- and trimetallic concentric tubing, multilayered foil and foil-mesh reinforced cylinders, collar clad pipe sections and transition joints, heat exchangers, multi-metallic arrays of concentric rods and tubes of specified geometrical patterns, and the tube-to-tube plate systems which incorporate, if and when necessary, explosive plugging of heat exchanger tubes.

Applications of these components range from the anti-corrosive and, possibly, heat resisting systems used in chemical industry in off-shore oil installations and nuclear plant, to light and strong heat exchangers, pressure containers—required in aircraft industry—and integrally bonded bimetallic tubular transition joints. The interest in the latter has increased considerably with the multiplication of the problems that arise in the operation of fluidised beds—associated with the combustion of coal—and superheated steam and cryogenic systems such as distillation columns, shafts and flow valves.

Extended area heat exchangers or, possibly remote control mechanisms for chemical and nuclear plant can be manufactured from the assemblies of coaxially welded arrays of rods and tubes, whereas direct cladding techniques are easily adapted to applications as diverse as the cladding of fuel and nozzle ducts, cladding of gun barrels, and the manufacture of honeycomb aircraft panels.

8.2. EXPLOSIVE AND IMPLOSIVE WELDING SYSTEMS AND BONDING PARAMETERS

8.2.1. Welding Systems

Although basic criteria of successful welding of axisymmetric components and flat plates are essentially the same, the geometry of the tubular and similar assemblies imposes certain limitations on a range of process parameters. The magnitude of the dynamic angle of collapse, as well as the acceleration of the flyer element are affected as a result of a relatively small stand-off distance. This limits, for instance, the number of ways in which the plastic hinge can develop and can also influence the workable size of the assembly itself. The characteristic features of the process can be influenced further by both the geometry adopted and the physical properties of the individual elements of the system. These various restrictions, common to both the relatively simple systems of, say, duplex or triplex cylinders and complex multilayered components, create additional problems in the latter case since the initial spacing of the individual layers is made particularly difficult.

Since the relative ease of manœuvrability enjoyed when welding large-area plates is totally lacking in this case, the elimination of the initial angled flyer-base configuration—favoured in the flat sandwich welding—is largely necessary. It is for this reason that with the notable exception of the tube-to-tube plate welding process, parallel surface systems have to be employed.

Because of their geometrical characteristics, the coaxial, axisymmetric assemblies can be welded using either of the two basically different techniques. These are known as the processes of explosive and implosive welding. Although both rely on the use of high explosives and depend on the generation of plane detonation waves at the respective points of initiation of the charge, their utilisation and the properties of the final products obtained differ widely.

In the explosive welding process, referred to as EWP, the coaxially aligned assembly is contained in a heavy metal die and carries a suitably cylindrical charge in the bore of the inner, flyer, tube. In the implosive (IWP) process, the coaxially positioned elements of the assembly are supported internally, where appropriate, in the bore of the base tube and the outer, flyer, tube or element remains in contact with a hollow cylindrical charge on its outer surface. The EWP suffers from the basic component size and tool cost limitations. The heavy die, required to contain high pressure products of explosion, on the one hand, and to maintain reasonable dimensional tolerances of the welded component, on the other, is subject to both serious length limitations and a high degree of progressive permanent deformation. The product itself is limited both in the length and in the minimum diameter of the bore which, in turn, must be compatible with the level of energy produced by the charge which it contains. The process is therefore primarily used for the manufacture of short length, larger diameter tubular components required in relatively small numbers.

Unlike the EWP, the IWP does not suffer from any of the above limitations. Apart from, possibly, an inner assembly support, it does not require any tooling and the only necessary requisite is a cylindrical charge container enclosing the flyer element of the system. The only size limitations imposed on the operation of IWP are either of ecological nature, i.e. the noise level created by large charges or the deterioration in the quality of the weld in very long tubular assemblies.

Types and levels of stresses and strains provide an additional means of comparison between the two processes. Whereas the EWP produces a triaxial residual stress system in the specimen, the IWP is basically a plane strain phenomenon which imparts different residual stress properties to the welded assembly.

8.2.2. Welding Mechanisms

Some of the phenomena associated with the propagation of stress waves through solid, thick metal plates are often obscured by the combined effects of the generated, high pressures and the mechanical collision of the welding components. The coaxiality of the tubular type of an assembly emphasises, however, the effect of the propagation of stress waves and may occasionally be responsible for so strong a degree of stress interference as to affect the welding process itself.

It is clear that an incident, initially detonation pressure wave—generated on the surface—will suffer reflections during its passage in the radial direction, and that these will result in changes in the type of stress induced at the respective interfaces. Although these effects are likely to be less noticeable when welding, say, thick-walled duplex or triplex cylinders, their influence on the mechanism of welding of multilayered thin-foil cylinders is pronounced. It is likely that in such cases reflection and refraction of a dilatational compressive or tensile wave, together with the effect of a transverse shearing, or distortional one, will lead to the formation of stress waves on the mating interfaces. This phenomenon, combined with the usual jetting, will affect materially the conditions of welding.

A stress wave mechanism of surface wave formation can be explained with reference to Fig. 8.1 which shows both the propagation and the successive reflections of a single compressive wave. One notes that the surface waves develop both in front of and behind the collision region. Ignoring, for the moment, the shear surface waves, it is noted that the reflections also meet the surface periodically at points ahead of and behind the collision point. Quite irrespective of whether a sandwich plate or an axisymmetric assembly is considered these in turn become sources of disturbance and continue as such as the collision point is moving forward. A continuous source of surface instability is thus created and has a bearing on the general geometry of the surface. Clearly, the formation of a hump, postulated by Bahrani et al.,[1] can be regarded as a small amplitude surface wave, that is when the velocity of the flyer is of the same order as that of the surface wave propagation. Interference between the hump and the jet can clearly occur, as described in Ref. 1 for instance. If, however, the velocity of the surface wave is much higher than that of the collision region, one has to consider two possibilities. Either

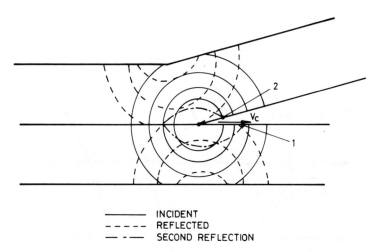

INCIDENT
----- REFLECTED
—·— SECOND REFLECTION

FIG. 8.1 Diagrammatic representation of pressure wave propagation.

the intensity is high enough to produce surface deformation at points 1 and 2 for example (Fig. 8.1), or the intensity is insufficient to cause deformation. In the former case, collision takes place between two surfaces which are already 'wavy'—the waves being either in or out of phase—and the character of the resulting interface is likely to be that defined by Stivers.[2] In the second case only elastic surface waves are formed and flat interfaces are likely to be observed.

The actual formation of the jet introduces a slight complication into this mechanism. The wave mechanism itself is unaffected of course, but the appearance of small vortices—at wave crests and troughs—of pockets of melt, or of continuous layers of it, can be attributed to the interaction of the jet with the base.

To assess critical conditions for a significant stress wave interference, it is sufficient to consider a multilayer, multimaterial assembly subjected to a single, incident compressive pressure wave[3] (Fig. 8.2). The intensity of the reflected wave depends, generally, on the differences between the acoustic impedances of the successive layers. If the particle velocity within a longitudinal wave is u, then the stress in direction, say, xx is

$$\sigma_x = \rho c_1 u \quad (8.1)$$

and the transverse stress τ is

$$\tau = \rho c_2 v \quad (8.2)$$

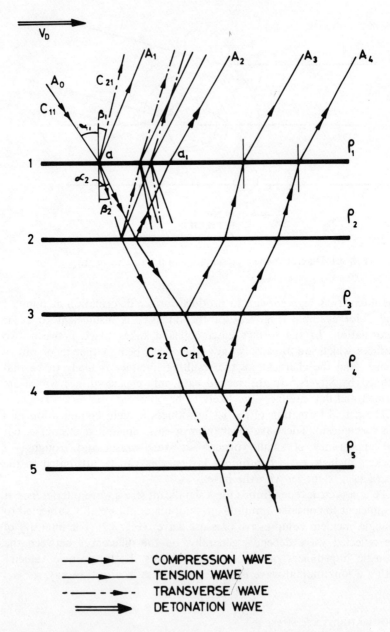

FIG. 8.2 Incidence and reflection of stress waves in a multilayer system.

where v is the particle velocity in the transverse direction and c_1 and c_2 are the velocities of the wave propagation in the two respective directions.

$$c_1 = \left[\frac{E(1-v)}{\rho(1+v)(1-2v)}\right]^{1/2} \tag{8.3}$$

and

$$c_2 = \left[\frac{E}{2\rho(1+v)}\right]^{1/2} \tag{8.4}$$

The relationship between any incident stress wave σ_i and a reflected wave σ_r is given by:

$$\sigma_r = R\sigma_i \tag{8.5}$$

whereas a shear stress wave τ is defined as

$$\tau_r = [(R+1)\cot 2\beta]\sigma_i \tag{8.6}$$

The coefficient of reflection, R, is given by

$$R = \frac{\tan\beta \tan^2 2\beta - \tan\alpha}{\tan\beta \tan^2 2\beta + \tan\alpha} \tag{8.7}$$

Further, $c_2 \sin\alpha = c_1 \sin\beta$, and consequently, with reference to Fig. 8.2, eqn. (8.1) takes the form

$$\sigma_r = \frac{\rho_2 c_{12} - \rho_1 c_{11}}{\rho_2 c_{12} + \rho_1 c_{11}} \sigma_i \tag{8.8}$$

The equation indicates that the reflected wave depends on the interplay between acoustic impedances, and that the coefficient of reflection is a function of the angle of incidence. It is, of course, obvious that the reflected wave will change signs and will be alternately compressive and tensile.

Considering, as an example, a compressive wave of intensity A_0—passing from a medium of density ρ_1 to that of ρ_2, where $\rho_1 c_{11} > \rho_2 c_{12}$—it is found that waves of intensities A_1 and A_2 are tensile and compressive respectively and that the angle of incidence constitutes the only factor which governs the displacement between the points of intersection with the surfaces. In a geometrically parallel welding configuration of Fig. 8.2, the point of collision between the successive layers is moving with a velocity $V_C = V_D$ and the distance between the points of incidence and reflection is

$$aa_1 = 2t \tan \alpha \qquad (8.9)$$

where t is the thickness of the layer. The velocity of the point of reflection is $c_1 \sin \alpha$.

Depending on the relative order of magnitude, two different situations can arise. If the collision point and the incident wave travel at the same speed, i.e. $V_D = c_1 \sin \alpha$, then interference will occur between the tensile reflection of wave A_0 and the compressive reflection of wave A_1. A similar interference—in reverse order—can be expected at the next surface if $\rho_2 c_{12} > \rho_1 c_{11}$. However, if $V_D > c_1 \sin \alpha$, a situation may arise in which a reflected wave, impinging on a newly welded surface, somewhere behind the point of collision, gives rise to high stress intensity. This can easily lead to destruction of the weld and separation of surfaces. It was consequently suggested by El-Sobky and Blazynski[3,4] that in general, the optimum condition is

$$V_D < c \sin \alpha \qquad (8.10)$$

It is evident that the number of points of interference of the reflected waves increases on those interfaces which are near to the explosive charge. A change in the sign of a reflected wave, which occurs alternately on each interface, is also associated with the formation of distortional, shear waves which can cause fracture of the already welded surface. The stress waves can, therefore, contribute to the total stress at the interface and can either reduce or increase the pressure and/or shear stress in the neighbourhood of the collision point. The nature of this intervention depends on the sign of the reflected or refracted wave.

8.3. WELDING OF DUPLEX AND TRIPLEX CYLINDERS

8.3.1. Development of Welding Techniques

Both the discovery of the welding process itself and its application to the welding of tubular components are attributed to Philipchuk,[5] who developed an explosive welding technique for the manufacture of bimetallic cylinders (Fig. 8.3(a)). The application of the EWP to this particular requirement was further pursued by a number of investigators among whom the earlier pioneers numbered Wright and Bayce,[6] Carlson,[7] Holzman and Cowan[8] and Dalrymple and Johnson.[9] Patents for the use of variants of these techniques were obtained by Philipchuk (as used by

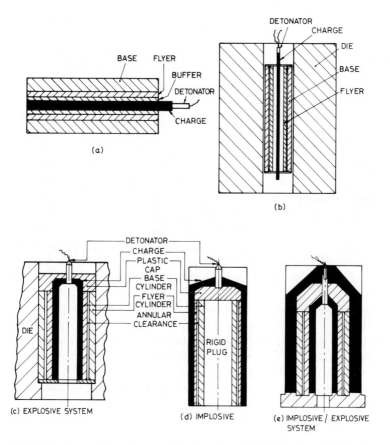

FIG. 8.3 Explosive and implosive welding systems. (a) Philipchuk, (b) Du Pont de Nemours, (c) explosive, (d) implosive, and (e) implosive/explosive (Blazynski and Dara).

the EFI[10]) and E. I. Du Pont de Nemours.[11] The latter technique, for instance, relies on the use of a heavy, containing die (Fig. 8.3(b)). Good results are obtained and the continuous bonding zone between the tubes is easily produced. The process imposes dimensional limitations and difficulties indicated already in the previous section. A somewhat modified EW system is that of Dalrymple and Johnson,[9] which, although including basic characteristics of Philipchuk's and Du Pont's, employs a tubular explosive charge whose axis of symmetry coincides with that of the assembly. A rubber lining is used to separate the explosive from the flyer cylinder and a thick outer tube serves as the die. Metabel (sheet) explosive and Cordtex detonating fuse were used in these experiments, with Metabel releasing excessive energy and consequently producing welds of poor quality. More satisfactory results were obtained with Cordtex, although microscopic examination revealed some variation in the quality of the weld along the cylinder. The unevenness in the weld properties can be attributed to the inherent difficulty of maintaining uniformly distributed charge throughout. Some of these difficulties are reduced in the EWP proposed by Blazynski and Dara[12,13] in which use is made of a low detonation velocity, powder explosive Trimonite No. 1— manufactured in the UK by ICI. In this case, the charge is made in the shape of a hollow cylinder and is positioned concentrically with the flyer tube (Fig. 8.3(c)). Satisfactory results are consistently obtained.

The IWP was investigated extensively in the late 1960s by Blazynski and Dara[12-15] with a view to establishing basic parameters of the operation and, if possible, the best welding conditions. A similar investigation, but on an industrial scale, was being conducted at that time by Shetky.[17]

The diagrammatic sketch of the Blazynski and Dara system is shown in Fig. 8.3(d). The bore of the base tube is supported by a plug made from a low melting point metal, but the use of solid, cylindrical plugs is inadvisable if initially straight cylinders are required. As the detonation front travels along the cylindrical surface of the flyer, an oblique pressure wave is transmitted through the tube wall, the passage of which creates high temperature and pressure within the material; the maximum effect of these being felt along the axis of the assembly since it is here that the interactions of the radially propagated waves are most pronounced. Melting of the core of the plug results and an uneven expansion and bulging of the cylinder are observed. However, these shape problems can be overcome by the provision of a plug with an initial axial hole of a sufficient diameter to accommodate the molten metal.[16]

The IWP proposed by Shetky[17] is similar to that described above, but employs wet sand as the plug and is conducted underground. Compound cylinders up to 1.2 m long have been satisfactorily produced by this method.

To simplify the welding system even further, it is possible to replace the bore supporting plug by another explosive charge which, naturally, must be detonated in phase with the 'implosive' charge. An implosive/explosive system (Fig. 8.3(e)) was designed in 1969 by Blazynski and Dara[16] and an industrially operated one, known as Dynaweld, is used in the manufacture of bimetallic tubing by the Explosive Fabricators Inc. of Louisville, Colorado.[10]

The success of a welding operation depends not only on the correctly designed geometry of the system, but also on the provision of the necessary level of energy. Estimation of the required charge is analytically complex and, at this stage, not particularly reliable and therefore recourse is made to dimensional analysis. This technique was successfully used by Carpenter et al.[18] and Kowalick and Hay[19] in connection with plate welding, by Blazynski and Dara in tube welding,[12] and by Ezra and Penning[20] and Blazynski[21] in explosive forming.

Carpenter's approach[18] is based on the energy balance between that dissipated during the explosion of the charge and the kinetic and elastoplastic deformation energies of the welded specimen. Starting with the assumption that the total (charge) energy $E_T \geq E_K + E_D$ (where E_K and E_D are the kinetic and deformation energies respectively), it is postulated that

$$F = K\theta^2 (\rho t Y/s) \qquad (8.11)$$

where F is the charge per unit area, θ is the dynamic angle of the plastic hinge, ρ is the density of the flyer, t is the thickness of the flyer, Y is the yield stress of the flyer and s is the stand-off distance. It is sometimes argued that eqn. (8.11) is valid only for compatible metal combinations defined by almost the same equations of state.

The Blazynski and Dara approach is concerned mainly with the geometry and mechanical properties of the system. If the diameter of the outer cylinder is D, the energy of the charge is E, the length of the assembly is L, the wall thickness of the inner cylinder is t', the detonation velocity is V_D; and with the other parameters being defined by eqn. (8.11), we have

$$E = f(\rho, Y, V_D, D, t, t', s, L)$$

This functional relationship can be expressed in the form of

$$E = A \rho^B Y^C V_D^E D^G t^H (t')^I s^J L^K$$

Using the concept of vectorial length, i.e. L_x, L_y and L_z, it is shown that

$$E = \rho^{8/5} Y^{-3/5} V_D^{16/5} t^3 f\left[\left(\frac{D}{t}\right)^\alpha \left(\frac{t'}{t}\right)^\beta \left(\frac{s}{t}\right)^\gamma \left(\frac{Y^{3/5}L}{\rho^{3/5} V_D^{6/5} t}\right)^\delta\right] \qquad (8.12)$$

An experimental investigation involving a range of engineering alloys and cylinder sizes[12] led to a simplification of eqn. (8.12) and to the postulation of semi-empirical expressions which define the required charge energy in the two basic welding systems.

Thus, for the EWP (in SI units) and short cylinders

$$E = 0.226 \rho V_D^2 L t^2 \left(\frac{D}{t}\right)^{3/2} \tag{8.13}$$

and for the IWP and cylinders over 300 mm in length,

$$E = 2.25 \rho V_D^2 L t \left(\frac{D}{t}\right)^{3/2} \tag{8.14}$$

For short cylinders ($L < 150$ mm)

$$E = 0.1 \rho^{1.1} Y^{-0.1} V_D^{2.2} t^3 (D/t)^2 (L/t)^{5/6} \tag{8.15}$$

These simplified expressions give only approximate values of charges, unless the mechanical properties of the welded materials happen to be similar.

8.3.2 Characteristics of the Welded Systems

The main application of duplex, bimetallic tubing lies in the field of heat exchangers and transport of chemicals, but occasionally marine requirements call for more specialised systems such as, for instance, flange bore cladding of mild steel with 70/30 cupro-nickel.[22] Mono-metallic duplex cylinders can of course, be manufactured more easily by explosive techniques than by conventional shrinking operations. The area of the triplex cylinder is relatively new and few applications have been recorded. Wylie and Crossland[23] reported a feasibility study of the welding of austenitic, titanium stabilised triplex tubing with axial leak detection channels, and Matin and Blazynski[24] investigated the welding and subsequent cold drawing of brass, copper and mild steel systems.

The three basic types of bonded interfaces, i.e. wavy with pockets of intermetallic alloy or phase change, and line metal-to-metal or line with a continuous layer, are present in duplex and triplex assemblies. Because of the relatively small stand-off distances and, hence, low jet velocities, the amplitudes and wavelengths observed are generally small as shown in Fig. 8.4. The presence of pockets of melt, produced by excessive charge energy, as well as that of a layer of intermetallic alloy, tends to weaken the strength of the weld by introducing an element of brittleness or, at least a reduction in ductility. The rigid support provided either by a plug or by a constraining die, limits the deformation of the elements of the

FIG. 8.4 Implosively welded trimetallic triplex tube.

system in the sense that only the flyer suffers a small amount of plastic strain, with the base remaining elastic. The assembly as a whole acquires, however, a system of in-built residual stresses. The plastic deformation manifests itself in the increase in the level of yield stress of the material and even when relatively small it affects the properties substantially (Fig. 8.5).

In a given material combination, the consistency of the weld quality can be checked, or a comparison between welding system and materials can be made, by means of a suitably adapted ASTM 264-44T proof test, originally devised for clad plate. Figures 8.6(a) and 8.6(b) show respectively a specimen machined out of a tube wall—with the respective coupons machined down to the level of the welded interface—in a testing machine, and the types of failure encountered.

The actual levels of strength in shear in duplex and triplex cylinders are given in Tables 1 and 2. Hot finished extruded or drawn seamless tubing was used to obtain these data. The respective material specifications were: aluminium BS1471 HT30WP, brass BS249, commercially pure copper, low carbon or mild steel BS3602, and stainless steel EN58. These materials are referred to throughout as Al, Br, Cu, MS and SS respectively.

In a satisfactorily welded combination, the uniformity of the weld strength along the specimen is well preserved. Failure is usually due to the shearing of the weld itself and the necking of the base. When a combination of 'strong' and 'weak' materials is used, the weak will shear

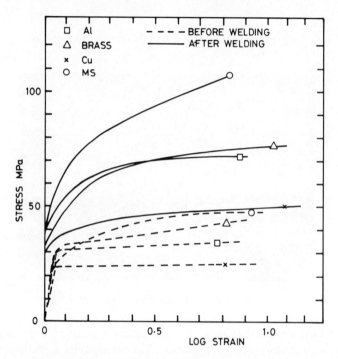

FIG. 8.5 Stress–strain curves before and after welding.

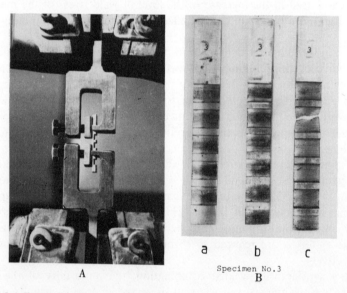

FIG. 8.6 Determination of the shear strength of weld. (A) Tensile test on coupons. (B) Modes of failure, (a) before loading, (b) failure in shear, (c) failure by necking.

TABLE 8.1
WELD STRENGTH IN SHEAR IN DUPLEX CYLINDERS

System	Material		Coupons (MPa)				Interface	Bond
	Flyer	Base						
Explosive	MS	AL	38	36	42	38	41	Continuous	b
	MS	Cu	236	197	189	196	189	Wave	b
	MS	MS	351	327	317	327	317	Wave	b
	MS	SS	347	331	336	331	336	Wave	a
Implosive	AL	AL	117	119	128	122	119	Wave	a
	AL	MS	40	39	32	42	38	Continuous and wave	a+b
	Br	Br	121	152	154	154	165	Wave	b
	Br	MS	185	189	210	248	220	Wave	b
	Cu	MS	123	167	178	177	179	Wave	a
	MS	AL	55	72	75	77	73	Wave	a
	MS	Br	120	138	172	169	158	Wave	a
	MS	MS	223	238	257	268	265	Wave	a
	MS	SS	375	411	454	458	450	Wave	a

a—metal-to-metal, b—intermetallic compound or phase change.

TABLE 8.2
IMPLOSIVE WELD STRENGTH IN SHEAR IN TRIPLEX CYLINDERS

Material	Flyer 1/Flyer 2			Flyer 2/Base			Interface	
	Coupons (MPa)			Coupons (MPa)			1/2	2/B
Br/Cu/MS	227	252	175	236	240	228	195	240	260	240	Wave	Wave
Br/Cu/MS	240	185	200	160	190	220	240	210	180	210	Wave	Wave
Br/Cu/MS	195	240	215	207	210	145	90	180	175	180	Wave	Wave+melt
MS/Br/Cu	185	210	0	60	40	210	145	195	180	200	Wave+melt	Wave

within itself leaving the actual weld intact. In the less satisfactory welds, i.e. when brittle elements are present on the interface, failure is sometimes caused by the machining operation. Generally, with the wave type bonds, the strength of the weld is at least that of the yield stress in shear of the weaker material.

A comparison between the EWP and the IWP shows that for smaller energy levels employed in the EWP, combined with larger stand-off distances, larger wavelengths and amplitudes are achieved.

A representative hardness distribution in implosive and explosive duplex systems is shown in Fig. 8.7. Figure 8.7(a) refers to bimetallic, implosive arrangements, whereas Fig. 8.7(b) gives a comparison between the EWP and IWP systems for mono- and bimetallic combinations. These figures, together with other experimental data,[13,14,25] lead to the following conclusions. An increase in hardness occurs on the outer surface of the flyer—caused by the shock wave of the detonation front—but the effect does not extend to any depth. Very little, if any, change in hardness occurs in the tube wall until the weld interface is reached. The hardness value attains its maximum level on the interface and the area affected depends on the presence, or otherwise, of an intermetallic alloy, change of phase or solidified melt. With direct metallic bonds, no discontinuity is observed on the interface of a mono-metallic combination, but a sharp difference is obvious in the bimetallic compounds.

8.3.3. Residual Stresses

Acceptance for industrial applications, of explosively welded compound cylinders depends to a degree on the in-built state of stress. This is particularly important of course, when the cylinder is to be used as a pressure container.

On detonation of the charge, the flyer undergoes a sequence of elastic, elastoplastic and plastic deformations, caused by the closing of the annular stand-off distance, and then welds to the base. In the absence of perfect rigidity, the welded assembly continues to deform elastically until the detonation front and the associated pressure pulse, have passed the section considered. Elastic relaxation follows, but is usually of lesser magnitude than the original deformation. This cycle, imposed on the assembly during the loading and unloading phases, produces longitudinal, hoop and radial residual stresses. Their pattern, type and levels will ultimately determine the suitability of the component for a given application.

The influence of the adopted welding system on the characteristics of

WELDING OF TUBULAR, ROD AND SPECIAL ASSEMBLIES 305

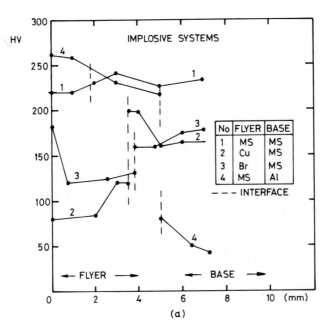

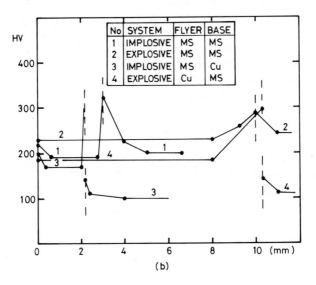

FIG. 8.7 Hardness distribution in welded systems.

the stresses was studied by Blazynski and Dara,[13,26] who, using electric resistance strain gauges in conjunction with a boring out/turning down technique, assessed residual stresses in MS/MS and Cu/MS duplex cylinders. A modified form of Sachs's equations,[27] checked by the Weiss[28] graphical method, was used. A comparison between the two systems is given in Fig. 8.8. Generally, the mechanism of the welding operation, irrespective of its nature, will give rise to a reasonably well defined residual stress pattern. It is observed that, in the case of hoop stress, σ_θ, the flyer is subjected to tensile stress. As the flyer is accelerated towards the base, it retains, and possibly increases, its tensile residual stress and passes into the plastic state. Depending on the level of the available energy, the flyer will either only just touch the base or it will impact it. In either case, the base will acquire a compressive stress which, in the case of impact, may be numerically significant. After the welding stage a spring-back is observed which results in a decrease in the level of the tensile stress in the flyer and in a possible increase in the compressive stress in the base. These tendencies are shown clearly in the figure.

With regard to the radial stress, σ_r, it is seen that the two welding systems produce, as expected, stresses varying in sign. Examination of the specimens obtained by either welding technique indicates that the longitudinal stress, σ_z, of the flyer will be tensile, whereas that of the base will remain essentially compressive.

Stress levels although low—especially for alike combinations—are significant and increase slightly for bimetallic duplex cylinders. These differences reflect the mismatch between the mechanical properties of the metals used. It should be noted, however, that for a given D/t ratio, exchanging the material of the flyer for that of the base does not affect the absolute values of the residual stresses in a bimetallic combination, and that, therefore, the effect of material properties is not very pronounced in thin-walled components. The highest stresses recorded in this instance[13] are of the order of 300 MPa and are considerably lower than the stresses generated in conventionally shrunk cylinders. It is clear that the general level of hoop stress will be affected by the strain suffered and therefore that the magnitude of the annular stand-off will play a considerable role.

A method of analytical prediction of the residual stress levels and distribution in implosive welding was proposed by Blazynski and Dara[26] in 1973. The method is based on Appleby's analysis of an internally pressurised cylinder,[29] but has been modified to account for the effects of the implosive welding technique.

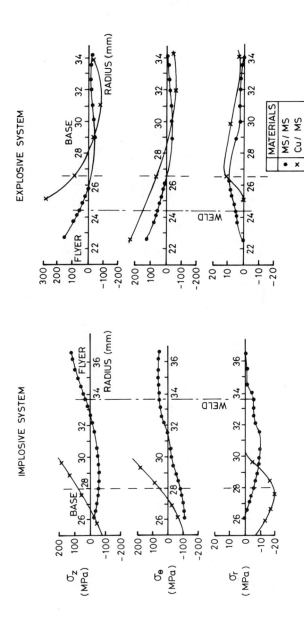

FIG. 8.8 Residual stress distribution. Comparison between EWP and IWP.

With reference to the notation of Fig. 8.9, and assuming $\phi(T)$ to be an arbitrary function of time, taking, also, p_1 as the internal pressure exerted by the bore supporting plug, and p_2 as the external pressure generated by the charge, the radial, σ_r, and hoop, σ_θ, stresses during the loading phase

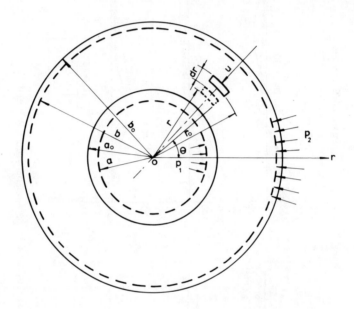

FIG. 8.9 Motion of an element of tube wall during the loading phase.

are given by:

$$\sigma_r = -p_1 + \ln(r/a)\left[2K + \rho\frac{d^2\phi}{dT^2}\right] + \tfrac{1}{2}\frac{d\phi}{dT}(1/a^2 - 1/r^2)\left[\lambda - \rho\frac{d\phi}{dT}\right] \quad (8.16)$$

$$\sigma_\theta = 2K + \sigma_r + (\lambda/r^2)\frac{d\phi}{dT} \quad (8.17)$$

and, with the plane strain conditions assumed

$$\sigma_Z = \tfrac{1}{2}(\sigma_\theta + \sigma_r) \quad (8.18)$$

where λ is the coefficient of viscosity, K is the dynamic yield stress in shear, and

$$a = [a_0 + 2\phi(T)]^{1/2} \quad (8.19)$$

and

$$b = [b_0 + 2\phi(T)]^{1/2} \qquad (8.20)$$

The relationship between p_1 and p_2 is given by

$$p_1 - p_2 = \rho \ln(k) \frac{d^2\phi}{dT^2} + \tfrac{1}{2}(1/a^2 - 1/b^2)\frac{d\phi}{dT}\left[\lambda - \rho\frac{d\phi}{dT}\right] + 2K \ln(k) \qquad (8.21)$$

The motion of an element of an implosively welded cylinder approximates closely to a sinusoidal function. With this assumption, $\phi(T)$ is given by:

$$\phi(T) = A^2 b_0^2/2\left[(1/A + 1/\omega) - \frac{\cos \omega T}{\omega}\right]^2 - b_0^2/2 \qquad (8.22)$$

where $A = \omega/2[b_1/b_0 - 1]$, ω is the frequency and $k = b/a$.

The loading phase is assumed to be contained within the elastic regime and consequently the classic relations for thick cylinders apply.

$$\sigma_r = \frac{(r^2 - b^2)}{r^2(k^2 - 1)}p_1 + \frac{(r^2 - a^2)}{r^2(1/k^2 - 1)}p_2 \qquad (8.23)$$

$$\sigma_\theta = \frac{(r^2 + b^2)}{r^2(k^2 - 1)}p_1 + \frac{(r^2 + a^2)}{r^2(1/k^2 - 1)}p_2 \qquad (8.24)$$

and

$$\sigma_z = 2\nu\left[\frac{p_1}{(k^2 - 1)} + \frac{p_2}{(1/k - 1)}\right] \qquad (8.25)$$

where ν is the Poisson's ratio.

If subscripts L, U and R refer to the loading, unloading and residual stresses respectively, the latter are defined by:

$$\sigma_{Rr} = \sigma_{Lr} - \sigma_{Ur} \qquad (8.26)$$

$$\sigma_{R\theta} = \sigma_{L\theta} - \sigma_{U\theta} \qquad (8.27)$$

and

$$\sigma_{Rz} = \sigma_{Lz} - \sigma_{Uz} \qquad (8.28)$$

The adoption of a sinusoidal time function results in the disappearance of the viscosity term in eqns. (8.16) to (8.20) and, therefore, calculations based on this assumption can only be approximate. Nevertheless, Fig. 8.10 indicates that the agreement between the calculated and experi-

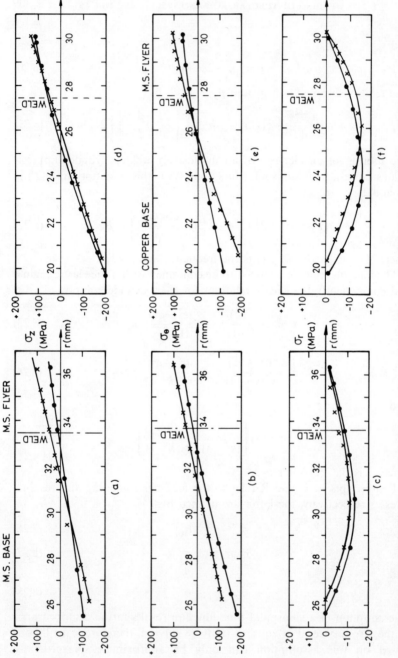

FIG. 8.10 Comparison between the calculated and experimentally assessed residual stresses in an implosive system. ×, Experimental; ●, Theoretical.

mentally assessed stresses is acceptable. It also follows from Figs. 8.8 and 8.10 that by producing an autofrettaging effect, the residual stresses present in an implosively welded duplex cylinder make it suitable for use as an internally loaded pressure vessel.

8.3.4. Conventional Processing

The relatively high levels of energy required, on the one hand, and natural deterioration in the quality of the weld in large and long specimens, on the other, limit the sizes of primary, welded components. In the circumstances, it is often necessary to manufacture short compound cylinders of larger diameters, and then to form them, either in extrusion or drawing, to the required final dimensions. The problems of differentiality in material properties of bi- or trimetallic combinations have to be considered when planning production, since marked differences in properties can, of course, affect the quality of the secondary, finished product.

The effect of the forming operation on the quality of the weld is, naturally, of interest. Some insight into this is afforded by the results of investigations of the process of cold plug drawing of implosively welded bimetallic tubing[25,30] which are summarised in Table 3. Although the actual correspondence between the same sections of a specimen before and after drawing cannot be achieved physically, tests carried out on

TABLE 8.3
COLD PLUG DRAWING OF BIMETALLIC TUBING IMPLOSIVE WELD STRENGTH IN SHEAR (MPa)

Material	After welding Coupons				After drawing Coupons			
Br/MS	129w	185w	234w	157w	102w	87w	163b	0m
Br/MS	190w	234f	230w	238f	135w	124w	35w	0m
Br/MS	190w	171f	177w	222f	187w	59w	231w	0m
Cu/MS	145f	81w	169f	161f	107w	130w	111w	0m
Cu/MS	137f	81w	161f	129f	122w	106w	91w	0m
Cu/MS	165f	173f	182f	161f	78w	120w	130w	0m
MS/Br	157b	161b	169b	157w	87w	46w	0m	0m
MS/Br	157b	161b	157b	166b	115w	46b	0m	0m
MS/Br	157b	97b	145b	157b	74w	96w	0m	0m
MS/Cu	133b	133b	105b	125b	139b	167f	93b	120b
MS/Cu	133b	129b	105b	97b	144b	213f	93b	120b
MS/Cu	133b	133b	105b	113b	204b	167f	93b	120b

Modes of failure: b—base by necking, f—shearing of flyer, m—when machining, w—weld.

specimens welded initially in a similar manner, provide a reasonable guide to the comparison. A general observation drawn is that a degree of degradation in the quality of the weld is invariably present. The weld strength in shear falls off as the reduction in the amplitude of a wave-profiled weld is observed. Line-type welds show a more rapid rate of deterioration. In addition to the weakening of the weld, the drawing operation also changes the mode of failure of the component. Whereas failure of the flyer by shearing appears to be a dominant feature before welding, failure of the weld itself is predominant after drawing. From the point of view of processing itself, even the total failure of the weld during drawing is not necessarily serious because in the condition of plane compressive strain—in which the drawing proceeds—shear stresses can still be supported on the interface. If, however, total failure does occur, small amounts of relative slipping may take place, as shown in Fig. 8.11 (B) and (C).

FIG. 8.11 Differential drawing of MS/Br/Cu triplex tubing. A—17%, B—21%, C—23%.

Theoretical assessments of the drawing loads involved, of the distribution of shearing stresses—particularly in the vicinity of the interface—and of the limits of drawability have been made by Blazynski and Townley[30,31] for implosively welded bimetallic tubing.

8.4. TUBE-TO-TUBEPLATE WELDING

8.4.1. Introduction

The difficulties experienced with joints fabricated either by mechanical expansion or by fusion welding are numerous. Mechanically expanded joints depend for their reliability on low operating temperatures and pressures since their otherwise relatively low integrity may be insufficient to maintain the required standard and quality. Relaxation of in-built residual stress systems may easily produce failure of the joint which can also be caused by thermal cycling or mechanical vibrations. Fusion welding, apart from being difficult to operate, cannot guarantee consistent quality of the weld and consequently, that of the joint. Explosive welding of tubes to tubeplates, on the other hand, introduces an element of consistency combined, from the practical point of view, with an ease of access and inspection, with the absence of preheating and of special, possibly, inert atmosphere. Relatively small limitation is imposed on the material combinations or for that matter, on the wall-thickness of tubes to be welded. Substantially longer welds are obtained than in fusion welding, the ratios of weld lengths to tube wall-thicknesses being of the order of three as compared with about one in the latter case.

The origin of the process is not quite clear, although indications of it are found in a paper published in 1952 by Holtzman and Rudershausen.[32] Crossland et al.[33] rendered an account of its possibilities in 1967, while patent applications for its exploitation were being made between 1964 and 1966.[34-37] Both exploratory and industrial scale work was carried out in the UK in the late 1960s and early 1970s by Chadwick et al.[38] and Cairns and Hardwick[39]—with an investigation into the use of titanium by the latter authors[40]—and by Crossland and Williams,[41,42] and Crossland et al.[43] The more recent papers deal with the evaluation of the parameters of welding,[44] improvements in the welding techniques[45] and the application to more unusual materials.[46]

Since 1966 the patented process has been exploited industrially by Yorkshire Imperial Metals Ltd., Leeds, UK, under the name of YIMpact, in which time some 200 000 tubes have been welded successfully. More recently the International Research and Development Co. Ltd., Newcastle upon Tyne, UK, started to use their IRDEX technique on a production basis.[47]

8.4.2. Geometry and Parameters

The two basic welding configurations of parallel and inclined flyer-base orientations are adopted in the tube-to-tubeplate welding process (Fig. 8.12(a) and (b)). Although the parallel system requires relatively little preparation of the tubeplate holes and is simple to operate, it does, nevertheless, depend on the use of a prefabricated explosive cartridge because the subsonic velocity of detonation needed can only be obtained with a powdered explosive. This limitation increases production costs and makes the technique less attractive than the inclined approach (Fig. 8.12(b)). In this case, a simple detonator—backed up if necessary by a pelleted cylindrical charge—can be used and the effect of supersonic detonation velocity of PETN—the explosive normally used—is counterbalanced by the obliquity of the assembly's geometry. The inclined system forms the basis for the YIMpact process.[36]

The essential parameters governing either operation, are the dynamic angle of collapse, β, the collison energy, IE, the tube velocity, V_t, V_w the velocity of the collision point S, and the detonation velocity V_D. Associated with this is the ratio R of the mass of explosive to the mass of the tube, or the modified ratio R' in which the mass of the buffer is added to that of the tube.

With reference to Fig. 8.12(b) it is seen that in the inclined system

$$V_w = \frac{V_D V_t}{V_t \cos \alpha + V_D \sin \alpha} \tag{8.29}$$

where α is the angle of the taper.

$$\beta = \tan^{-1}\left[V_t \cos \alpha / (V_w - V_t \sin \alpha)\right] \tag{8.30}$$

and the velocity of the tubes relative to the collision point

$$V_F = V_t \cos \alpha / \sin \beta \tag{8.31}$$

In a parallel arrangement (Fig. 8.12(a)), $\alpha = 0$ and the above equations simplify to

$$V_w = V_D \tag{8.32}$$

$$\beta = \tan^{-1}(V_t/V_D) \tag{8.33}$$

and

$$V_F = V_t / \sin \beta \tag{8.34}$$

It is clear that if the tube velocity is determined, other parameters become assessable. An expression giving the terminal tube velocity in terms of R and E (the experimentally determined energy per unit mass of explosive converted to mechanical work) was proposed by Gurney[48] in

WELDING OF TUBULAR, ROD AND SPECIAL ASSEMBLIES 315

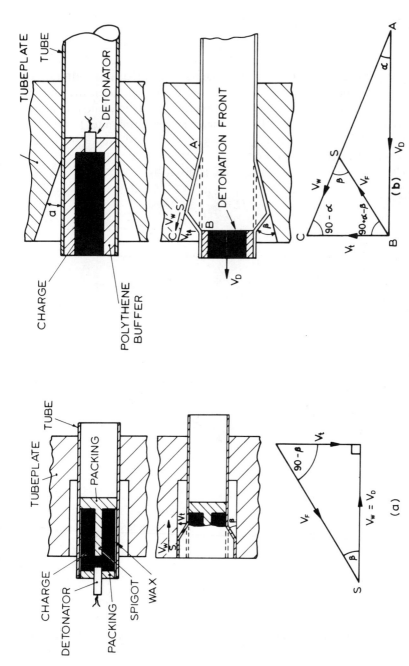

FIG. 8.12 Parallel (a) and inclined (b) tube-to-tubeplate welding geometries.

the form

$$V_t = \sqrt{(2E)[R/(1+\tfrac{1}{2}R)]^{1/2}} \qquad (8.35)$$

It was pointed out by Crossland,[41] among others, that a substitution of R' for R in eqn. (8.35) will give a better agreement between the Gurney equation and experimentally established tube velocities. These relationships are shown in Fig. 8.13 for Metabel and Trimonite 1 (powder) explosives.[41] The correlation is satisfactory and affords, therefore, means of theoretical determination of V_t.

The collision, or impact energy IE, is given by

$$IE = \frac{mV_t^2}{2\pi D} \qquad (8.36)$$

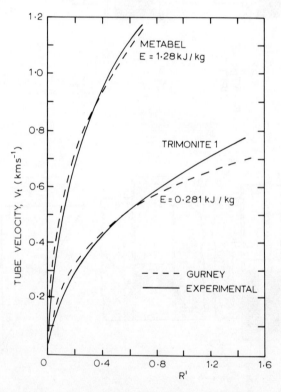

FIG. 8.13 Comparison between the theoretical and experimental values of tube velocity for Metabel and Trimonite 1 (after Crossland[41]).

where m is the mass of the tube and buffer per unit length of the assembly, and D is the mean diameter of the tube at collision.

Conditions necessary for the appearance of an acceptable weld are determined by the critical values of the impact velocity between the tube and tubeplate, and by the collision energy IE. Although some doubt exists as to which of these parameters is truly critical for the welding process, experimental evidence presented to date seems to point to the minimum tube velocity V_t as being of significance. It is clear, however, that once this value is exceeded for a particular physical situation defined either by the angle β or the collision point velocity V_w, the success of welding will depend on the interplay between V_t and IE. This is borne out by experimental data which indicates that, for instance, an increase in V_t, carried out when IE remains unchanged, produces front and rear vortices in the welding wave, whereas an increase in IE—linked to an unchanged angle β and constant V_t—increases the size of these vortices.

The levels of the impact velocity and/or collision energy necessary to produce welding in a given material and size combination are related to the type of flow generated and, hence, to the geometry of the system. When the flow is laminar, i.e. $V_t < V_{t\ crit}$, the required levels of the two parameters are very much higher than those in the turbulent flow. It is in the latter case that wavy-type welds are obtained. The critical, or transition, velocity V_w between the laminar and turbulent modes can be assessed by means of an expression derived by Cowan et al.[49] for an ideal elastoplastic material.

An illustration of the use of the concepts which defined the lower limits of welding is afforded by Fig. 8.14, which shows experimental data referring to an all steel system of 30 mm diameter tubing, 1 mm wall-thickness, welded to the tubeplate by means of Trimonite 1 explosive. The boundary between the laminar and turbulent flows is defined by $V_t = 2.08$ km/s and the failure to weld is indicated.

The minimum value of the collision angle β is difficult to specify since it may well depend on the degree of roughness of the surfaces to be welded and, therefore, can easily vary from tube to tube. The problems of surface contamination and of the ability of the jet to remove it also arise. It was shown by Shribman,[50] for instance, that the surface asperities must be appreciably less than the interface wave amplitude. From the point of view of the geometry of the system, however, β depends principally on the stand-off distance and the initial angle of inclination, α. It is generally agreed that the stand-off distance $s > 0.5\,t$ (where t is the tube wall-thickness). The angle, α, should be such that the minimum

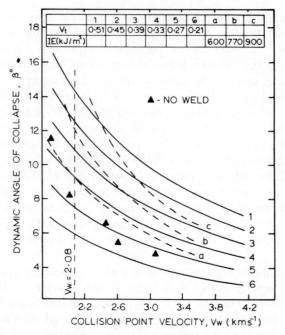

FIG. 8.14 Assessment of the lower limits for welding (after Christensen et al.[44]).

value of β is exceeded, and also the following two conditions are satisfied:

$$V_c/C_{tp} < 1.25 \quad \text{and} \quad V_F/C_t < 1.25$$

where C_{tp} and C_t are the velocities of sound in the tubeplate and tube materials respectively.

8.4.3. System Characteristics

The preparation of the tube and tubeplate as well as the properties of the metals to be welded decide finally the characteristics of the joint produced.

The distortion of the plate during the welding of successive holes can easily occur if the critical thickness of the ligament is not established *a priori*. Further complications arise if the surface finish and cleanliness are not of a high order and if the pressure to be generated is not correctly related to the yield stress.

Estimation of the minimum required thickness of the ligament can be made on the basis of the expression proposed by Crossland and

Williams.[42] Their experiments led to the postulation of an expression derived by means of dimensional analysis,

$$\Delta R = f(\rho, V_t, D, t, L, \sigma)$$

which, when developed, gives

$$\frac{\Delta R}{L} = K \left(\frac{D}{L}\right)^a \left(\frac{\rho V_t}{\sigma}\right)^b \left(\frac{t}{L}\right)^c \quad (8.37)$$

where ΔR is the radial displacement of the adjoining hole, σ is the dynamic yield stress of the tube, L is the ligament thickness, ρ is the density of the tube, D and t are the tube diameter and wall-thickness respectively and K is a material constant.

For mild (low carbon) steel tubeplates, the equation reduces to

$$\Delta R = K' \rho V_t^2 D^2 t^{1/2} L^{-3/2} \quad (8.38)$$

and $K' = K/\sigma$. The validity of this expression is confirmed by Fig. 8.15 which shows a good correlation between the theoretical and experimental data. In the case of non-ferrous combinations, it seems that an expression incorporating the density of the tubeplate, ρ_{tp}, is preferable. The following relation has been suggested:

$$\Delta R = 3.06 \times 10^{-5} \rho V_t^2 D^2 t^{1/2} \sigma^{-1} \rho_{tp}^{-1} L^{-3/2} \quad (8.39)$$

In a parallel welding system a ligament distortion $\Delta R = 0.2t$ can be accepted. Using this value in conjunction with the data provided by, say, Fig. 8.13, the value of L can be calculated.

The entrapment and resolidification of the jet generated are, generally, unacceptable in this particular application, because the shock transmitted during the welding of the other holes in the plate, is likely to damage the welds already made. To minimise the entrapment of the jet, the surface roughness should be kept low, certainly below $2.54 \mu m$ and preferably less than $1.25 \mu m$.

Again, to minimise the charge and hence to reduce the permissible ligament thickness, contamination of surfaces should be avoided. Clean, free from scale and grease surfaces are recommended.

With regard to the yield stresses of the materials concerned, the interface pressure required should exceed the higher yield stress by several times. The higher the latter, the bigger the charge required and consequently the higher the value of L. The problem can be eased by the use of annealed tubing when thick-walled components are to be welded.

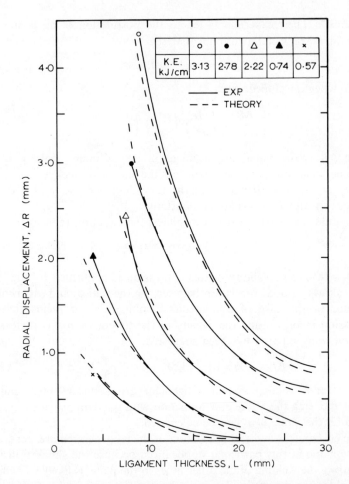

FIG. 8.15 Variation of radial displacement ΔR with the ligament thickness—comparison between the theoretical and experimental results. (after Crossland[42]).

8.4.4. Properties of Joints

The selection of the particular method of testing the properties of the joints depends naturally on whether a production run or an experimental situation is considered. For production purposes non-destructive techniques have to be used, whereas in the experimental development-type tests destructive methods can be employed.

The non-destructive techniques include hydraulic pneumatic leak detection, and ultrasonic weld presence and quality checks. Unlike the fusion welding where the latter test is difficult to operate, the length of the explosive weld—up to 12 mm—is sufficient to allow the use of an ultrasonic technique. Basically, the test should differentiate between the areas of true welding and mere mechanical expansion, but cases have been reported in which wear cohesion, but not welding, of wavy surfaces was confused with the presence of a weld.

Destructive testing helps to quantitatively evaluate various parameters of the system. The most common test employed is the peel test, in which a strip is cut through the tube and into the tubeplate. The strip is then bent through 180° and the tube is peeled off. In the case of a good weld, the tube breaks first. Differing degrees of response can, of course, be expected and can be assessed by means of a tensometer.

If more sophisticated information about the likely performance of the joint is required, thermocycling and mechanical fatigue tests can be employed.

The strength of the bond is determined also from the consideration of the stress required to shear off the interface or to fail the weaker material of the combination. The test is carried out in the manner described in Section 8.3.2. The shear strength of the bond is influenced by IE and, generally, increases with an increase in the energy dissipated. A reduction in the strength is observed in the laminar zone of the flow.

8.5. EXPLOSIVE PLUGGING OF TUBES IN TUBEPLATES

8.5.1. Applications

Tube plugging techniques have been developed mainly for use in feed water heaters and heat exchangers of modern power stations, in which the possibility of tube failure in, say, a generator containing up to 10 000 elements is high. In the operating conditions of power stations, leaking and damaged tubes are very difficult to repair and require the application of either fusion welding or of mechanical plugs designed to produce interference fit. These techniques are often unsatisfactory and even unacceptable, especially so in the nuclear power plant where front crevices in plugged tubes are not tolerated, and where strict helium tests would fail most of the conventionally produced plugging. An additional problem present is that of poor accessibility which, combined with radiation hazard, calls for the development of remotely controlled and operated processes.

The tube-to-tubeplate welding methods, discussed in Section 8.4, offer an obvious solution to the problem and lead naturally to the design of an explosive tube plugging technique. In this, a suitably shaped, hollow plug is explosively welded to the bore of the damaged tube and thus provides an effective and lasting leak-proof seal.

The development of these techniques has been carried out by the Yorkshire Imperial Metals, Leeds (YIM), International Research and Development Co., Newcastle upon Tyne (IRD), Babcock & Wilcox, USA, Atomic Energy of Canada Co. and the Queen's University, Belfast.

The application of explosive welding to plugging was first reported in 1971 by Johnson[51] and has since been described in detail by various groups of investigators. The plugging of elements of nuclear power plant, developed mainly by Crossland and his collaborators in conjunction with their work on the Prototype Fast Reactor (PFR) and the Advanced Gas Cooled Reactor (AGR) is reported in, for instance, references 52 to 56. The YIM method, applied to power stations, particularly to those of the Central Electricity Generating Board system in the UK, is described by Hardwick,[57,58] and the IRD technique, used in plugging feed water heaters and PFR heat exchangers, is reported by Jackson.[47,59]

8.5.2. Plugging Systems

The development of the plugging systems followed closely on the lines of that of tube-to-tubeplate welding in that both parallel and inclined geometries have been adopted. These are shown in Figs. 8.16 and 8.17 respectively.

Since the plugging operation applies usually to an already existing and working set-up, the provision of inclined, conical surfaces in the affected tubeplate may prove impractical or expensive. A parallel welding system, basically established already in the existing tube-plate configuration is therefore likely to be adopted and the charge has to be designed so as to conform to this geometry. As can be seen from the diagrammatic representation of welding in Fig. 8.16, the dynamic angle of collapse depends on the stand-off distance between the outer surface of the plug and the bore of the tube, and on the shape of the charge. To obtain sufficiently large values of β, two geometries of the charge—which consists of a detonator and a pellet of either Trimonite 1 or Detasheet—can be used.[52-56] Larger tube bores call for the use of enhanced values of β and consequently ring or annular shaped charges are employed (Fig. 8.16 (a)–(d)). The sequence of welding is shown in the figure, but the important point of this technique is that the plug end, which initially

WELDING OF TUBULAR, ROD AND SPECIAL ASSEMBLIES

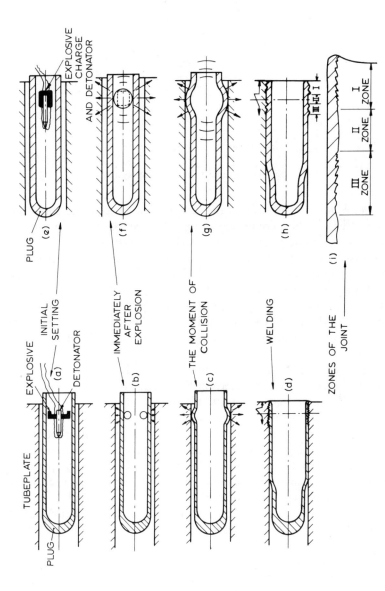

FIG. 8.16 Parallel technique of tube plugging. Ring or annular charge (e–d), point charge (d–h) (after Crossland[42]).

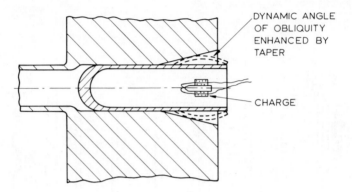

FIG. 8.17 An inclined tube plugging system.

protrudes beyond the tail end of the tube, is sheared off at the entry and the tube itself is welded completely at the entry section. The formation of front crevices—prone to corrosion— is thus avoided.

For smaller diameter tubes, of up to 25 mm bore, a point charge—generating a spherical pressure wave—which produces a spherical expansion of the plug (Fig. 8.16 (e)–(h)) is sufficient. The welded zone is subdivided, in this case, into three sections (Fig. 8.16 (i)), two of which correspond to the actually welded surfaces, i.e. Zones I and III, and Zone II which experiences normal impact and is, therefore, subject to expansion only.

If a large enhancement of β is required, the inclined system, based on the YIMpact technique and shown in Fig. 8.17, is introduced. Naturally the taper present will affect the thickness of the ligament.

The parallel and inclined systems do not compete with each other but are complementary. In addition to catering for somewhat different ranges of β, they also display their own characteristic advantages and disadvantages. Whereas the inclined system for example, facilitates preparatory inspection and cleaning of the tube, the parallel one lends itself more easily to the remote control operation of the type required in atomic power plant, where repairs guided by TV monitors are carried out at some distance from the access area.[56]

8.5.3. Plugs, Materials and Testing

The shape of the plug itself is of considerable importance because it determines the stand-off distance and, taking into account the expansion of the plug, the length of bonded and unbonded zones. Operational problems that are likely to arise can be discussed with reference to Fig. 8.18, which shows a parallel plugging system used by Crossland[56] for the inlet of the reheater of an AGR. The ligament distortion in the area adjoining the plugged hole is often unacceptable and has to be prevented by insertion of bore supporting bungs. Bungs made from metal, say mild steel, are inserted into the holes surrounding the one to be plugged, but non-metallic, perhaps nylon, bungs are used in farther holes. The design of a bung must incorporate a slight taper which will facilitate the withdrawal after plugging. More recently Stanko[60] has introduced a ribbed surface plug, incorporating several rings on its outer surface, with an idea of spanning pitted or otherwise contaminated areas and thus ensuring good welded zones. Good results are claimed, particularly where ligament distortion is concerned.

The early experiments of Bahrani[53] indicated the preferred choice of plug materials. This particular work was concerned with plugging 20 mm diameter tubes—in stainless steel tubeplate—of a PFR system, using

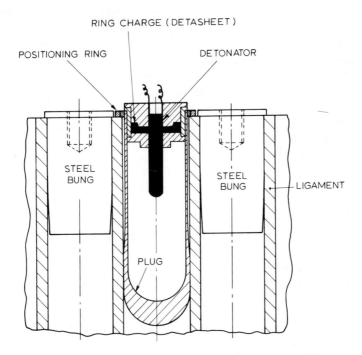

FIG. 8.18 Design of a plugging system (after Crossland[56]).

point charges and vacuum remelted stainless steel plugs. Similar work carried out on the outlet side of an AGR superheater, involving 29 mm tubes in stainless steel plate, established the fact that vacuum remelted SS316 steel is a suitable plug material. The plugging of the inlet side of a superheater is better accomplished with Incoloy 800 plugs. Typical lengths of zones obtained were as follows: (i) on the outlet side—Zone I 4 mm, Zone II 4.5 mm, Zone III 2.5 mm; (ii) on the inlet side, Zone I 3.2 mm, Zone II 5.9 mm and Zone III 2.1 mm. Similar weld lengths are normally achieved in these operations. More recent work, concerned with the plugging of inlet and outlet sides of a SS316H reheater, inlet side of an MS feed system, and a SS316H outlet side of a superheater in the AGR,[56] has confirmed that vacuum remelted stainless steel and Incoloy 800 are the most suitable materials.

The quality of welds obtained in individual cases is normally checked by simulated pressure tests—with pressure up to 50% in excess of the normal working one—and by vacuum helium checks for leaks. Ultrasonic tests are employed when the lengths of the bonded zones are sufficiently long, and dye penetrant checks for the presence of a front crevice are carried out.

Apart from the actual geometry of the plug-tube system, the quality of the weld is influenced to a considerable degree by surface cleanliness and finish. Removal of contaminated surfaces, whether grease, oxides or films of substances originating from previous handling, should be thorough to ensure good welds. Equally, the surface finish should be appreciably better than the amplitude of waves formed, which, of course, has a direct bearing on the magnitude of β and consequently on the selection of the welding system. It should be appreciated that a direct link exists between roughness of the surface and its cleanliness, since rougher surfaces will provide natural catchment areas for contaminants.

8.6. MULTILAYER FOIL REINFORCED CYLINDERS

8.6.1. Introduction
The need for a variety of light, but strong, metallic composites, in the aircraft, chemical and atomic power industries respectively, has been growing steadily, and with the advent of sophisticated material combinations, often difficult to manufacture by conventional methods, the implosive welding technique has found a ready application.

Earlier investigations of the welding techniques, and of the consequential material properties, were concerned with the explosive welding of layered, reinforced flat sheet components. Of particular interest in this respect is the work of, among others, McClelland and Otto[61] on tungsten reinforced steel matrix, Jarvis and Slate[62] on copper– and aluminium–tungsten combinations, Reece[63] on the filament reinforced laminated columbium matrix–molybdenum wire composites, and Bouckaert[64] on tantalum–copper–steel sandwich plates. The problems involved in the manufacture of multilayered cylindrical components have been investigated by Blazynski and El-Sobky,[4,65,66,67] who have suggested that the jetting indentation mechanism needs to be supplemented with a stress wave mechanism (see Chapter 6) in the case of thin foils.

8.6.2. Implosive Welding System

The IWP is employed as the simplest solution to the manufacturing problems. The operation can be regarded as a multiple cladding process which results in a compound cylinder of two or more metallic constituent elements which are uniformly distributed in both the radial and axial directions. Although a suitable combination of materials can provide strengthening of the weaker one by the interleaving foil sheets of the stronger, this particular arrangement is more useful in those industrial applications in which the physical properties of the system, i.e. electrical and heat conductivity, are of greater importance than the mass to strength ratio.

The cylinder is prepared by building up a multilayered flat sandwich which is then wound on to a central, bore supporting tube, to form the unwelded cylinder. The wall-thickness is determined by the number of layers of foil and by the gauge of the individual layers. Normally, the explosive charge is separated from the outer foil by a buffer and, occasionally, by a tubular metallic driver which is used to attenuate the pressure wave generated by the detonation.

On detonating the charge, the outermost foil welds to the neighbouring layer and then both act as the new flyer with respect to the next or 'base' foil element. The process continues radially inwards until the bore tube is reached. Reflections of the stress waves from the respective foil interfaces occur in the manner discussed in Section 8.2.2.

Generally, the same or very similar foil thickness gauge is used in bi- or multimetallic combinations; with the respective stand-off distances maintained at the gauge order of magnitude. Suitable foil thickness is of the order of 0.1 mm and most of the experimentation reported has been made on cylinders of about 70 mm in diameter.

The magnitude of the charge required depends, naturally, on the material combination used, the number of layers and on the stand-off distances. In these calculations the number of layers is of particular importance since the energy of the charge has to be of a sufficient level to allow the welding of the innermost layer while not exceeding its critical value on the surface which could easily suffer damage through excessive melting, cracking or even disintegration. An assessment of the optimum number of foil layers is provided by, for instance, El-Sobky and Blazynski[65,68] (see Chapter 6 for details).

The weldability index N is defined as

$$N \doteq V_{t\ max}/V_{t\ min} \tag{8.40}$$

where V_t can be obtained from, say, eqn. (8.35). If, n, is the number of layers, t, is the thickness of a layer, subscript, e, refers to the charge, and K is defined as

$$K = \frac{\rho_e t_e}{\rho} \tag{8.41}$$

then

$$N \geq n[(K+4nt)(K+nt)/(K+4t)(K+t)]^{1/2} \tag{8.42}$$

For each value of K, the number of layers, n, can be determined. However the number can be increased by employing an energy 'storer' in the form of a driver. This is of course, equivalent to increasing, albeit artificially, the thickness of the flyer. If the equivalent thickness of the driver is, say mt, then approximately

$$N \geq [(m+n_1)(m+1)]^2 \tag{8.43}$$

8.6.3. Pressures and Stresses

A method of estimating magnitudes of the pressures and stresses developed on the interfaces was proposed by, among others, Persson,[69] who suggested the following expressions:

(i) For a small angle of impact β and a metal displaying 'fluid-like' behaviour, the pressure p_β related to the properties of the two media, a, and b, is given by

$$p_\beta = p_0[1 + (U_s\beta/V_f)^2] \tag{8.44}$$

where p_0 (at $\beta=0$) is defined as

$$p_0 = \rho_b U_b V_f / (1 + \rho_b U_b / \rho_a U_a) \tag{8.45}$$

where U is the shock velocity.

(ii) For linear, elastic materials,

$$p_\beta = p_0 (1 + A\beta^2) \tag{8.46}$$

and, in the usual notation,

$$p_0 = (1 + v_b)\rho_b U_b V_f / 3(1 - v_b)(1 + \rho_b)(1 + \rho_b U_b / \rho_a U_a) \tag{8.47}$$

and $A = f(v_a, v_b, U_a, U_b, \rho_a, \rho_b)$.

The shear stress, τ, on any interface has been estimated by Chou,[70] and is given by

$$\tau = V_m (K_m v/c - \rho_m cv) \tag{8.48}$$

where V_m is the volume fraction of the matrix, v is the particle velocity, c is the wave velocity in the matrix and K_m is the stiffness of the matrix.

8.6.4. Structural Properties

Although the mechanism of wave formation in a multilayered system remains unchanged, the pattern of 'waviness' alternates between interfaces. This is caused partly by the continual increase in the thickness of the flyer—as it forms during the successive collisions—and partly by the stress wave pattern which develops during the collisions. Additional difficulties, and hence variations in structural properties, are created by either a less suitable material combination or the use of very thin, (0.025 mm) and therefore highly strain-hardened, foils. Cold rolled foils show considerable differences in their initial properties—related to thickness—and consequently can display very low strain to fracture. Differences in acoustic impedances of the welded metals will also influence their response to the passage of stress and pressure waves, affecting the weldability of the combination. This is of particular significance in the case of combinations containing aluminium and metals like copper, brass and steel where acoustic impedance ratios are 0.420, 0.375 and 0.339 respectively. The difficulties of practical nature, encountered in these situations, are solved by using driver/buffer cylinders.

An example of well welded surfaces is afforded by the copper–brass combination shown in Fig. 8.19(a). The previously described build-up of the wave amplitude on the surfaces nearer the bore can be seen clearly. The wave formation is associated with those interfaces on which the incident compressive stress wave passes from the medium of the lower

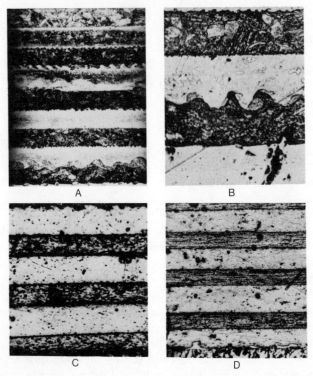

FIG. 8.19 Structural phenomena in implosively welded foil–foil cylinders. (a) Cu/Br × 100, (b) Cu/Br × 160, (c) AL/Cu × 100, (d) AL/MS × 100.

acoustic impedance to one of the higher one. The variation in the amplitude and wavelength of the welding wave is most likely to be caused by the increasing number of disturbances which are produced by the transverse stress developing on each surface.

In the case illustrated here, the build-up of stress waves and of the associated energy level near the base leads to the welding of the latter to the innermost foil and also to the melting, and then resolidification of the melt, on the interface. To prevent this from happening, the base tube is usually coated with a layer of grease.

The variation in the welding conditions in a similar copper–brass system is shown in Fig. 8.19(b). The influence of the driver tube on the pattern of waves is illustrated in Figs. 8.19(c) and (d). The former shows, again, the variation in the amplitude and wavelength in the radial direction in an AL–Cu system welded without recourse to a driver, whereas the latter shows line welds in an AL–MS system welded with the help of a metallic driver. In this case, again, the unprotected MS base tube produces a layer of solidified melt.

8.7. INTERFACE WIRE MESH REINFORCEMENT

8.7.1. Welding Systems
General information regarding wire mesh reinforced cylinders can be found in references 4, 60, 66 and 71.

The basic welding geometry is the same as in the case of the foil–foil systems. The reinforced cylinder is made up of a number of orginally flat, foil sheets interleaved with a suitable metal mesh. The latter is intended to give strength to the assembly. The 'sandwich' is wound on to a base tube to form a cylinder. This arrangement results in a non-circular transverse section which shows a spiral disposition of one material with respect to the other. Radial clearances are small and depend on the relative thicknesses of the component materials. In all instances, the clearances have to be artificially maintained. A variety of foil–mesh combinations are possible which involve one or more foil materials and possibly, gauges. The smaller the latter, i.e. the higher the volume fraction V_m of the mesh, the more concentric a transverse section of the cylinder will be.

Four basic foil matrix-reinforcing wire mesh systems can be distinguished. These are indicated in Fig. 8.20(a). System A represents the standard type of the sandwich; System B, provided with an additional layer, is particularly useful for widely spaced low gauge mesh because it gives improved penetration of the foil; System C can be employed in low volume fraction, relatively thick foil situations; and System D is well suited for applications that require bi- or multimaterial matrices. The post-weld integrity of any system depends largely on the effective meshthickness, since, for a given foil gauge, it is this factor that determines whether or not the interstrand voids are filled and distortions avoided. Various possible mesh arrangements are indicated in Fig. 8.20(b), from which it becomes clear that the ratio of the angles θ_1 and θ_2 is of importance. As a further illustration of the material and foil/mesh gauge combinations used normally, Table 8.4 is included.

The equations of Sections 8.6.2 and 8.6.3 are applicable to these cases.

8.7.2 Metallurgical Characteristics
The quality of the interfacial welds is affected by three basic parameters,

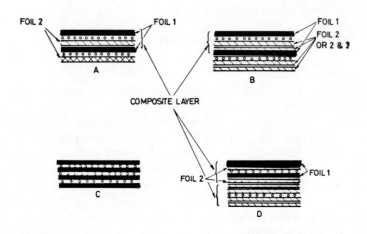

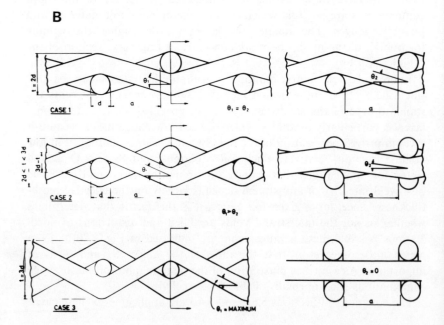

FIG. 8.20 Foil (matrix)–wire mesh (reinforcement) systems. (A) Basic foil–mesh combinations, (B) types of wire mesh.

TABLE 8.4
WIRE MESH REINFORCED CYLINDERS

Test	Matrix material	Welding system (Fig. 8.3(d))	Wire dia. (mm)	Mesh thickness (mm)	Foil thickness (mm)		Vol. fraction (V_w)
					1	2	
1		D	0.254	0.55	0.1524	0.1524	11.8
2	AL	A	0.213	0.41	0.1016	0.1016	21.9
3		A	0.234	0.55	0.1016	0.1016	19.1
4		A	0.213	0.41	0.1524	0.1524	15.8
5		A	0.213	0.41	0.0762	0.0762	27.3
6		A	0.254	0.55	0.1016	0.1016	28.6
7	Cu	A	0.254	0.55	0.0762	0.1270	28.6
8		A	0.416	1.00	0.1778	0.2032	37.2
9		B	0.254	0.55	0.1524	0.0762	21.9
10		C	0.213	0.41	0.2540	—	18.3
11		A	0.254	0.55	0.1016	0.1016	28.6
12	Brass	A	0.254	0.55	0.1524	0.1524	21.1
13		A	0.254	0.55	0.2540	0.2540	13.8
14		A	0.294	0.66	0.2540	0.2540	12.9
15	Steel	A	0.294	0.66	0.1270	0.1270	24.3

viz. the level of energy available, volume fraction of the reinforcement (V_w) and the acoustic impedances of the component materials. Particularly important is the question of the correct level of energy delivered. Underestimation of energy requirement will lead to incomplete welding and, therefore, to a structure characterised by a large number of interfacial voids and unwelded zones. Correct level of energy will give an integral structure (Fig. 8.21(a)) in which welding of the individual wires to the neighbouring foil will take place (Fig. 8.21(b)) and, as the experience shows, the initially free elements of the mesh itself will weld to each other. On the other hand, an overestimate will produce melting on the interfaces (Fig. 8.21(c)) and will tend to increase the effect of stress wave reflections. The amplitudes of the waves are generally small—this reflects small standoff distances—but vorticity of the weld is of a very low order.

The matching or otherwise of the volume fraction of the mesh and foil also affects the integrity of the structure. For instance, when welding a relatively thick mesh to thin matrix, the degree of deformation of the foil required to give perfect contact between two successive layers in any mesh opening, can cause so high a loss of kinetic energy that the

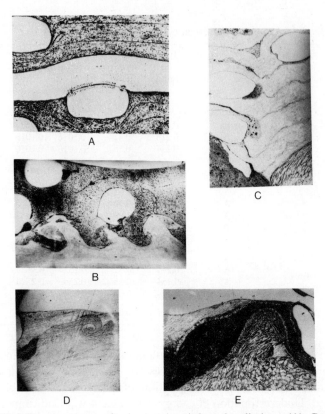

FIG. 8.21 Microstructures of wire mesh reinforced cylinders. (A) Cu foil/SS mesh × 100, (B) welding of mesh to matrix, (C) distortion of structure, (D) rotational flow, (E) formation of large wave in the base.

available energy level may prove insufficient to effect complete welding. Large voids in the structure will be present.

Large mismatch of acoustic impedances of the welded materials may result in either the failure to weld or in the separation of the successive layers after welding.

A characteristic feature of these composites, and, for that matter, of the foil–foil systems, is the occasional appearance of the rotational flow, associated with large angular movements caused by the discussed stress wave reflection and refraction. Since the build-up of the stress wave effects is biggest nearer the base, the rotational flow is more likely to occur in that region. The amplitude of its wave is often in excess of the thickness of the foil (Fig. 8.21(d)) and sometimes even in excess of the diameter of the wire, in which case it can damage the mesh if the wave is formed near a wire (Fig. 8.21(e)).

8.7.3. Mechanical Properties

As indicated in reference 67, the mechanical properties of the foil–mesh–foil system are difficult to assess and only approximate values can be arrived at, either experimentally or by calculation. The geometry of the wire mesh is such that the actual length of a wire element in a unit length of the system is longer than unity by a factor of $1/\cos\theta_1$ longitudinally and $1/\cos\theta_2$ transversely. For this reason, when loaded, the wire of the mesh will undergo an initial extension before actually becoming a load-bearing element of the composite. This is illustrated in Fig. 8.22, which refers to a load–extension curve of a Cu–SS mesh, welding System A composite.

With a 'soft' matrix and 'tough' reinforcement, the increase in the load increases the elastic deformation of both the matrix and the mesh to a comparatively high level of some 0.4 per cent for a load of 3 kN (Zone I). A discontinuity in the slope of the curve, at that level of stressing, is accompanied by the elastic failure of the matrix, but also by the continuing elastic extension of the mesh in Zone II. Eventually both materials become plastic in Zone III, with the load reaching its maximum value of 6 kN and the elongation increasing to some 1.6 per cent. As the wires continue to straighten, the mesh finally fails in Zone IV, but the plastic matrix continues to support the load—at a reduced level—until it fails structurally in Zone V.

The two basic elastic constants for a foil–mesh–foil composite can be assessed by means of simple tensile tests and also by calculation, using the rule of mixtures.[67]

The effective Young's modulus E is given by

$$E = 0.5 E_L + 0.5 E_T \quad (8.48)$$

where subscripts L and T refer to the longitudinal and transverse directions, and

$$E_L = V_w + V_m E_m \quad (8.49)$$

and, further

$$\frac{1}{E_T} = \frac{V_w}{E_w} + \frac{V_m}{E_m} \quad (8.50)$$

Figure 8.23 serves as an example of the types of experimental elastic

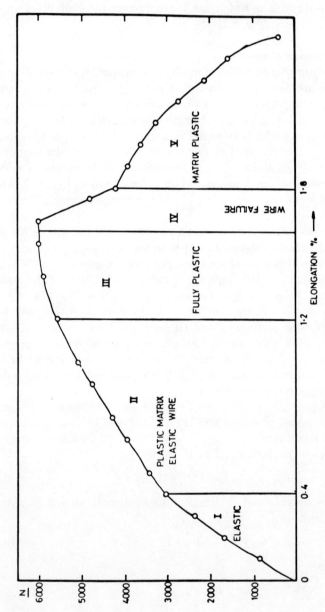

FIG. 8.22 Load–extension curve for a Cu/SS mesh cylinder (welding system A).

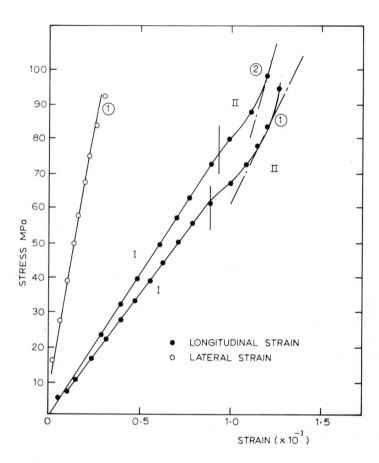

FIG. 8.23 Elastic stress–strain curves for AL–SS mesh cylinders.

stress–strain curves normally observed, and Table 8.5 provides a comparison between the calculated and experimental values in a selection of SS-mesh reinforced material combinations. The Poisson's ratio of the composite is usually lower than that of the constituent materials and the values of the modulus differ slightly from those predicted. The discrepancy is mostly due to the fact that eqns. (8.48) to (8.50) do not account for the effects of the geometry of the mesh on the properties of the welded composites.

TABLE 8.5
TENSILE TESTS OF SPECIMENS MACHINED FROM REINFORCED COMPOSITE CYLINDERS

Test no.	Matrix material	V_w	Modulus E GNm^{-2}		Poisson's ratio (measured)
			Measured	Calculated	
1	AL	11.77	67.7	79.1	0.30
2	AL	21.94	80.01	88.24	0.22
3	Br	21.07	96.8	118.63	0.17
4	Br	13.8	101.0	113.32	0.20
5	Cu	15.78	106.0	120.84	0.30
6	Cu	28.59	148.0	129.84	0.28
7	Cu	21.97	93.3	125.13	0.28

8.8. TRANSITION JOINTS

8.8.1. Applications and Systems

From the industrial point of view, the importance of bi- or even multimetallic transition joints has been increasing steadily and particularly so in the areas of heat transfer—in chemical and atomic plant—ultra high vacuum and cryogenic systems, and to a lesser degree in high frequency electronic equipment.

The interest is centred mainly on the situations in which the transport of fluids at differential temperatures and often of corrosive properties is involved, and in which, therefore, differential properties of the conveying systems themselves are required. Typical examples of these are provided by hot water-superheated steam systems in boilers, by fluidised beds used in conjunction with the combustion of coal, and generally by chemical apparatus. On the cryogenic site, aluminium–stainless steel combinations are required in, for instance, distillation columns, in shafts, in the extended parts of flow valves and similar apparatus. On the other hand, quite a different problem is posed by the requirements imposed by the construction of long vacuum systems in which intersecting storage rings must be able to contain pressures as low as 10^{-12} Torr. Again, difficulties are often encountered in the production of electronic equipment, particularly in the area of high frequency microwave systems. The explosive/implosive welding techniques developed to solve simpler design and manufacturing problems are usefully employed in this field.

Two basic manufacturing approaches have been adopted. A transition joint is either machined out of a multilayered, explosively welded plate, or a bimetallic tubular component is obtained by implosive or explosive end-to-end welding of two tubes with or without a collar over the joint itself. The machining technique has been discussed by, for instance, Kameishi et al.[72] and Izuma et al.,[73] whereas the actual tube welding has been investigated by Grollo,[74] Chadwick and Evans[75] and Wylie and Crossland.[76]

8.8.2 The Machined Joint

The machined joints are intended to operate mainly in cryogenic temperatures and are sometimes used in preference to the truly 'tubular' ones because of metallurgical and economical reasons. Direct explosive welding of aluminium to stainless steel can generate brittle intermetallic compounds on the interface and thus impair the all important bore quality of the weld. Also, the manufacture of special size tubing is very expensive, whereas explosive welding of a multilayered plate of suitable thickness provides means of separating the less compatible materials and of reducing cost of the starting elements of the assembly.

A typical example of such a combination is afforded by the Cryocoup Joint[73] (Fig. 8.24), which is machined out of a five-layer plate of aluminium, titanium, nickel and stainless steel alloys. It is claimed that joints of up to 0.7 m in outer diameter can be manufactured.

Mass spectrometer helium leaks tests showed that no leaking, at a sensitivity of 10^{-10} cm^3/s, took place in the tested range of joint sizes between 25 mm o.d. and 450 mm o.d. The effects of internal pressurisation

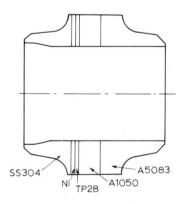

FIG. 8.24 Cryocoup joint.

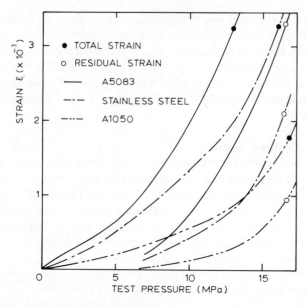

FIG. 8.25 Stress–strain curve for pure AL, AL-alloy and SS cryocoup joint.

of this type of joint can be gauged from Fig. 8.25, which shows the relationship between strain and pressure for an aluminium alloy, pure aluminium, stainless steel joint, designed for a working pressure of about 4 MPa.

8.8.3 Welding of Tubular Joints

Although welding of aluminium–stainless steel tubular joints for cryogenic operations has been reported,[74] the real importance of the process is seen in its adaptability to pipeline building *in situ*.

Since the pipes to be joined are normally of the same material and certainly of the same diameter, simplicity of design of the welding system is the main objective. Either implosive or explosive techniques are employed, but the geometry of the adopted system is often modified to suit the particular requirement.[75] Basically, irrespective of the process chosen, the actual welding is carried out either by using an external joining sleeve, made of suitably selected material, or by overlapping one tube by another. The use of the sleeve—which, of course, must be welded to both tubes—requires that the tubes to be joined be tapered, or scarfed beforehand; with the taper on one tube differing by 1° or so from that on the other. The cost of machining can be high and will increase with the size of the pipe. Clearly the method is more suitable therefore for the welding of small diameter tubing. Although the tapering technique applies primarily to the explosive sleeve-type system, it can be used to advantage for small diameter tubes if the implosive process is employed. In this case no sleeve is required, but the bore must be supported.

The overlapping, which involves belling out of one end of the tube, is also expensive, but at least removes the need for both the tapering operation and the provision of a sleeve. Internal or external support has to be provided depending on the process selected.

All these operations are usually employed in the building of mild steel (low carbon) pipelines for transmission of either gas or oil. Leak proofing is therefore essential, particularly with gas, but is easily achieved even with considerable internal pressures. Satisfactory joints have been obtained in tube sizes ranging from 25 mm to 200 mm in diameter, and 2.3 mm to 9.5 mm in wall-thickness.

8.9. SOLID AND HOLLOW AXISYMMETRIC COMPONENTS

8.9.1 Introduction
The manufacture of multistrand, coaxial arrays of rods and/or tube-rod composites, to be used as remote control elements working in toxic systems or more sophisticated heat-exchangers, is of considerable interest in the chemical and atomic power industries. For the latter case, in particular, the use of refractory metals is fairly common and since these are often available only in the form of short rods, means of joining them together and processing to the required length have to be devised. The differentiality in the response of metals differing in their mechanical properties to processing, calls for the manufacture, in the first instance, of an integral structure such as a welded assembly.

Studies of welding techniques and parameters influencing the implosive welding of rod and rod-tube arrays have been carried out by Blazynski and Bedroud,[77,78] and the application of hydrostatic extrusion to the processing of prewelded composites has been described by Blazynski and Matin.[79]

8.9.2. Welding Systems
The basic system is that of Fig. 8.3(d) and consists of a central rod, or a tubular element (the core), surrounded by one or more layer(s) of flyer rods, possibly of a different material (Fig. 8.26). Depending on the intended use of the composite, its elements are either confined initially in

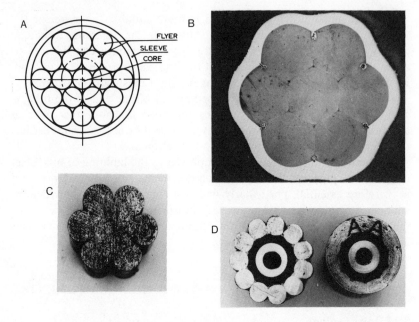

FIG. 8.26 Implosive welding of arrays of rods. (A) General arrangement, (B) Metal sleeve system after welding, (C) Plastic sleeve system after welding, (D) Before and after hydrostatic extrusion.

a metal sleeve (gas shield) or are held in position in a plastic container tube. In the first case, on detonation of the charge, the tube deforms plastically to conform to the profile of the outer layer of the flyer rods (Fig. 8.26(A,B)). This is really equivalent to the application of a shaped charge and results in the melting of the sleeve material and its subsequent jetting into the inter-rod voids. This first stage of the operation is followed by the collapse of the individual rods, and of their respective layers, on to each other and finally on to the core. Plastic deformation and welding of the individual strands to each other follow. The presence of the solidified melt in the now filled voids introduces an element of brittleness and may prove undesirable. One possible solution consists in using 'filler' rods which would eliminate the original interstrand gaps, and the other is to adopt the second welding system, in which plastic container tube is employed. A clean, extended area surface is then obtained (Fig. 8.26(C)).

8.9.3. Hydrostatic Extrusion

Since conventional extrusion of prewelded rod arrays results in total disintegration of the structure, hydrostatic extrusion must be used for processing.[79]

Irrespective of the welding system adopted, a certain amount of machining of the specimen is required to enable effective sealing of the assembly in the container of the extrusion press. Final results are generally satisfactory as indicated, e.g. in Fig. 8.26(D). This shows a stainless steel core tube—copper tube—mild steel rod assembly, prewelded in a plastic container.

Assemblies in stainless and mild steels, copper, brass, aluminium and titanium have been processed.

REFERENCES

1. BAHRANI, A. S., BLACK, T. J. and CROSSLAND, B. *Proc. Roy. Soc. A* **296** (1967), 123.
2. STIVERS, S. MS Thesis (1974), University of Denver, Colorado.
3. EL-SOBKY, H. and BLAZNSKI, T. Z. *Proc. 15th Int. M.T.D.R. Conf.*, 1975, Macmillan Press Ltd., 399–400.
4. BLAZYNSKI, T. Z. and EL-SOBKY, H. *Proc. III Int. Symp. on Explosive Working of Metals*, 1976, Czechoslovak Scientific and Technical Society, Prague, 351–356.
5. PHILIPCHUK, V. *Creative Manufacturing Seminar*, 1965, ASTME, Paper SP65-100.
6. WRIGHT, E. S. and BAYCE, A. E. *Proc. Conference on High Energy Rate Working of Metals*, 1964, NATO Advanced Study Institute, Oslo, 448.
7. CARLSON, R. J. *Western Metal Congress*, 1965, ASM Tech. Rep. W6-3-65.
8. HOLZMAN, A. H. and COWAN, G. R. *Welding Res. Council*, Bull (1965), No. 104.
9. DALRYMPLE, D. G. and JOHNSON, W. *Int. J. M.T.D.R.* **7** (1967), 257.
10. YOBLIN, J. A. and MOTE, J. D. As Ref. 4, 161–180.
11. E. I. Du Pont de Nemours, British Patent No. 945452.
12. BLAZYNSKI, T. Z. and DARA, A. R. *Proc. 11th Int. MTDR Conference*, 1970, Pergamon Press, Oxford, 940–965.
13. BLAZYNSKI, T. Z. and DARA, A. R. *Proc. 3rd Int. Conf. of the Center for High Energy Forming*, 1971, University of Denver, Colorado, Paper 8.3.
14. BLAZYNSKI, T. Z. and DARA, A. R. *Metals & Materials* **6** (1972), 258–262.
15. BLAZYNSKI, T. Z. *Proc. Int. Conf. on the Welding and Fabrication of Non-Ferrous Metals*, 1972, The Welding Institute, 62–71.
16. DARA, A. R. and BLAZYNSKI, T. Z. *Proc. Int. Conf. on the Use of High Energy Rate Methods for Forming, Welding and Compaction*, 1973, University of Leeds, Paper 10.
17. SCHETKY, L. M. *J. Inst. Met.*, **98** (1970), 364–367.

18. CARPENTER, S. H., WITTMAN, R. H. and CARLSON, R. J. *Proc. 1st Int. Conf. of the Center for High Energy Forming*, 1967, University of Denver, Colorado.
19. KOWALICK, J. F. and HAY, R. *Proc. 2nd Int. Conf. of the Center for High Energy Forming*, 1969, University of Denver, Colorado.
20. EZRA, A. A. and PENNING, F. A. *Experimental Mechanics* (*1962*), No. 8, 234.
21. BLAZYNSKI, T. Z. *Proc. 10th Int. MTDR Conf.*, 1969, Pergamon Press, Oxford, 511–523.
22. JACKSON, P. W. *Proc. 6th Int. Conf. on High Energy Rate Fabrication*, 1977, Haus der Technik, Essen, West Germany, Paper 1.5.
23. WYLIE, H. K. and CROSSLAND, B. *Ibid*, Paper 2.4.
24. BLAZYNSKI, T. Z. and MATIN, M. *Proc. 7th Int. Conf. on High Energy Rate Fabrication*, 1981, University of Leeds, U.K., 164–172.
25. BEDROUD, Y. and BLAZYNSKI, T. Z. *J. Mech. Working Techn.* **1** (1978), 311–324.
26. BLAZYNSKI, T. Z. and DARA, A. R. *Proc. 4th Int. Conf. of the Center for High Energy Forming*, 1973, University of Denver, Colorado, Paper 2.3.
27. DENTON, A. A. *Met. Rev.* **11** (1966), 1.
28. WEISS, V. *Proc. S.E.S.A.*, 1957, **XV**, 53.
29. APPLEBY, E. J. *J. Appl. Mech.* 1964, Ser. E, **31**, 654.
30. TOWNLEY, S. and BLAZYNSKI, T. Z. *Proc. 17th Int. MTDR Conf.*, 1976, Macmillan, London, 467–473.
31. BLAZYNSKI, T. Z. and TOWNLEY, S. *Int. J. Mech. Sci.*, **20** (1978), 785–797.
32. HOLTZMAN, A. H. and RUDERSHAUSEN, C. G. *Sheet Met. Ind.*, **39** (1962), 399–410.
33. CROSSLAND, B., BAHRANI, A. S., WILLIAMS, J. D. and SHRIBMAN, V. *Weld. and Met. Fab.*, **35** (1967), 88–94.
34. GIBBON, R. B. and WHITEMAN, P. British Patent No. 1029494.
35. WEC, British Patent No. 1124891.
36. BERRY, D. J. and HARDWICK, R. British Patent No. 1149387.
37. ROBINSON, J. A. and GASKELL, P. D. British Patent No. 1179107.
38. CHADWICK, M. D., HOWD, D. WILDSMITH, G. and CAIRNS, J. H. *Brit. Weld. J.*, **15** (1968), 480–492.
39. CAIRNS, J. H. and HARDWICK, R. *Select Conf. 'Explosive Welding'*, Weld. Inst. 1969, 67–72.
40. CAIRNS, J. H., HARDWICK, R. and TELFORD, D. G. As Ref. 13, Paper 2.3.
41. CROSSLAND, B. and WILLIAMS, P. E. G. As Ref. 26, Paper 7.3.
42. CROSSLAND, B. and WILLIAMS, P. E. G. As Ref. 16, Paper 9.
43. WILLIAMS, P. E. G., WYLIE, H. K. and CROSSLAND, B. As Ref. 15, Paper 15.
44. CHRISTENSEN, T. B., EGLY, N. S. and ALTING, L. As Ref. 26, Paper 4.3.
45. PRÜMMER, R. and JAHN, E. As Ref. 22, Paper 2.7.
46. OTTO, H. E., CARPENTER, S. H. and PFLUGER, A. R. As Ref. 22, Paper 2.8.
47. CHADWICK, M. D. and JACKSON, P. W. *Developments in Pressure Vessel Technology 3*, 1980, Applied Science Publ. Ltd., Barking, England, 245.
48. GURNEY, R. V., Ballistic Research Report No. 648, 1947, Aberdeen Proving Ground, Maryland, U.S.A.
49. COWAN, G. R., BERGMAN, P. R. and HOLTZMAN, A. H. *Met. Trans.* **2** (1971), 3145–3155.

50. SHRIBMAN, V., WILLIAMS, J. D. and CROSSLAND, B. *Select Conf. on Explosive Welding*, 1968, The Welding Institute, 47.
51. JOHNSON, W. R. *Weld. J.* **50** (1971), 22–32.
52. CROSSLAND, B., HALIBURTON, R. F. and BAHRANI, A. S. As Ref. 26, Paper 4.4.
53. BAHRANI, A. S., HALIBURTON, R. F. and CROSSLAND, B. *Pres. Ves. & Piping* **1** (1973), 17–35.
54. CROSSLAND, B. As Ref. 16, Paper 21.
55. CROSSLAND, B., BAHRANI, A. S. and TOWNSLEY, W. J. As Ref. 4, 59–98.
56. TOWNSLEY, W. J. and CROSSLAND, B. As Ref. 22, Paper 2.5.
57. HARDWICK, R. As ref. 16, Paper 20.
58. HARDWICK, R. *Weld. J.*, 1975, 238–244.
59. JACKSON, P. W. Explosive Welding—A National Seminar—1975, The Welding Institute, 25–28.
60. STANKO, G. *Proc. The Pressure Vessels and Piping Conf. 'Explosive Welding, Forming, Plugging and Compaction'*, 1980, ASME PVP–44, 13–24.
61. MCCLELLAND, H. T. and OTTO, H. E. As Ref. 26, Paper 9.1.
62. JARVIS, C. V. and SLATE, B. M. H. *Nature* **200** (1968).
63. REECE, O. Y. As Ref. 13, Paper 2.1.
64. BOUCKAERT, G., HIX, H. and CHELIUS, J. *Proc. 5th Int. Conf. on High Energy Rate Fabrication*, 1975, University of Denver, Colorado, Paper 4.4.
65. EL-SOBKY, H. Ph.D. Thesis 1979, University of Leeds.
66. BLAZYNSKI, T. Z. and EL-SOBKY, H. *Metals Technology* **7** (1980), 107–113.
67. BLAZYNSKI, T. Z. and EL-SOBKY, H. As Ref. 60, 69–86.
68. EL-SOBKY, H. and BLAZYNSKI, T. Z. As Ref. 24, 100–112.
69. AKE-PERSSON, C. *J. Appl. Mech. Trans. ASME* 1974.
70. CHOU, P. C. and ROSE, J. L. *J. Comp. Matls.* **5** (1975), 405.
71. BLAZYNSKI, T. Z. and EL-SOBKY, H. As Ref. 64, Paper 4.6.
72. KAMEISHI, M., BABA, N. and NIWATSUKINO, T. As Ref. 4, 205–223.
73. IZUMA, T. and BABA, N. As Ref. 22, Paper 2.14.
74. GROLLO, R. P. As Ref. 13, Paper 8.4.
75. CHADWICK, M. D. and EVANS, N. H. As Ref. 16, Paper 13.
76. WYLIE, H. K. and CROSSLAND, B. As Ref. 26, Paper 4.2.
77. BEDROUD, Y., EL-SOBKY, H., and BLAZYNSKI, T. Z. *Metals Techn.* **3** (1976), 21–28.
78. BLAZYNSKI, T. Z. and BEDROUD, Y. As Ref. 22, Paper 2.12.
79. BLAZYNSKI, T. Z. and MATIN, M. *Proc. 4th Int. Conf. on Production Engineering* 1980, The Japan Society of Precision Engrg. and the Japan Soc. for Technology of Plasticity, Tokyo, 731–735.

Chapter 9

EXPLOSIVE FORMING

J. W. Schroeder

Foster Wheeler Development Corporation, Livingston, New Jersey, USA

9.1. INTRODUCTION

The beginning of the use of explosives for forming of metals dates to 1888 when the earliest recorded application is described as the engraving of iron plates by imprinting a design on a block of explosives or by interposing a stencil between the explosive and the plate.[1] With the exception of a few isolated applications, very little explosive forming was done until the mid-1950s and the advent of the Space Age, which, in turn, brought the need for a low volume of parts with complex shapes. Often the parts were large and formed from materials that were difficult to work by conventional means.

Explosive forming allows large, shaped, parts to be formed without an investment in machine tools and die sets, both of which are costly and time-consuming to produce. Thus for a decade, explosive forming experienced a rapid growth. With diminishing activity in aerospace, the use of these techniques has gradually decreased in recent years, although a few companies and military establishments continue to use explosive forming as a regular manufacturing process.

This chapter discusses the practical, everyday aspects of explosive forming and points out its potential.

9.2. FORMABILITY OF ENGINEERING ALLOYS

Studies have shown that the same factors—ductility and toughness—are critical for forming metals explosively or mechanically. The elongation determined by uniaxial testing must not be exceeded during explosive forming or the part will fail. Because of the multiaxial stresses thay may occur during explosive forming, uniaxial test data can be used only as a relative value. This value, when combined with experience, die design, and the forming technique used, will yield a comparison for formability. Such a comparison of materials, using annealed Type 1100 aluminum as the basis, is shown in Fig. 9.1.[2]

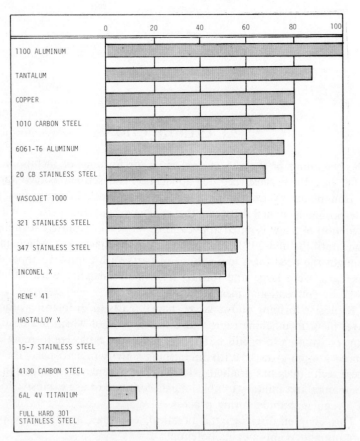

FIG. 9.1 Relative formability of metals when explosively formed.

9.3. MECHANICAL PROPERTIES OF EXPLOSIVELY FORMED COMPONENTS

The mechanical properties of materials after explosive forming cannot be described in one general statement. Many materials exhibit increases in strength much like those exhibited by materials evidencing equivalent deformations produced by conventional methods. Most common carbon and stainless steels behave in this way. This increased strength, along with the uniform stress patterns normally characteristic of explosively formed components, generally produces shapes superior to their mechanically formed counterparts.

Aluminum alloys remain virtually unchanged when explosively formed. Their ultimate strength remains almost the same and the increase in yield strength is very small.

Precipitation hardening of stainless steels, which are austenitic at room temperature and martensitic when deformed, has greatly increased ultimate and yield strengths after explosive forming. This increase in strength, discussed in more detail in Section 9.10, is further enhanced when the deformation is performed at cryogenic temperatures.

A comparison of the effects of explosive forming and conventional forming was made for three forming materials, low-carbon steel, AISI 304-L, and HY100, by F. E. Van Wely.[3] Large-diameter domes were formed from 12 mm thick sheets of each material using cold pressing and explosive forming. Van Wely arrived at the following conclusions as a result of extensive tests of the three materials after forming:

(i) The outside fibres of the explosively formed products showed compressive stresses; tensile stresses were present in these same locations for the pressed specimens.
(ii) Extreme work hardening was present in skirted portions of the pressed domes; work hardening was evenly distributed over the entire formed part of the explosively formed domes.
(iii) Toughness was essentially the same for both cold-pressed and explosively formed parts.
(iv) Corrosion resistance and hydrogen embrittlement were approximately the same for both methods.

The results of this study indicate that (for these materials) explosive forming produces a formed part mechanically equal and possibly superior to the cold-pressed part because of the more uniform strain.

9.4. AIR AND UNDERWATER FORMING SYSTEMS

The ideal explosive forming system would combine the simplicity and ease of operation inherent in an air system with the advantages of energy distribution, safety, and sound control found in underwater systems.

Air forming systems are generally used when the size of the explosive charge is small, the operation is in a remote area, or the nature of the operation excludes the presence of water. Typically, an air system is a blast chamber into which the entire forming operation is placed. This chamber must be designed to withstand the force of the explosive blast, be insulated to contain the sound waves of the detonation, and be fitted with a means of removing the gases generated by the explosives. These chambers can be as simple as a bench-top blast booth, or they may be above-ground bunkers and, in some cases, underground caverns. The primary consideration when designing these chambers is that they must withstand the pressure pulse released by the explosive charge. The work by Loving[4] yielded the formula:

$$P = WK/V \qquad (9.1)$$

where P is pressure pulse in psi (6.894 kPa) W is weight of explosive in lb (0.4536 kg) V is volume of a sphere surrounding the charge in ft^3 (2.83×10^{-2} m^3) and K is the constant for the explosive used (15 000 for PETN).

(When the chamber is not spherical, use the volume of a sphere having a radius equal the nearest wall.) The formula has been a reliable method of establishing the requirements for chamber pressure integrity.

If the manufacturing facility permits, it is often advantageous to locate the explosive forming operation at a remote outside site. The disadvantage of this procedure is that the operation is exposed to the weather and time can be lost when the weather is inclement. This type of facility can only be justified when detonations are few.

Finally, some explosive forming applications have a built-in shock-wave and sound-containment system. A prime example is explosive expansion of tubes into the tubesheets of heat exchangers. If the expansion operation is performed after the exchanger channel closure is in place, the closure itself provides a very efficient blast chamber.

Underwater forming systems require a large amount of water (preferably contained in a tank) able to withstand the repeated impact of the explosive shock. The use of an expendable water containment system made from thin plastic sheets or bags placed over the part to be formed and then filled with water is also possible. This method works very well for many types of operations but has the disadvantages of high water consumption and less effective sound abatement than if the explosive were completely submerged in a pool of water.

The design of a water tank or forming pond to be used for explosive forming must be carefully considered. Only a fraction of the energy of the explosive is directed to the part being formed; the remainder of the energy is released into the water as a shock wave, which travels radially outward from the charge. When designing tanks for underwater explosive forming, a safety factor of four should be used. The most common material used for tank fabrication is mild steel.

An 8.6 kg charge of TNT will generate a 13 765 kPa pressure at 6.1 m. A vertical-walled tank strong enough to sustain repeated shocks of this intensity would be extremely costly. To lower the stresses in the tank sides and bottom, several approaches are recommended:

(i) First, use a tank with sloping walls. Analysis of a tank reveals that the stress is reduced by a function of the angle of the slope.
(ii) Second, use a cushioning medium to moderate the shock waves. By lining the inner tank walls with inflated rubber tubing, a reduction in stress of 83 per cent has been realized.[5]
(iii) Third, release controlled amounts of air through holes in tubes located at the base of the tank. The air cushion of bubbles produced has been found beneficial in lowering the blast-caused stress on the tank walls. A 6.1 m dia, 6.1 m deep, 2.54 cm thick steel, cylindrical tank with a bubble cushion has withstood repeated 4.3 kg explosive charges with no ill effects.[6]

9.5. DIE AND DIELESS FORMING

When explosive forming is discussed, the immediate concept that comes most often to mind in the use of an explosive charge to drive a metal sheet or blank into intimate contact with a preformed die. This type of forming has been popular and probably is the most used method; however, the concept of free-forming or dieless forming has also enjoyed popularity and is used whenever extreme accuracy of the finished part is not a prerequisite.

Die forming is generally performed in the manner shown in Fig. 9.2. A die is machined to the exact shape of the part to be formed. A metal

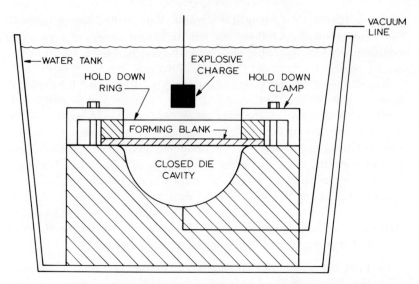

FIG. 9.2 Explosive forming in a closed die.

blank is placed over the die and its edges are firmly clamped to the upper die surface. The die cavity under the forming blank is evacuated to prevent adiabatic compression, which could result in damage to the formed part and the die. The die and blank are submerged in an energy-transfer medium, usually water, and the explosive charge is positioned at a predetermined distance over the assembly. The charge, when detonated, produces a shock wave and a gas bubble. The shock wave, which is the major energy source, produces a pressure pulse of several million MPa for 5 to 10 μsec. The pressure pulse imparts a loading on the blank, driving it into intimate contact with the die cavity. Photographic observations have shown that the blank is formed within one to two milliseconds after detonation.

A hold-down ring is needed to prevent edge wrinkling and deformation as the blank is driven into the die. (A more detailed discussion devoted to die design will follow in Section 9.7 of this chapter.) An evacuation port must also be located within the die cavity. Since the formed part takes on every detail of the die cavity, the port must be placed in an area of the die where the imprint can be removed or does not cause a deleterious effect.

Forming into closed dies results in extremely accurate symmetrical parts. It is generally accepted that the dimensional tolerances of parts produced by explosive forming are equal to or better than the dimensional tolerances of similar conventionally formed parts.

Dieless forming with stand-off charges, commonly called 'free-forming', enjoys much success where extreme accuracy is not required in symmetrical parts. With dieless forming the final shape of the formed part is determined by the size and placement of the explosive charge along with the resistance of the forming blank to deformation. The details of the free-forming system are shown in Fig. 9.3. The forming is done over a

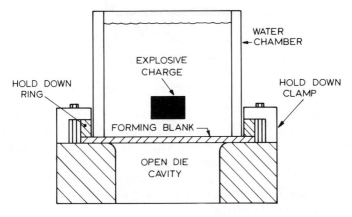

FIG. 9.3 Explosive free forming in an open die.

deep cavity, the function of which is merely to control the edge shape of the formed part. Again, a hold-down device must be employed to prevent wrinkling and deformation at the edges where the blank is drawn into the cavity. This type of forming has been very successful for elliptical, spherical, and hyperbolic heads. It has also been used to preform heavy plate prior to machining, thereby decreasing the amount of metal that must be removed to achieve a desired shape.

A second method of dieless forming or free-forming using near-contact charges has been described in a paper by Berman and Schroeder.[7] They define a method whereby the amount, shape, type, and distribution of a charge, in conjunction with the same factors for the media, determine the magnitude, distribution, and duration of the energy input to the workpiece.

Figure 9.4 shows the details of free-forming using near-contact charges. As with free-forming using stand off charges, a simple edge support is included. The hold-down device is eliminated; an explosive

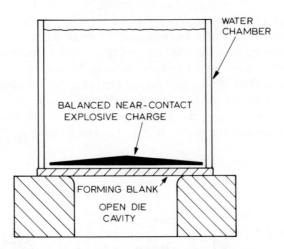

FIG. 9.4 Explosive free forming using near-contact charges.

charge placed over the edge performing this function. The charge is distributed over the workpiece and varied in amount and distance to achieve the specific energy input to produce the desired deformation. The media—the substances chosen to direct the energy from the explosive to the workpiece–can be varied if necessary for local control of the magnitude of the energy.

When explosively free-forming a homogeneous workpiece of uniform thickness, the final shape is determined by five factors: (i) amount of explosive, (ii) type of explosive, (iii) shape of explosive, (iv) explosive placement and (v) media used to transmit the energy.

The amount of explosive, in conjunction with the media, determines the amount of total energy that will be imparted to the workpiece. Because of the existence of the five variable factors, no reliable formulas are available to aid the first-time explosive former in precisely sizing an explosive to free-form any given part correctly. Instead, carefully controlled tests must be conducted to establish the proper parameters to achieve the desired expansion. Thus the amount of explosive is directly dependent upon the other four factors. The relative power and velocity of the explosives used are determined by the type. The common higher explosives have a rate of detonation ranging from 2600 to 8060 m/s and produce pressures ranging from 12 to 25 MPa.

The explosive shape helps to determine the profile of the energy front as it strikes the workpiece, thus contributing to the final formed shape. The placement, which includes the stand-off distance of the charge, also determines the shape and magnitude of the energy front as it strikes the workpiece. This variable, more than any other, can be used to effect the desired shape in an explosively free-formed part.

The medium used to transfer the explosive energy can also be used to direct it. Conversely, the medium can be altered to attenuate the energy in local areas if desired. Water is the most common medium used in explosive free-forming; but air, sand, and other liquids and solids, such as wax and plastic, have also been used successfully.

9.6. ANALYSIS OF FINAL SHAPES IN FREE-FORMING

As stated in the previous section, the final shape in explosive free-forming is dependent upon five factors; the most critical being the charge placement. Many tests have been conducted to study this forming parameter and classic observations are shown in Fig. 9.5. The profiles of

12.7 mm
STAND-OFF

25.4 mm
STAND-OFF

38.1 mm
STAND-OFF

FIG. 9.5 Free formed head shapes using identical disc charges but varying stand-off distances.

free-formed heads are distinctly different when the stand-off distance of a flat disc charge is varied. I have observed this forming characteristic many times and have used it successfully when forming prescribed shapes. An example of control of the final shape of a formed part is demonstrated by a dished plate free-formed in a single shot, using the near-contact charge technique. The dished shape was formed from SA-516 Gr 70 carbon steel plate, 2.5 m long and 0.8 m wide. The plate was 25 mm thick at the center before forming and was tapered by machining to 6 mm at the narrow ends. The depth of the dish was 114 mm at the deepest center and 22 mm in the shallowest corners. The dish itself was to be spherical at all points, with the radius of the spherical sections varying from 34 to 310 mm.

To achieve the desired shape, the charge was positioned over the forming blank as shown in Fig. 9.6. The charge was unbalanced to follow the contour of the desired form, and the stand-off distance varied from 50 mm at the center to 100 mm at the ends. Figure 9.7 shows how closely the final free-formed shape conformed to the designed shape of the dished plate. Four identical plates were formed during this program and the final shapes of all plates were nearly identical.

FIG. 9.6 Near-contact explosive charge positioned over dished pass partition plate forming blank.

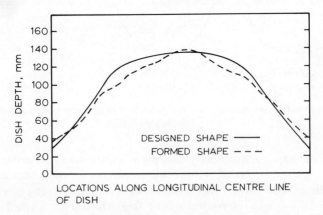

FIG. 9.7 Conformity of free formed dished plate to the design shape.

The strain distribution within explosively free-formed parts is very uniform, primarily because there is no constricting force during forming caused by die interference. Because of this desirable feature, some explosive forming fabricators have free-formed parts to their approximate shape and then explosively sized these same parts in a closed die to achieve greater accuracy.

9.7. PARAMETERS AND ANALYSIS OF DIE DESIGN

One factor that makes explosive forming economically attractive is that only one die component—usually the female half—is needed. This die component must be carefully designed to withstand the type of loading experienced by the part to be formed. In addition, because the energy wave associated with explosive forming leads to unusual stress patterns within the die material, sharp corners or notches should be eliminated wherever possible. A safety factor of four is usually sufficient to prevent shock-induced failures. Material selection is also important; the use of materials with high compressive or tensile strengths will optimize die performance and minimize costs.

When determining the strength required of a die, one must first calculate or estimate the peak pressure needed to form the component. Then, using this peak pressure, conventional methods of stress analysis can be used to size the die. Normally, the yield strength of the material is taken as the working strength, with the safety factor of four included to obtain the design stress level.

Important in die design and analysis is the type of loading the die will experience. The loading is influenced by the shape of the component to be formed and the method of charge placement. For instance, the die for a head-drawing operation can be analyzed by realizing that the die acts as an end closure of a high-pressure vessel. It is apparent that the maximum stress occurs near the upper edge of the die; thus the formulas for stress concentrations in a heavy-walled cylinder can be used. For proper stress distribution under shock loading, the base of the die should be as thick as the die wall.

To determine peak pressure, the forces generated by an explosive charge can be estimated from the basic laws of physics. The problem with the application of these laws is that short time intervals make difficult the measurement of the parameters that determine the applicable laws. Strohecker et al.,[5] noted that experiments have shown that peak pressures, developed by detonating an explosive underwater, decrease in a regular manner with detonation velocity and distance. This relationship, for point charges, can be expressed by the following equation:

$$P = \sqrt{(6.9 \times V)} \times \left(\frac{W^{1/3}}{R}\right)^{1.13} \times 10^2 \qquad (9.2)$$

where P is peak pressure in psi (6.894 MPa), V is detonation velocity in m/s, W is weight of explosive in lb (0.4536 kg), and R is the stand-off distance in ft (0.3048 m).

When a particular type of explosive and charge shape is to be used, it is frequently useful to construct a nomogram, using the above equation and substituting the detonation velocity of the explosive to be used. A typical nomogram, which can be used for line charges of PETN explosive, is shown in Fig. 9.8. This nomogram has two scales for stand-off distances—one for line charges located along the axis of a cylinder and

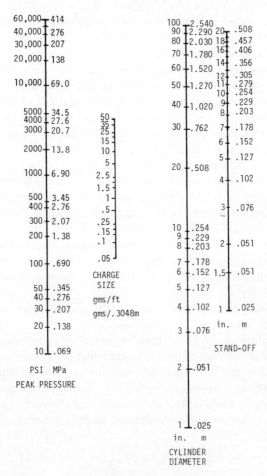

FIG. 9.8 Nomograph showing peak underwater shock pressure for line charges of PETN explosives.

the other for determining the peak pressure of line charges to form flat blanks.

While many methods are available for determining initial peak pressures based on energy requirements, the following equation by Strohecker et al.,[5] provides a fair correlation with experience in free-forming operations:

$$d = D(Eu/2St)^{1/2} \qquad (9.3)$$

where d is depth of draw in in. (0.0254 m), D is cup diameter in in. (0.0254 m), Eu is energy required per unit area in psi (6.894 KPa), S is dynamic yield strength of material in psi (6.894 KPa), and t is the material thickness in in. (0.0254 m).

This equation will give the die designer and forming engineer a basis from which to begin. The precise charge sizing will have to be determined by testing. Bench tests can be useful and the scaling laws apply in predicting the forming blank response.

The die surfaces should be as smooth as the surface desired on the finished part, because the surface of the die will be reproduced in every detail on the finished part. Where split dies are used, the parting line will show because the large loading conditions will result in extrusion of material into the crack between the two parts.

Where a hold-down ring is required, it must be sufficiently sturdy to prevent wrinkling of the blank edges as they are being drawn into the die. A prime consideration in designing the hold-down is the time required for its clamping and removal. Bolts are an excellent attachment method but are inefficient because of the time required to load and unload them. If the production rate of the operation is expected to be high, a hydraulic clamping system can be an efficient time saver.

The final parameter of die design is the material selected. Important in this selection is the die requirement in terms of accuracy, finish, durability, and handling ease. Where long life, high accuracy, and good surface finish are required, heat-treated alloy steel provides the best results. These alloy steel dies are generally heat treated to a maximum hardness of 50 Rockwell C.

Kirksite, a castable zinc-based alloy which melts at 380 °C, has been widely used for explosive forming dies. This material has an ultimate tensile strength of 241 MPa and an ultimate compressive strength of 517 MPa and is used when the loading levels on the die are low and production does not exceed 100 pieces. The material is popular because of its low-temperature casting properties.

Other materials that have been used successfully are: reinforced concrete for low-volume, low-accuracy parts; plaster for one-shot dies where only one piece is needed; and ice (but the cost of the auxiliary equipment required to keep the dies in their frozen state offsets the material costs and ease of shaping). Composite dies such as epoxy facings on concrete and alloy steel and epoxy facings on kirksite have also been successful.

9.8. FORMING OF DOMES AND OF ELEMENTS OF SPHERICAL VESSELS

Much work has been directed toward forming domes and spherical vessels, mainly because the symmetry of the parts lends itself so well to the process. Among the numerous approaches that have been established, the most common are:

(i) Forming from a circular blank into a closed die.
(ii) Free-forming from a circular blank using a simple edge support.
(iii) Sizing of a prefabricated preform into a spherical shape.

The method chosen is dependent upon three factors: (a) the dimensional accuracy required, (b) the formability of the material and (c) the facilities available for performing the operation.

Shallow domes of elliptical, spherical, or other shapes with very accurate dimensional tolerances can be formed into a closed die. This method was discussed in Section 9.5; the die system was shown in Fig. 9.2. The stand-off method is normally used for charge placement. The forming can be completed in a single shot if the total strain to be introduced to the blank is not excessive to the point of failure. If the strain is excessive, the dome can be formed in numerous steps, with the blanks stress relieved or annealed between steps.

The closed die method is used most often where numerous parts are to be made with a high degree of dimensional accuracy. The large number of parts produced offsets the cost of preparing the closed die.

Free-forming using a simple edge support, discussed in Section 9.5 and illustrated in Fig. 9.3 is the most common method of explosively forming elliptical and spherical domes. This method will not produce a part with extreme dimensional accuracy. It will, however, produce adequate shapes for many applications. This method has also been used to preform very thick domes that are finally machined to the desired tolerances. As stated earlier free-formed domes are also often explosively sized into a closed die to their final dimensions. As in forming with closed dies, the part can be formed in a single step or in numerous steps if required.

The charge used and its placement are, of course, critical when free-forming domes. As discussed in Section 9.5, the final formed shape is dependent upon the charge shape and stand-off distance.

The near-contact charge method can probably produce the most accurate free-formed shape. The resultant shape can be controlled by placing explicit amounts of the explosive precisely where needed. This method also eliminates the need for a mechanical hold-down at the edge of the blank. If an acceptable free-formed part can be made, this method should be used, because a considerable saving is realized by the elimination of a closed die and evacuation system.

A third method of explosively forming deep spherical domes is to fabricate a preform by joining cylindrical, conical, and flat members, as shown in Fig. 9.9, and to form it explosively to the desired shape by the

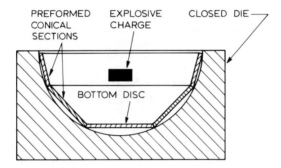

FIG. 9.9 Dome forming with fabricated preforms to limit the amount of strain.

closed-die method, as shown, or by free-forming using a balanced charge suspended at the centerline of the preform. The advantage of this method is the ability to produce deep domes while controlling the amount of strain introduced in the material.

A fourth method, which was developed by Patel,[8] forms a hemispherical or spherical tank from individual gore sections having equal curvature in all directions. These gore sections can be fastened together by welding or riveting to form a complete sphere (Fig. 9.10). This method of fabrication allows very large spherical domes to be explosively formed in a single, simple die. The strain introduced into the individual gore segments is very low. The results of tests by Patel indicated that a spherical liquid–natural-gas tank with an 18 m radius can be fabricated from 6061–T6 aluminum by this method.

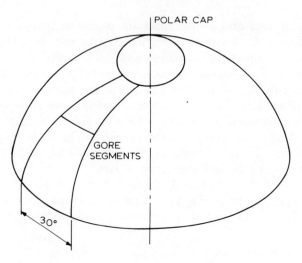

FIG. 9.10 Dome formed from 24 gore segments having equal curvature and one polar cap.

9.9. FORMING AND PUNCHING OF TUBULAR COMPONENTS

Almost as prevalent as dome forming has been the forming of tubes into unique shapes by putting a bulge in selected areas of the tube. This work is generally done in a split closed die; thus close tolerances can be achieved. The split die facilitates removal of the tubular part after forming.

When explosively forming tubes, the charge is placed along the centerline of the tube. This charge alignment is fairly critical; if the charge is long, spacers of a thin material are usually placed at close intervals to ensure centering of the charge within the tube. Once the charge is centered, the tube is filled with the selected energy-transmitting medium (usually water). The water is contained by placing a plastic bag or a plug at one end of the tube or by sealing the cavity between the die and the tube and submerging the entire set-up in water. Normally, the extreme ends of the tube will expand less than the remainder because of the portion of the energy wave directed axially during the forming. This problem can be overcome by adding to the length of the tube to be cut off after forming or by placing additional charge at the ends to balance the expansion.

This process of forming unique shapes lends itself very well to high-quantity production because it can replace with one operation many handforming and fastening steps and yet produce a part with better tolerances.

Another application of tube forming that has been used extensively is the expansion of tubes into tubesheets of heat exchangers. This is done for two major reasons:

(i) To close the annulus between the tube and its hole, preventing vibration and corrosion and improving the heat transfer into the tubesheet.
(ii) To provide the seal between the tube and shellsides of the heat exchanger, preventing mixture of the products within the system.

Tube expansion into tubesheets is described by Berman et al.,[9] as an example of manufacturing by means of an explosive contained in a solid substance and used as a package (Fig. 9.11). Berman and Schroeder[7]

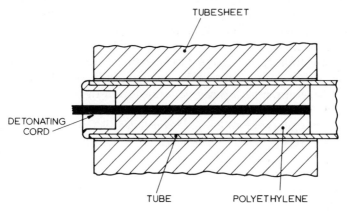

FIG. 9.11 Explosive insert positioned in a tube and tube sheet assembly.

categorized this process as further application of near-contact forming because the resultant expansion can be carefully controlled by the charge placement, the shape of the media, or both. The medium most often used is polyethylene; the explosive is detonating cord. The explosive imparts its energy locally to each area of the energy-transmitting media, which then locally imparts that energy to the tube. Polyethylene was chosen as the medium for this type of forming because it leaves little residue, is flexible and can accommodate large strains without rupturing, alleviates shock waves readily, avoiding tubesheet fracture, and has a high density to conduct the pressure to the tube from the explosive. Detonating cord is used as the explosive because of its consistency, low cost, and safe handling characteristics.

This process has been used successfully to expand several million tube ends into many heat exchangers. The process has produced a metallurgically superior expansion mainly because of the minimal cold work imparted to the tube. Explosive tube expansion also compares favorably in cost with the mechanical expansion methods also being used.

Explosive forming of conical shapes for butt-welded pipe reducers was discussed by Johnson.[10] These reducers are made in the shape of a frustum; thus the cross-section of flow is gradually changed over a length of one pipe diameter. They are readily available in the common pipe materials. However, modern piping systems use new corrosion-resistant materials, and fittings of mating materials are not always readily available. In addition, the conventional method of hot-forging these reducers usually includes multiple heating and forging operations—at added cost. Explosive forming of these parts provides an economical alternative that is particularly suited to corrosion-resistant alloys, where grain growth and constituent segregation are potential problems.

Explosive forming of these reducers is similar to other tubular part forming, with the exception that the die need not be split. Because of the conical shape of the part, it can be pressed from the die by a moderate force.

The tests conducted by Johnson[10] concluded that explosive forming of butt-welded pipe reducers could offer an economic advantage where quantities are relatively low and raw material costs are high. Many materials were included in this program; and those materials with sufficient dynamic ductility are apparently successful candidates for conical forming.

The variety of configurations that can be formed from tubular preforms is almost limitless. The equipment required for this type of explosive forming can be very basic, thus making it a very viable process for short-run productions.

The force from the detonated explosives can be used to punch holes in metal blanks as well as to form the blanks themselves. Very often the punching and forming operations occur simultaneously—the desired shapes for the holes are placed in the cavity of the forming die before the explosive forming operation.

A classic example of simultaneous forming and punching was described by Peak and Grollo.[11] A 130 mm I.D., 1 mm wall, 2.3 m long, Type 316 SS tube was sized, formed, and punched in a single shot. It had a circumference alignment bead, 27 countersunk and perforated rivet dimples, two circular holes (19 and 12.5 mm dia), and an elliptical slot 15.7 mm wide and 59.5 mm long. The final sized inside diameter was 132 mm. The tube was formed and punched in a split die submerged in water during detonation. A vacuum was applied between the die and the tube to eliminate air compression indentations and air burning. Zinc chromate putty and rubber tape were used to seal the cavity between the die and the tube. During this program 24 tubes were successfully formed.

9.10. MISCELLANEOUS FORMING OPERATIONS

Numerous other explosive forming operations have been performed that were not discussed earlier in this chapter. These operations, while not used frequently, have sufficient merit for mention.

One of these operations is explosive autofrettage. Autofrettage is described as the state of a tubular component whose inside diameter is maintained in residual compression by modification of its structure or stress pattern. Grollo[12] describes how naval gun barrels were autofrettaged using a radial piston method. This method used the energy of an explosive to expand a steel tube (i.e., radial piston) centered inside the gun barrel. Water between the piston and the gun barrel was the pressure-pulse transfer medium. The explosive charge was centered within the radial piston. For the work described by Grollo, a slurry explosive contained in a glass tube was used. An air space was maintained between the explosive and radial piston I.D.; end plugs were fitted to contain the water between the barrel and piston. These plugs were restrained to maintain the water pressure for a short time to help prevent reyielding of the inside surface of the gun barrel. Using the explosive method, gun barrels were successfully autofrettaged with residual compressive stresses at the base ranging from 309 to 653 MPa.

A somewhat similar application of explosive forming was the successful fabrication of a large-diameter duplex tube. A single 127 mm O.D. duplex tube assembly of Incoloy 800H was required in a nuclear test facility. The tubes in the assembly were each about 9.5 mm thick. The completed tube was to be 12 m long, and the required interface radial gap after expansion was to be between 0.000 and 0.076 mm for the entire length. Explosive fabrication was chosen because of the minimal cold work produced by this method, thus preserving the characteristics of the material, and also because the cost of alternative methods to make this individual duplex assembly was much greater. The duplex tube assembly was formed by inserting the inner tube into the outer tube and standing the assembly vertically. The inner tube was filled with water. Because both tubes had caps welded on the lower ends before assembly, the water was easily contained. An explosive charge of detonating cord was centered along the entire length of the inner tube I.D. and detonated from the upper end.

Explosive reshaping of metal is a unique operation having unlimited application. One of these applications is the reshaping of worn teeth on large gears. The gear to be reshaped is fitted with a driver plate of thin steel equal in width to the gear. The driver plate completely encircles the outside gear diameter. An explosive is placed at the outside surface of the driver plate and detonated. The force of the explosive pushes the driver plate against the outside of the gear, permanently deforming each tooth. The resultant gear is slightly decreased in diameter, which is not detrimental to its function, with a corresponding increase in individual tooth width. This reshaped gear can then be ground in the same manner as a new gear. The proper application of metal reshaping can result in substantial savings where large, costly parts can be simply reshaped when worn and then refinished to their exact original configuration.

Finally, there is the technique of transforming the austenitic crystal structure of Type 301 SS to a high-strength martensitic material by straining with explosive forming. Type 301 SS is a ductile material in its annealed condition and is easily welded and formed. The deformation required to transform metastable austenite to high-strength martensite is dependent upon the chemistry of the material and its temperature at transformation. Berman et al.,[9] describes tests where thin cylinders were fabricated of Type 301 SS about 150 mm in diameter. These cylinders were explosively free expanded while at cryogenic temperatures. After free forming, the cylinders were sized into a die at ambient temperatures. The total strain during various tests ranged from 12.8 to 20.3 per cent. The cylinders were tested to failure by hydraulic pressure, under open-ended conditions. The engineering stress to fracture was determined to be 1410 MPa for the specimen strained 12.8 per cent and 2070 MPa for the cylinder strained 20.3 per cent. An unstrained cylinder tested in the same manner exhibited an engineering stress to fracture of 818 MPa.

This method is feasible for the fabrication of lightweight high-strength cylinders or pressure vessels. It is also possible to form flat plate into high-strength super panels using this technique.

9.11. CONCLUSION

The subjects discussed in this chapter are only the tip of the iceberg of the possibilities existing for future uses of explosive forming. Here, man has at his command nearly unlimited energy—deterred only by his lack of imagination and innovation.

As the future needs of the world increase, so will the need for unique manufacturing techniques. Explosive forming cannot be expected to become a dominant fabrication method, but will continue to fill the gap when a limited number of otherwise hard-to-form components are needed.

REFERENCES

1. ZERNOW, L. *High-Velocity Forming of Metals*, Englewood Cliffs, New Jersey, Prentice-Hall, Inc., 1964.
2. LOCKWOOD, E. E. and GLYMAN, J. The explosive formability of metals, ASTME Report No. SP62-09, *Creative Manufacturing Seminar*, 1961–62.
3. VAN WELY, F. E. Comparison between the influence of cold pressing and explosive forming on material properties of steel. *4th International Conference of the Center for High-Energy Forming*, Vail, Colòrado, July 9– 13, 1973.
4. LOVING, F. A. Sound-reducing explosions testing facility, US Patent No. 2,940,300, granted June 14, 1960, US Patent Office, Washington, D.C.
5. STROHECKER, D. E. et al. Explosive forming of metals. Battelle Memorial Institute, Defense Metals Information Center, *Report No. 203*, 1964.
6. PEAKE, T. A. and GROLLO, R. Development of HERF facilities and capabilities at U.S. naval ordnance station, Louisville. US Naval Ordnance Station, Louisville, Kentucky, 1972.
7. BERMAN, I. and SCHROEDER, J. W. Near-contact explosive forming, in *Explosive Welding, Forming, Plugging, and Compaction—PVP-44*, I. Berman and J. W. Schroeder, ed. (New York: American Society of Mechanical Engineers, 1980).
8. PATEL, B. C. Explosive forming of spherical gore segments. Denver Research Institute, Army Materials and Mechanics Research Center, *Report No. AMMRC CTR 74-69*, 1978.
9. BERMAN, I. et al. Development and use of explosive forming for commercial

application. *1st International Conference for High-Energy Forming*, Estes Park, Colorado, June 1967.
10. JOHNSON, M. W. Explosive forming of butt-welded pipe reducers, US Naval Ordnance Station, Louisville, Kentucky, *Final Report No. MI–052*, 1979.
11. PEAKE, T. A. and GROLLO, R. Explosive forming and punching of 316 SS tubes: a case history. US Naval Ordnance Station, Louisville, Kentucky.
12. GROLLO, R. Explosive autofrettage of 5 in./54 mark 18 mod. 3 gun barrel Liners. US Naval Ordnance Station, Louisville, Kentucky, *Report No. MT–032*, 1975.

Chapter 10

POWDER COMPACTION

R. PRÜMMER

Fraunhofer-Institut für Werkstoffmechanik,
Freiburg, Federal Republic of Germany

10.1. INTRODUCTION

For a long time explosive compaction was considered an unsuitable tool for industrial application. Usually powder compaction takes place in heavy equipment at very low loading rates. The very high loading rates during explosive compaction were difficult to control. On the other hand no data were available about the explosive's parameters and their dependence on the method used and the type of powder to be compacted.

Early experiments were undertaken with indirect methods of explosive compaction. La Rocca and Pearson[1,2] described a single piston and also a double piston press for explosive compaction of powders. The single piston device is described in Fig. 10.1: an upper plate together with a piston is explosively accelerated against the sample, which is supported by the lower plate.

In the two-piston arrangement two pistons are simultaneously accelerated against the powder.

Another apparatus for the compaction of powders with the use of a ram was also developed at the Battelle Memorial Institute.[3]

A similar method, described by Brejcha and McGee,[4] consists of a

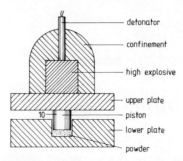

FIG. 10.1 Single piston arrangement for explosive compaction of powders.

modified 0.38 calibre gun barrel. A ram is accelerated with the effect of driving a punch against the powder to be compacted (Fig. 10.2).

A further impact press of this type was built earlier by Hagemeyer and Regalbuto,[5] their system, with a punch accelerated by compressed air, is described in reference 6.

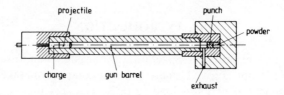

FIG. 10.2 Modified gun barrel used for explosive powder compaction.

Other indirect methods of explosive compaction make use of liquid-transmitting media. The powder is contained in a pliable rubber bag, placed in a water chamber. The explosive then either propels a piston into the water[7] or is detonated in the water itself. An electric discharge in the water chamber instead of the detonation of the explosive was also proposed, the pressure acting isostatically for a longer period which is an advantage of this method. However, heavy equipment is required and it was also shown that with indirect compaction procedures a certain ductility of the test material is required to make the material adhere. For the compaction of ceramic materials such methods appear to be unsuitable. Further limiting factors are the die materials.[3,5]

The direct method of explosive compaction[8-15] requires practically no capital investment and leads to high densities of the compact. The arrangement is very simple: a tube of mild steel or aluminium with end plugs is filled with powder and mantled with a layer of a proper explosive of uniform thickness and density (Fig. 10.3). After ignition of

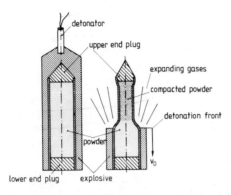

FIG. 10.3 Assembly for direct explosive compaction of powders.

the detonator the detonation proceeds down along the tube wall at a velocity v_D (Fig. 10.3, right), leading to a compression of the tube and the contained powder. With detonation velocities ranging from 1700 m/s to 8400 m/s the acting pressures range from 7 to about 300 kbar, depending on the type of explosive used.

10.2. DYNAMIC COMPRESSIBILITY OF POWDERS

The main feature of the method of explosive compaction compared to isostatic methods is that compaction does not occur simultaneously across the whole volume of the sample. Furthermore, a shock wave travels through the powder, leaving compacted material behind. The pressures in the shock front are several times greater than the shear stress of the material to be compacted. The high loading rate associated with a very short shock pulse duration also causes temperature rises. Transient temperature rises in the order of the melting point of tungsten are possible as will be shown below. Therefore, the state of the material under the high pressure shock loading can be considered as hydrostatic loading. Under these conditions the compression of matter by strong shock waves can be described by the Rankine–Hugoniot[16,17] equations which make use of the conservation of mass, momentum and energy:

$$(U_s - U_p) \cdot \rho = U_s \cdot \rho_0 \tag{10.1}$$

$$(P_1 - P_0) = \rho_0 \cdot U_s \cdot U_p \tag{10.2}$$

$$(E_1 - E_0) = \frac{1}{2}(P_1 + P_0)(V_0 - V_1) \tag{10.3}$$

where U_s is the shock velocity, U_p the particle velocity, ρ_0 the density before, ρ the density after the passage of the shock wavefront and V, P and E the specific volume, pressure and internal energy (with subscript 0 referring to the initial state before passage of the shock wavefront.

The shock wave velocity U_s and the particle velocity U_p are directly measurable; consequently it is possible to establish a pressure–volume relationship for different materials and set up the 'Hugoniot' curve. The pressure–volume relationship for porous materials, of course, differs quantitatively from the corresponding curve for solid materials. Qualitatively the shock wave behaviour of a porous material can be described as follows (Fig. 10.4). Starting from the specific volume V_0, powder compaction takes place along the Hugoniot curve A up to the state $(V_1 P_1)$. The release adiabate curve B describes the unloading up to the end volume V_E at ambient pressure. The end volume V_E is a measure for the achieved compaction.

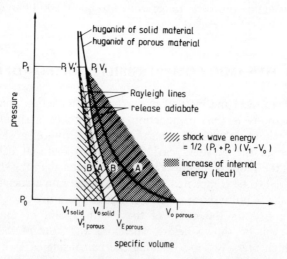

FIG. 10.4 Pressure–volume relations for the explosive compaction of powdered substances.

In the case of a solid material the Hugoniot and the release adiabate curves are approximately the same, providing the pressures are not too high. The total work necessary for compaction of the powder is represented by the triangle $V_0-P_1-V_1$. A part of it is released during unloading, area $V_E-P_1-V_1$. The difference (dotted area) of both represents the increase of internal energy due to one cycle (loading and unloading). As can be seen from the schematic diagram the increase of internal energy is much higher in powdered substances than in solid materials. Its increase is higher the greater the starting volume $V_{0\ porous}$. Most of the internal irreversible energy is dissipated as heat.

It is easy to understand that the Hugoniot shock curve not only depends on the specific starting volume (its inverse is the starting density) but also on parameters such as size and shape of grain and state of powder (work hardened or annealed particles).

Alternatively, a shock wave propagating in a porous medium can hardly be considered a continuous wave. In the early stage of compaction the shock wave is only transmitted at points of contact of individual particles, and local pressures exceeding the pressure given by the Hugoniot to a greater extent, are likely to occur. Due to different orientations of the individual particles and the effect of elastic anisotropy, of yield strength and work-hardening of the individual particles the wavefront becomes 'Wavy Wave Front' to a much greater extent than, e.g. observed in polycrystalline solid nickel.[18] More detailed information on theoretical calculation and the experimental set-up of Hugoniots for porous substances is given by Butcher and co-workers.[19,20] Their theory is based on time-dependent pore closure during shock wave transmission through a ductile material with spherical voids. Hugoniot determinations for foamed graphite and other porous substances have also been established by experiment.[21-24] There is lack of information, however, on experiments taking into account not only the degree of porosity of the powder, but also grain size, grain shape and their distributions. Undoubtedly, such information would be very useful for the application of explosive compaction to certain industrial powders.

10.3. TYPE OF SHOCK WAVE AND DENSITY DISTRIBUTION

The shape of the shock wave during its propagation in the powder in the course of explosive compaction is of main importance and decisive, whether a homogeneous compaction is achieved over the cross-section of the cylindrical sample or not. The container wall just acts as a transmitting medium of the shock wave to the powder. As the shock wave proceeds towards the centre of the cylindrical specimen peak shock wave pressure and shock wave velocity can vary continuously. There are two main factors governing this process:

(a) During the propagation of the shock wave in the powder material energy is consumed. The compaction work done by the shock wave is represented by the dotted area in Fig. 10.4. Plastic deformation in ductile materials and crushing occurring in brittle materials are necessary in order to stack the particles in a more dense manner.

(b) As the shock wave proceeding towards the centre of the cylindrical sample is a converging one, its pressure and velocity are increasing towards the centre. Therefore different shock wave configurations are possible.

A decreasing pressure and hence a decreasing shock wave velocity during propagation of the shock wave towards the centre lead to a shape of the shock front as described in Fig. 10.5(a). In this case it can be attenuated to such an extent that uncompacted material is left in the centre.

An increasing pressure and with that an increasing shock wave velocity leads to a parabolic shock wavefront as shown in Fig. 10.5(c). In

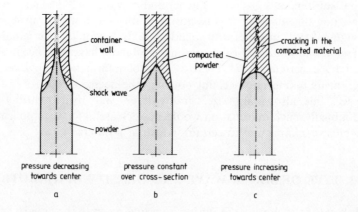

FIG. 10.5 Shape of the shock wave when propagating in an axial direction and towards the centre of a cylindrical specimen (stable configuration).

this case cracks are formed in the sample due to the intense release wave. At a very high intensity of the converging shock wave even a 'Mach-disk' can be formed[25,26] resulting in a hole in the centre of the cylindrical sample.

The different configurations of shock wavefronts were experimentally investigated by means of X-ray flash patterns[27-30] and by introducing thin foils of leads.[29,30]

The ideal case of a shape of the shock wavefront leading to a uniform compaction over the cross-section is shown in Fig. 10.5(b) with its conical shape. It is achieved when the absorption and convergence effects of the shock wave compensate each other. The shape of the shock wave is then conical. Figure 10.6 shows an X-ray flash picture of such an explosive compaction of aluminium powder. The exposure time was 120 ns.

Figure 10.7 shows the results of the three cases of different shock waves discussed. The cross-sections were taken from cylindrical samples which were under-compacted (a), correctly compacted (b) and over-compacted (c). The latter shows the above mentioned hole in the centre which is approximately 0.5 mm diameter.

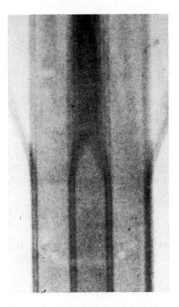

FIG. 10.6 X-ray flash pattern of an explosive compaction of Al-powder, exposure time 120 ns.

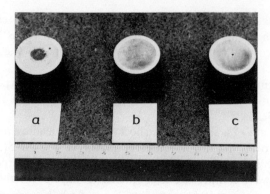

FIG. 10.7 Examples of under-compacted (a), correctly compacted (b), and over-compacted (c) specimens of Ferrotic-material (50% steel − +50% TiC powder) 3 μm grain size, compacted in a steel tube of 25 mm diameter and 1.5 mm thickness of wall.

In order to achieve an optimum compaction in most investigations[27,28,31-37] the E/M ratio, where E is the mass of explosive and M the mass of powder to be compacted, was the main parameter considered. It has been shown that the rate of compaction, whether an under-compacted or over-compacted specimen is obtained, is a matter of explosive loading. The densities achieved amount to more than 90% of theoretical density depending on the type of powder and often reach 99% or more. A linear relationship seems to exist for the E/M ratio, which is necessary for a uniform compaction over the cross-section of the sample with the compressive yield strength of the powder material, the range being $E/M=0.2$ for aluminium up to $E/M=1.22$ for nickel powder.[27]

In early experiments high explosives were used with the intention of obtaining high densities in the compact. In a further extensive study, however, it could be shown that maximum density of the compact largely depends on the detonation velocity of the explosive used for compaction.[28,38-40] In an E/M versus the square of detonation velocity plot (upper part of Fig. 10.8) a curve is established experimentally which gives the parameters for correct compaction with uniform density over the cross-section of the sample and no cracking. It separates the region of under-compacted and over-compacted samples. The higher the detonation velocity v_D (the higher the pressure) the lower is the amount of explosive E/M necessary for compaction. The detonation pressure is related to the detonation velocity and the specific weight by the following

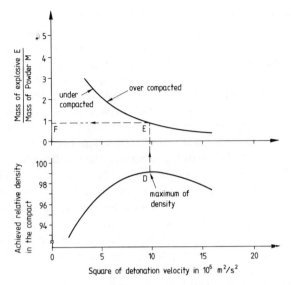

FIG. 10.8 Determination of the parameters for explosive compaction.

equation

$$p = \left(1 - \frac{\rho_0}{\rho_s}\right) \cdot \rho_0 \cdot v_D^2 \approx \frac{1}{4}\rho_0 v_D^2$$

where ρ_0 is the density of the crystalline explosive and ρ_s is the density of the explosive in the reaction zone, which is approximately 4/3 of the density of the crystalline explosive. Therefore the abscissa in Fig. 10.8 is linearly related to the detonation pressure. The lower part of Fig. 10.8 shows the achieved density (relative density) versus the square of the detonation velocity. The most important feature of this curve consists of a maximum of density at a certain value of the square of the detonation velocity (pressure).

This feature represents the effect of over-compaction: due to a strong release wave the interparticle locking which was achieved by the compressive wave is interrupted. This results in localised debonding of the material and thus to a lower density of the compact. This effect is the more pronounced the higher the (excess) detonation pressure. In order to achieve optimum results an explosive has to be chosen with a detonation velocity where maximum density is observed (point D). The appropriate E/M value can be taken from the upper graph in Fig. 10.8, following the dashed line up to point F.

Investigations with a variety of different powders lead to similar shapes of the E/M–v_D^2—and density–v_D^2—curves. Depending on the type of material to be compacted, the detonation velocity, where maximum density of the compact is observed, is shifted to higher or lower values. Aluminium powder, for instance, only requires low pressure (explosive with a low detonation velocity) whereas f.i., a highly alloyed steel powder, needs a high pressure (high explosive) in order to achieve a compact with optimum density. Different explosive compaction curves for a variety of materials and mixtures (metal and ceramic powders) were established.[28,38–40] Table 10.1 summarises the results of explosive compaction of different powders, which are difficult to be consolidated.

A shift towards smaller pressures for the same material to be compacted is possible by heating the powder during compaction. As the strength of the material decreases, lower pressures (explosives with lower values of detonation velocity) and less amounts of explosive are required. Hot explosive compaction especially can be applied to high strength materials such as highly alloyed steel powder or chromium- or nickel-based superalloys. The metal capsule containing the powder is heated in a furnace located at a secure distance from the explosive and after reaching the desired temperature is rapidly moved into the cylindrical hole of the explosive, which is then detonated with the detonation wave proceeding in the axial direction of the metal capsule.[41] A similar arrangement can also be applied to the compaction of heated flat porous plates.[42]

An investigation of the explosive's parameter for compaction of iron, nickel, copper and aluminium powders in tubes with varying wall thicknesses makes use of the kinetic energy which is imparted to the container wall by the detonating surrounding explosive.[36] The velocities are taken from measurements of the acceleration of the container wall in radial direction in a configuration with the same dimensions but without powder filling. Therefore, the shock wave shape in the powder is not taken into account. An interesting approach[43] making use of the shock adiabatic curves of the powdered substances also accounts for thermal heating and increase of internal energy during the explosive compaction. The initial packing density of the powder is also considered.

10.4. TEMPERATURE AND STRAIN RATE EFFECTS

A part of the total energy applied to a powdered substance by means of a shock wave is necessary in order to overcome the elastic repulsive forces

TABLE 10.1
EXPLOSIVELY COMPACTED POWDERS

Powdered material	Grain size	Theoretical density (g/cm³)	Packing density (g/cm³)	(TD%)	Density of explosive compact (g/cm³)	(TD%)
Tungsten	2–10 μm	19.3	8.76	45.6	18.80	97.4
Aluminium—fine	10–160 μm	2.7	1.54	57.0	2.708	99.9
Aluminium—coarse	10μ–2 mm	2.7	1.59	59.0	2.69	99.8
Iron	1 μm	7.86	3.45	44.0	7.76	98.7
Ferrotic*	3 μm	6.55	3.65	55.8	6.26	95.6
Al_2O_3	5 μm	3.94	2.32	59.0	3.75	95.5
ZrO_2-Si_3N_4, ZrO_2+CaO +2^w/$_o$ MgO	3 μm	5.4	2.85	53	5.3	98
B_4C—fine	0–90 μm	3.18	1.66	52.3	3.06	96.3
B_4C—coarse	0–300 μm	2.51	1.56	62.4	2.45	97.7
alloyed steel	0.03–1.0 mm	2.51	1.75	70.0	2.50	99.6
		8.00	6.11	76.4	7.95	99.4

* Mixture of 50^w/$_o$ steel powder + 50^w/$_o$ TiC powder.

of the atomic lattice with an amount represented by the area $V_E - P_1 V_1 - V_0$. It is reversible and causes fracturing of the compacted sample, if the compaction pressure chosen was too high in magnitude and if the release rate is very high.

The dotted area $V_0 - P_1 V_1 - V_E$ in Fig. 10.4 is the work irreversibly applied to the powder. It leads to an increase of the internal energy of the powder (compact) and is considerably greater than that for a solid material. Most of it is dissipated as heat in the material leading to a temperature increase. Whereas residual temperature increases achieved in solids after passage of a shock wave of a magnitude of a few hundred kbars are in the order of a few hundred degrees[44,45] the residual temperatures in compacted powders can be much higher and easily reach melting temperature.

Figure 10.9 shows the centre of a cylindrical sample of 17 mm diameter

FIG. 10.9 Centre area showing recrystallised tungsten after explosive compaction of 5 μm tungsten powder.

of tungsten fabricated by explosive compaction of tungsten powder of a grain size of 5 μm by means of an explosive with a detonation velocity of 3500 m/s and a pressure of about 50 kbar.[40] Due to the fact that the shock wave was converging, the melting point of the material ($\sim 3000\ °C$) was reached, subsequently leading to a recrystallised area of 0.12 mm in the centre of the rod.

But even in most cases where no gross melting occurs during compaction, it does take place locally. The shock wave during compaction is not of a plane nature, like propagating in a solid, but is wavy, consisting of a series of shock waves in the individual grains. The shock wave crosses particles and gaps in sequence leading to local shock jumps. The discontinuity of pressure and of velocity is a very important feature of explosive compaction[40] since it launches the relative movement of the individual particles and therefore friction with neighbouring particles. Local temperature spots, created this way, enable friction welding of individual particles to one another.

Individual particles on the other hand can be accelerated to velocities of the magnitude of the theoretical particle velocity, similar to the acceleration of a pellet fixed at the free end of a shock's solid material. Collisions with other particles are likely to occur and lead to high local dynamic pressures greater than the compaction pressure of the explosive. Subsequently hydrodynamic flow is initiated in regions with oblique collision under similar conditions as in explosive welding. Therefore, jetting is very likely to occur between the particles during compaction. In this case higher temperatures are also created at local areas of compact.

Several authors have investigated heating during explosive compaction. The necessary measurements can be performed by means of thermoreactive materials[46] which are embedded in the powder, by means of thermocouples[30,47-50] or by optical methods.[51] A unique method is the direct registration of the thermoelectric effect during the passage of the shock wave through the interphase of a porous copper – nickel sample during compaction.[30,52]

It could be experimentally verified to show that shock loading of porous substances under conditions of explosive compaction leads to temperatures far higher than those obtained in solids during the same treatment. Especially, it was shown[48] that the tap density of the powder to be compacted is of great importance. No data are yet available, however, taking into account the grain size and shape and the distribution of the powder.

10.5 PHASE TRANSITIONS IN SHOCK LOADING MIXTURES

High pressures and elevated temperatures during the explosive compaction and the high strain rate associated with the rapid rise time of shock waves and its short duration can cause chemical reactions in mixtures of reactive substances or phase transitions in pure substances. The first observed chemical synthesis was that of the formation of zinc-ferrite with its constituents[53] and shortly after that of titanium carbide through explosive compression of a stoichiometric mixture of titanium and carbon powder.[54] Starting with mixtures of acetylene black and tungsten or aluminium powder, metal carbides were formed when tetryl was applied.[55] The E/M ratio was ~ 5 for the formation of tungsten carbide at a yield of 90% tungsten carbide (α-W_2C and WC) in the pellet and about 16 for the formation of Al_4C_3 at a yield of 42.5%.

No yield at all or only a low one of the carbides was obtained, when the E/M ratio was too small. But in this case it has been reported that after subsequent heating of the compact an increased reaction rate can be observed, compared to the known reaction rate of the unshocked material. From this it was concluded, that the preshock treatment leads to activation of the chemical reaction leading to the carbide formation.[56]

Many efforts were made in order to quantitatively describe the synthesis of a great variety of components.[50] Of special interest should be the findings that the amount of explosive for initiation of the synthesis is related to the reaction temperature under ordinary conditions.[57,58] It has been shown that the initiation of the synthesis can also be accompanied by an exothermic reaction. So after explosive compaction of a mixture of tin and sulphur a temperature of 1100 °C was observed, whereas the explosive compaction of SnS powder only resulted in a temperature increase of 130 °C.[59] It can therefore be concluded that the impulse $I = \int p \, dt$ as imparted to the capsule is a measure for the amount of chemical reaction. The explosive synthesis was also applied for the formation of super-conductive compounds like Nb_3Sn[60,61] and Nb_3Si[62-64] from the constituents. The pressures required range up to about 1 Mbar. A technique is used where an outer driver-tube is accelerated against the metal cylindrical capsule containing the mixture. There is evidence for a lower transition temperature in explosively fabricated super-conductive alloys as compared to alloys produced by conventional methods. Also the range where transition takes place is slightly wider for explosively synthesised super-conductive materials. Barium titanate which is widely used for pressure sensors was synthesised with a mixture of titanium oxide (TiO_2) and barium carbonate ($BaCO_3$) using RDX as an explosive.[64]

In comparison with conventionally produced materials explosively synthesised materials reveal different behaviour. So the hardness of explosively synthesised tungsten carbide particles with a size of approximately 1 mm reveal a hardness of about 3500 HV compared to 2000 HV of conventionally fabricated tungsten carbide.[65] After explosive synthesis of CuBr from a mixture of $CuBr_2$ and Cu a lattice constant of $a = 5643$ Å was observed.[66] The general value is 5690 Å. The density and dielectric properties also appeared to be greater and the transition temperature for the sphalerite – wurtzite transition of CuBr was lower: it was 375 °C compared to the known transition temperature of 396 °C. A subsequent heat treatment of 1/2 h at 400° leads to the properties of conventionally fabricated CuBr. Different lattice constants were also observed with BN, CaF_2, CdF_2 fabricated by explosive synthesis. Annealing leads to physical parameters known for these substances.[50,66]

A significant discovery has been the possibility of making diamond from graphite by means of shock compression.[67,68] Figure 10.10 shows

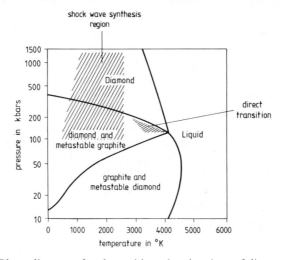

FIG. 10.10 Phase diagram of carbon with explored regions of diamond synthesis.

the phase diagram of carbon, with the triple point at 125 kbar and 4100 °C. The minimum pressure required for a substantial synthesis of diamond is about 150 kbar. The yield of carbon which under-went the transition to diamond is a few per cent and the particle size of the diamond powder is a few μm less. Also thin platelets of polycrystalline diamond are obtained with a size of almost 1 mm.

When the chosen shock pressure was too high and/or the starting density of the graphite powder was too low the temperatures in the shocked graphite were considerable.

As after the release of the pressure wave the shocked graphite cools down too slowly, the shock wave synthesised diamond might be graphitised. The best yield is obtained with a graphite starting density of about 1.8 g/cm^3 and at compaction pressures between 100 and 700 kbar. Several methods of the procedure of making diamond are described in the De Carli patent.[69]

A different method of shock wave treatment and rapid cooling exists in a high velocity collision of graphite with water at a velocity of about 6.5 km/s. The collision pressure is about 450 kbar. The acceleration to that high velocity is achieved in a hollow charge arrangement with graphite as a liner material. The yield of diamond powder collected from the water is about 3%.[70]

A better yield is achieved when a mixture of crystallised graphite and a metal powder is subjected to a shock wave treatment,[71,72] or when the graphite particles are embedded in a metal matrix which serves as a shock plate and as a heat sink after passage of the shock wave.[72,73] Today large charges, about 1 m in diameter and several metres long are detonated in order to produce diamond powders.[74] To obtain a large pressure and duration of pressurisation a thick driver tube is accelerated towards the container tube.

The diamond powders produced by this method are used for grinding and surface finishing of metals and ceramic materials and for cutting purposes. They are superior to natural diamond powder because of the irregular shape of the particles, with particle size ranging from 0.2 to about 10 μm. There also exists the possibility of using the explosively synthesised diamond powder to form polycrystalline bodies by means of dynamic compaction of the powders.[75]

A similar development took place with boron nitride. BN occurs at three modifications and the graphite like form is stable at ambient temperatures. The diamond like forms, cubic sphalerite type BN and a hexagonal wurtzite type BN are stable at high pressures but can also exist in a metastable form at ambient conditions. Their hardness is approximately as high as that of diamond and can be used in the same way as diamond powder. Shock wave transformation can be performed in the same way as explosive compaction.[76-80] The transformation during shock loading was also observed by means of X-ray flash diffraction.[78] The yield of wurtzite like BN amounts to above 80%, is dependent on the density of the green compact, and is the higher the more perfect the crystal structure of the starting material.[79,81] Because of its application potential the high density form of BN is also of great interest for industrial compaction.[82,83]

Both diamond powder and cubic boron nitride fabricated by shock wave synthesis reveal a severe lattice distortion, being dependent for its amount upon cooling rate after passage of the shock wave.[68]

During explosive compaction of silicon nitride, which is a material currently of great interest for high temperature applications (turbine blades and wheels in gas turbines) phase transitions are also observed.[84] When α-Si_3N_4 is subjected to shock pressures of 400 kbar, a transition to the β-modification can be observed. It is expected that this form of β-Si_3N_4 also contains a severe lattice distortion so that a subsequent activation of the sinter treatment can be expected.

Transformation can also occur during explosive compaction of metallic glasses. These are produced in thin strips of thickness of 40 μm or as a powder by rapid cooling at a rate of 10^{60}/s. After explosive compaction[85-87] in certain cases as increase of hardness was observed[86] which to some extent might be due to recrystallisation as a result of thermal effects at high shock pressures.

Explosive synthesis or shock wave synthesis is a new tool in materials science with a number of possibilities of developing new substances and studying high pressure phenomena in matter.

10.6. GENERAL MECHANICAL PROPERTIES OF COMPACTED POWDERS

A great number of investigations exist about shock wave parameters in explosive compaction, however, little information is available about the ultimate properties of the compacts. As the achieved densities are approximately or exactly equal to theoretical density, post-sinter treatment is often said to be unnecessary or even leading to a softening of the green compact. On the other hand, a sinter treatment is necessary in most cases in order to give the green compact its final strength. This certainly depends on the material's properties as well as the chosen explosive parameters. Low pressures merely lead to a dense packing of the particles; high pressures also give the compact a certain strength, however with the implication of difficulty in obtaining a uniform density in the compact over the cross-section of the cylindrical sample. As materials with high compressive strength or melting points (like high strength steel powder, refractory metals and alloys and Co- or Ni-based superalloys) require high pressures in order to obtain a sufficient density, the actual problems arise with them. The compaction pressures being extremely high, the pressure release wave is very likely to disrupt the bonding which was achieved by the compressive wave.

The material properties after explosive compaction are in the first place governed by the same metallurgical effects which are known from shock wave hardening. A work hardening of the material is mainly due to: (a) dislocation multiplication, (b) introduction of large quantities of point defects, (c) twinning, and (d) phase transitions.

These points have already been mentioned in numerous review articles and are also treated in Chapters 3 and 4. In effect, they have to do with the appropriate hardening of the volume of the particles. In contrast to solid materials in the shock loading of powdered substances, the wavy nature of the shock wave and the relative movement of the individual particles to one another, are further mechanisms which have to be taken into account, mostly relating to the surface areas of the particles. The surface of the particles is affected by the friction and hydrodynamic flow.

Friction leads to an increase of temperature at the surface of the particles which can be so high that local melting may occur. Hydrodynamic flow will be initiated during oblique collision of surface areas of individual particles and jetting can occur under the same conditions as in explosive welding.

Both mechanisms are restricted to appropriately oriented surface areas with respect to the direction of the shock wave and are only of short duration during the passage of the shock wave. Friction welding, fusion welding and explosive welding therefore are bonding mechanisms which are restricted to certain areas and can hardly be expected to cover the total area between the particles of the compact, with the exception of very high pressures and high explosive loads (E/M ratios) applied for compaction.

The third mechanism which, however, does not very much contribute to the work-hardening of the material is the crushing of brittle materials, necessary for obtaining a better packing density.

Experimental evidence exists for this scheme. Spherical grains can serve as a model of a porous substance when packed to a body. Spheres from ball bearings (steel with 1.6% Cr) were made the subject of explosive compaction in a flat arrangement at a pressure of 200–250 kbar.[88] The metallographic investigation revealed interparticle melting and temperatures in excess of the melting temperature (1460 °C) while the interior of the particles was only slightly heated and remained without any visible changes of the microstructure. It is to be concluded that the time required for solidification in the surface areas is between 10^{-6} and 10^{-5} s. No significant hardness increase was observed.

Experiments with a highly alloyed steel powder[28] with Udimet 700[28,89] and with Mar M-200 powders[90] in a cylindrical arrangement show that compaction can be performed with or without interparticle melting, depending on the conditions of compaction. These powders consist mainly of spheres with sizes from 40 μm up to about 1000 μm. Good mechanical properties are only observed when interparticle melting occurs. Without interparticle melting the strength of the compact in some cases is just good enough for a sample to be machinable by turning or grinding. In other cases individual particles break off during the preparation of a sample for micrographic investigations.[89] Post-sinter treatment leads to a strength comparable to otherwise produced parts of the same material or higher, especially when interparticle melting has occurred in sufficient quantities. In Mar M 200 powder a hardness increase from 357 HV to 700 HV was observed when explosive compaction was performed at E/M values of 10 and 14. Also a high density of lattice defects was detected by transmission electron microscopy.[90]

Similar results are obtained with the dynamic compaction of stainless steel powder (SAE 304 L) using a dynamic compaction press propelled with compressed air.[91] Interparticle melting can occur and a strength is obtained which increases with the strain rate during compaction and a magnification of hardness almost three times the value of annealed powder is measured.[92] Interparticle jetting with wave formation is observed in this case and also in a mixture of copper and iron powder after explosive compaction.[30] Similar jetting was also observed during the compaction of cylindrical rods[92] or filaments during the fabrication of fibre reinforced metal matrix composites.[93]

If temperatures, created in the compact during explosive loading are too high, possibly because the starting density was too low or the compaction pressure too high, an annealing of the sample can result in a lower hardness and strength.[94,95]

As a result, no general rule can be established for the strength, the hardness and other mechanical properties of materials after explosive compaction from their powders, these vary with the parameters applied for compaction. Explosive compaction at low pressures has the advantage of producing samples with uniform density over the cross-section, however, without considerable strength, whereas high pressures lead to dense samples with interparticle melting and a strength which is often superior to the strength of commonly produced solid materials. There exists good potential in development of a new type of material of metastable state.

10.7. X-RAY AND OTHER METHODS OF EVALUATING RESIDUAL STRESS DISTRIBUTION

Part of the irreversible energy required for compaction is dissipated as heat, while another part is stored in the solid increasing its internal energy. Often this is hoped to increase the material's chemical or sinter activities. In the case of carbon powder it was clearly demonstrated how its chemical reactivity with an oxygen atom is increased by shock treatment.

A great deal of the stored energy is found in the crystal lattice and can be analysed by X-ray diffraction methods. The procedure consists of X-ray line broadening analysis. The broadening of diffraction patterns is caused by lattice distortions and also subgrain crystallite size.

Several investigations were performed about the structural changes after explosive compaction in MgO single crystals[97,98] and in polycrystalline samples of Al_2O_3, ZrO_2, SiC and B_4C[99] and further substances.[50,100,101] Relatively high values of stored energy are found after explosive compaction, ranging up to 0.8 cal/g for ceramic materials. All the mentioned investigations, however, are limited to the application of one type of explosive with perhaps the wrong detonation velocity or compaction pressure. Further investigations, taking into account the pressure during compaction and the amount of the explosive (as a measure of pressure duration) should certainly reveal the relationship between pressure, density and lattice distortion and subgrain size for Al_2O_3,[38] Mo and Ti[39] and tungsten.[40,102] It has also been shown that the internal energy in Al_2O_3 compacts is influenced by the type (grain size) of starting powder and was also interpreted with crushing mechanisms during compaction. The density of dislocations is found comparable to the one usually obtained in heavily deformed metals. Scanning electron microscopy investigations support these findings. Transmission electron microscopy investigations with similar samples[103] of Al_2O_3 even revealed indications of melting at grain boundaries.

Transmission electron microscopic investigations performed on explosively compacted super-alloy powders[90] in interparticle molten areas revealed a microcrystalline structure which can only be due to a high cooling rate. The liquid phase between the particles during compaction serves as a grease and enables further relative movement necessary for densification. Interparticle grease during the explosive compaction of tungsten powder (achieved by adding of a few per cent of Fe and Cr powder) decreases interparticle friction during explosive compaction, as is indicated by a lower amount of lattice distortions.[40]

Further investigation will be necessary to gain more insight into the microstructural processes underlying the explosive compaction.

10.8. BASIC PROBLEMS IN FABRICATING SEMI-FINISHED PARTS

The fact, that porous materials attenuate shock waves rapidly is the reason why flat plates of porous substances of greater thicknesses require large amounts of explosive. When applying the oblique detonation technique, problems arise with the occurrence of shear waves which limit this method to ductile materials in order to obtain crack-free compacts. Plane wave techniques are difficult to handle because of the release wave leading to a disruption of the compact.

The converging nature of the shock wave during compaction of cylindrical samples is a good compensation for the attenuation of the shock wave. Therefore, explosive compaction methods are practically limited to parts with a rotational symmetry. Large rods of several metres in length were explosively compacted using tungsten powder. Other parts include tubes, hollow cones, nozzles for rockets, hemispheres and also cylindrical parts with corrugated surfaces. Figure 10.11 shows parts which were explosively compacted and the subject of post-sinter treatment.[28]

Large diameter rods were successfully explosively fabricated from Ni-base superalloy powders (115 mm diameter) and aluminium powder

FIG. 10.11. Hollow shapes with a wall thickness of 2 mm fabricated by explosive compaction from tungsten powder.[104]

(350 mm diameter). Practically there are no limitations to compact metal powders at large diameter except environmental ones. The larger the diameter, the higher will be the E/M ratio necessary for a uniform compaction. As the compaction wave proceeds towards the centre of the sample, the compacted outside layers have to undergo plastic deformation in order to enable further compaction. Therefore, relative to the compacted powder a greater duration of the compaction pressure is required. To the best knowledge of the author the largest charge ever used for compaction is 1100 mm.

10.9. STATIC AND DYNAMIC COMPACTION: A COMPARISON OF MATERIAL PROPERTIES

The mechanics of explosive compaction and the underlying mechanisms are different from static compaction. As outlined above, there is a greater increase of internal energy during explosive compaction (Fig. 4) in comparison to static compaction. This is especially the case with the effect of local heating as a support for rearrangement of the particles for better packing and for bonding. Therefore, in some given cases, a post-sinter treatment can be omitted after explosive compaction and in some other cases, which have to be investigated through future research, superior mechanical properties can be obtained.[5,94,105] Shock hardening and rapid temperature rises and decay are responsible for that complex. Alternatively, a uniform density over the cross-section of the compact, which in static compaction is achievable with little difficulty, is a problem in explosive compaction. Supposedly this problem can only be solved by means of extensive parameter studies, including type and amount of explosive, being adjustable to different kinds of material under investigation. Improper compaction with the result of density variations over the cross-section would cause deformations and possibly void formation during sinter treatment. The advantage of obtaining superior densities in the compact without heavy equipment will economically compensate for the required and sometimes complicated parameter studies in explosive welding.

Despite the great differences in static and explosive compaction, however, similar density–pressure relations are obtained. Comparative investigations were performed with tungsten, titanium[102] and stainless steel[91] powders. The resultant densities are shown as a function of compaction pressure in Fig. 10.12.

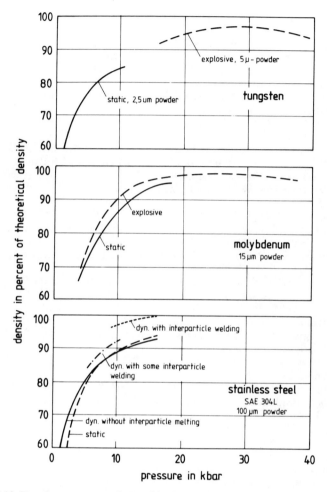

FIG. 10.12 Density–pressure relationship for explosive and static compaction, a comparison.

The explosively obtained results with tungsten form a continuation of the static curve to higher pressures with a maximum density occurring at ~30 kbar. With titanium, explosive compaction yields higher values in comparison to static compaction. The dynamic (gas-gun) investigations with stainless steel powder show lower values of density at pressures up to ~7 kbar, while at higher pressures, as soon as interparticle melting occurs, the dynamically obtained densities are larger than the isostatically obtained values for corresponding pressures.

REFERENCES

1. LA ROCCA, E. W. and PEARSON, J. *Rev. Sci. Instr.*, **29** (1958), 848.
2. LA ROCCA, E. W. and PEARSON, J. US Patent 2,948,923, June 4, 1958.
3. STEIN, E. M., VAN ORSDEL, J. R. and SCHNEIDER, P. V. *Metal Progress*, **83** (April 1964), No. 4, 83.
4. BREJCHA, R. J. and MCGEE, S. W. *American Machinist*, **106** (1962), 63–65.
5. HAGEMEYER, J. W. and REGALBUTO, J. A. *International Journal of Powder Metallurgy*, **4** (1968) No. 3, 19; see also vol. I of the Institute of Metals (1971), 209.
6. RAYBOULD, D. *15th Int. MTDR Conf.*, Macmillan, London, 1975.
7. MCKENNA, P. M., REDMOND, J. C. and SMITH, E. N. US Patent 2,648,125.
8. POREMBKA, S. W. and SIMONS, C. C. *Powder Metallurgy*, **6** (1960), 125.
9. MONTGOMERY, W. T. and THOMAS, H. *Powder Metallurgy* (1960), No 6, 125.
10. PEARSON, J. High energy rate forming, *ASTME SP*, 60–150, Dec. 1961.
11. PAPROCKI, S. J., SIMONS, C. and CARLSON, R. J. Advanced high energy rate forming II, *ASTME SP*, 62–29, 1962.
12. LIEBERMANN, I., KNOP, L. H. and ZERNOW, L. High energy rate forming, *ASTME SP*, 23–29, 1963.
13. RINEHART, J. S. and PEARSON, J. *Explosive Working of Metals*, 1963, Pergamon Press, Oxford, London, New York.
14. LEONARD, R. W. *Battelle Techn. Review* **17** (1968), No. 10, 13.
15. DERIBAS, A. A. et al. *Fizika Gorenija i Vzryva* **9** (1973), 883.
16. RANKINE, W. J. M. *Phil. Trans. Roy. Soc., London*, **160** (1870).
17. HUGONIOT, H. *J. ecole polytech*, **58** (1889).
18. MEYERS, M. A. *Proc. 5th Int. Conf. HERF*, University of Denver, Colorado, June 1975.
19. BUTCHER, B. M. and KARNES, C. H. *J. appl. Phys.*, **40**, No. 7 (1969), 2967.
20. BUTCHER, B. M., CARROLL, M. M. and HOLT, A. C. *J. appl. Phys.*, **45**, No. 9 (1974), 3864.
21. BOADE, R. R. *J. appl. Phys.*, **39**, No. 3 (1968), 1609.
22. BOADE, R. R. *J. appl. Phys.*, **39**, No. 12 (1968), 5693.
23. BOADE, R. R. *J. appl. Phys.*, **40**, No. 9 (1969), 3781.
24. HERMANN, W. *J. appl. Phys.*, **40**, No. 6 (1969), 2490.
25. WALSH, I. M. and RICE, H. M. *J. Chem. Phys.*, **26** (1957), 815.
26. HOLLENBERG, K. and MILLER, F. *BMVg WT* 72–3 (1971).
27. LEONARD, R. W., LABER, D., LINSE, V. D. *Proc. 2nd Int. Conf. HERF*, Estes Park, Co., USA, 1969.
28. PRÜMMER, R. *Ber. Deutsche Keramische Gesellschaft* **50** (1973), 75. See also *Proc. 4th Int. Conf. HERF*, Estes Park, Co., USA, 1973.
29. DERIBAS, A. A. and STAVER, A. M. *Fizika Gorenija i Vzryva*, **10** (1974), 568.
30. STAVER, A. M. In *Shock Waves and High Strain Rate Phenomena in Metals*, eds. M. A. Meyers and L. E. Murr, Plenum Press, New York and London, 1981. See also *Proc. 4th Intl. Conf. HERF* (1975), Denver, Co., USA.
31. MONTGOMERY, W. T. and THOMAS, H. *Powder Metallurgy* (1960), No. 6, 125.

32. PAPROCKI, S. J. et al. *Paper Nr. SPA 2–29*, Creative Manufacturing Seminars, Detroit: ASTME (1961/62).
33. STROHECKER, D. E. High velocity compaction of metal powders, in *High Energy Rate Working of Metals*, Oslo: Central Inst. Ind. Res. (1964), 481.
34. LABER, R. W. *Batelle Techn. Rev.*, Nov/Dec. (1968), 13.
35. DERIBAS, A. A. and STAVER, A. M. *Fizika Gorenija i Vzryva*, **10** (1974), No. 4, 568.
36. LENNON, A. R. C., BALLA, A. K. and WILLIAMS, J. D. *Proc. 6th. Int. Conf. HERF*, Sept. 1977, Essen, Germany.
37. MIKHAILOV, A. N. and DREMIN, A. N. *Fizika Gorenija i Vzryva* **13** (1977), No. 1, 115.
38. PRÜMMER, R. and ZIEGLER, G. *Proc. 5th Intl. Conf. on High Energy Rate Fabrication*, 1975, Denver, Colorado, USA. See also: *Powders Metallurgy International No. 1* (1977), 14.
39. PRÜMMER, R. and ZIEGLER, G. *Proc. 6th Intl. Conf. on High Energy Rate Fabrication*, 1977, Essen, Germany.
40. PRÜMMER, R. *Proc. 6th AIRAPT Conf. High Pressure Science and Technology*, eds. K. D. Timmerhaus and M. S. Barber (1979), Plenum Press, New York, p. 814.
41. PRÜMMER, R. Germ. Pat. No. 2436951 (1974).
42. ROMAN, O. V. and GOROBTSOV, V. G. *Int. J. Powder Metallurgy and Powder Technology* **11** (1975) 55.
43. ROMAN, O. V. In *Shock Waves and High Strain Rate Phenomena in Metals*, Meyers, M. A. and Murr, L. E. eds., 1981, Plenum Press, New York.
44. MCQUEEN, R. J. and MARSH, S. P. *J. appl. Phys.*, **31** (1960), 1253.
45. WALSH, J. M. and CHRISTIAN, R. H. *Phys. Rev.*, **97** (1955), No. 6, 1544.
46. ATROSHENKO, E. S. and KOSOVICH, V. A. *Technologija Masinostroenija* (Russ.) (1971), 67.
47. KRUPNIKOV, K. K., BRAZHNIK, M. I. and KRUPNIKOVA, V. P. *Soviet Physics JETP* **15** (1962), No. 3, 470.
48. PIKUS, I. M. and ROMAN, O. V. *Fizika Gorenija i Vzryva*, **10** (1974), 10, 782.
49. KUNIN, N. F., YURCHENKO, B. D. and OVICHINNIKOV, O. P. *Poroshkovaja Metallurgija*, **10** (1971), 106, 19.
50. BATSANOV, S. S. In *Preparative Methods in Solid State Chemistry*, Academic Press Inc., New York, London (1972), pp. 133–146.
51. BELJAKOV, G. V., LIVSHITS, L. D. and RODIONOV, V. N. *Izv. Earth Physics*, **10**, (1974), 92–94.
52. NESTERENKO, V. E. Dissertation Novosibirsk, 1974.
53. KIMURA, Y. *Japan J. of Appl. Phys.*, **2** (1963), 312.
54. HORIGUCHI, Y. and NOMURA, Y. *Bull. Chem. Soc. Japan* **36** (1963), 486.
55. HORIGUCHI, Y. and NOMURA, Y. *J. Less-Common Metals*, **11** (1966), 378.
56. HORIGUCHI, Y., NOMURA, Y. and KATAJAMA, S. *Kogyo Kogaku Zassi* **69** (1966), 1007.
57. BATSANOV, S. S. and ZOLOTOVA, E. S. *Dokl. Akad. Nauk SSR* **180** (1968), 93.
58. BATSANOV, S. S. et al. *Dokl. Akad. Nauk SSR* **185** (1969), 330.
59. BATSANOV, S. S. et al. *Phys. Gorenija Vzryva*, **5** (1969), 283.

60. OTTO, G., REECE, O. Y. and ROY, U. *Applied Physics Letters*, **18** (1971), 418.
61. BARSKI, I. M., DIKOVSKII, V. Y. and MATYTSIN, A. I. *Fizika Gorenija i Vzriva*, **8** (1972) 474.
62. PAN, V. M. et al. *IETP Lett.*, **21** (1975), No. 8, 228.
63. DEW-HUGHES, D. and LINSE, V. D. *J. Appl. Phys.*, **50** (1979), 3500.
64. DERIBAS, A. A., STAVER, A. M. *Fizika Gorenija i Vz'ryva*, **6** (1970), 122.
65. DERIBAS, A. A. et al. *Proc. Sympos. Behavior of Dense Media Under High Dynamic Pressures*, 1967, Paris.
66. BATSANOV, S. S. *J. Struct. Chim.*, **11** (1970), 156.
67. DE CARLI, P. S. and JAMIESON, J. C. *Science*, **133** (1961), 1821.
68. DE CARLI, P. S. In *Science and Technology of Industrial Diamonds* ed. J. Burls, 1967, Industrial Diamond Information Bureau, London.
69. DE CARLI, P. S. US Patent 3238019 (1966).
70. KIYOTO, K. et al. *J. Industrial Explosive Society Japan*, **37** (1976), Nr. 3, 152.
71. SEKATA, N. and SEKIKAWA, Y. *J. of Materials Science*, **16** (1981), 1730.
72. HANNEMAN, R. E., STRONG, H. M. and BUNDY, F. P. *Science*, **155** (1967), 995.
73. PING-HUANG, S. et al. *Proc. II Meeting on explosive working of materials*, Novosibirsk, 1981.
74. BERGMAN, O. R. *Proc. 7th Intl. Conf. HERF*, Leeds, England, 1981.
75. BALCHAN, A. S. and COWAN, G. R. US Patent 3851027 (1974).
76. AL'TSHULER, L. V. *Usp. Fiz Nauk*, **85** (1965), 198.
77. ADADUROV, G. A. and ALIEZ, Z. G. et al. *Dokl. Akad. Nauk SSR*, **172** (1967), 1066.
78. JOHNSON, Q. and MITCHELL, A. C. *Phys. Rev. Letters*, **29** (1972), 1361.
79. DULIN, I. N. and AL'TSHULER, L. V. et al. *Fiz. Tverd. Tela*, **11** (1969), 1262.
80. COLEBURN, N. L. and FORBES, J. V. *J. Chem. Phys.*, **48** (1968), 555.
81. BATSANOV, S. S. and BATSANOVA, L. R. et al. *Zh. Strukt. Khim*, **9** (1968), 1024.
82. VERESHTSHAGIN, L. F., et al. German Patent claim 2 235 240.8 (1972).
83. DRJOMIN, A. N. et al. German Patent claim 2219394.8 (1972).
84. MITOMO, M. and SETAKA, N. *J. of Mat. Science-Letters*, **16** (1981), 852.
85. MORRIS, D. G. *Metal Science*, July 1980, 215–220.
86. ROMAN, O. V. et al. *Proc. 4th Intl. Conf. on Rapidly Quenched Metals*, Sendai, Japan 1981.
87. CLINE, C. F. *Proc. 4th Intl. Conf. on Rapidly Quenched Metals*, Sendai, Japan 1981.
88. BELYAKOV, G. V., LIVSHITS, L. D. and RODINOV, V. N. *Izv. Earth Physics*, 1974, No. 10, 92.
89. PRÜMMER, R.: *J. Mat. Techn.* **4** (1973), No. 5, 236.
90. MEYERS, M. A., GUPTA, B. B. and MURR, L. E. In *Shock Waves and High Strain Rate Phenomena in Metals*, eds. M. A. Meyers and L. E. Murr, 1981, Plenum Press, New York.
91. REYBOLD, D. *J. Mat. Sci.*, **16** (1981), 589.
92. BLAZYNSKI, T. Z. and EL-SOBKY, H. *Metals Technology*, March 1980, 107.

93. WOLFF, E. and PRÜMMER, R. *Raumfahrtforsch*, **17** (1973) 16.
94. BÖHLE, P. and ERDMAN-JESNITZER, F. *Aluminium*, **44** (1968), 683.
95. WITTKOWSKI, D. S. and OTTO, H. *Proc. 4th Intl. Conf. HERF* (1973), Denver.
96. HORIGUCHI, Y. and NOMURA, Y. *Kogyo Kagaku Zassi*, **68** (1965), 910.
97. KLEIN, M. J., ROUGH, F. A. and SIMONS, C. C. *J. Am. Ceram. Soc.*, **46** (1963), 356.
98. KLEIN, M. J. and RUDMAN, P. S. *Phil. Mag.*, **14** (1966), 1199.
99. BERGMAN, O. R. and BARRINGTON, J. *J. Am. Ceram. Soc.*, **49** (1966), 502.
100. HECKEL, R. W. and YOUNGBLOOD, J. L. *J. Am. Ceram. Soc.*, **51** (1968), 398.
101. SAMSONOV, G. V. *Poroshkaya Metallurgiya*, **109** (1972), 93.
102. PRÜMMER, R. and ZIEGLER, G. *Proc. 7th Intl. Conf. HERF* Sept. 1981, Leeds, England.
103. HOENIG, C. L. and YUST, C. S. *Ceramic Bulletin*, **60** (1981), 1175.
104. PRÜMMER, R. *Europa industrie revue*, **1** (1975), 23.
105. BOGINSKI, L. *Planseeberichte für Pulvermetallurgie*, **17** (1969), 225.

INDEX

Adiabatic shear
  compaction, 37
  forming, 37
  mechanism, 88
  welding, 244
Aluminium
  2024, 44
  forming, 349
  impact in 6061-T6, 35
  strain rate
    fcc, 139, 143
    6061-T6, 85, 143
    7075-T6, 143
    testing, 130
  welding, 205, 220, 267, 278, 301
  yield stress, 148
Angle of collision, 195, 202, 229, 231, 234, 242, 290, 314
Arrhenius, equation of, 142
Ashby, equation of, 97

Beryllium
  dislocations, 65
  strain rate, 143
Blazynski and Dara, equation of, 201, 299
Bond
  drawing, in, 311
  interface, 166, 192
  parameters, 165, 200, 300
  strength, 159, 300
Bonding. See Explosive Welding
Brass
  flow stress, 148
  welding, 205, 301

Carpenter, equation of, 299
Christensen, equation of, 245
Chromium
  twinning, 72
Cladding. See Explosive Welding
Collision
  angle, 194, 200, 202, 229, 232, 234, 242, 314
  parameters, 195
  particles, of, 381
  point, 242, 296
  region, 210, 213, 292
  subsonic, 200, 229
  supersonic, 200, 229
  velocity, 197, 200, 223, 227, 242
Compressibility of powders, 371
Compression
  impact, 126
  shock, 90
  testing, 123, 146
  wave, 292
Copper
  dislocations, 61
  fcc, 143
  flow stress, 61, 148
  nozzles, 184
  plate, 50
  shear strain, 89
  welding, 205, 220, 275, 301
Crossland and Williams, equation of, 318

Defect
  dislocation generation, 61, 62, 65, 68, 94, 386

Defect—*contd.*
  phase change, 386
  point, 61, 69, 106, 386
  spalling, 5
  strain energy, 90
  transformation
    diffusionless, 61, 72
    displacive, 61, 72
    twinning, 5, 61, 64, 68, 95, 102, 106
    welding distortion, 318
Deformation
  dynamic, 17
  grain, of, 192
  Mar M-200, of, 103
  plastic, 300
  shock, 83, 88, 104
  strain-rate, at low, 104
  tube-plate, of, 318
  twins, 103
Deribas, equation of, 199
Detonation
  front, 229
  pressure, 230, 292, 371, 381
  velocity, 165, 371
Duval and Erkman, equation of, 199, 241

Explosions
  air, in, 5, 350
  underwater, 5, 350
Explosive
  charge, 175, 354, 361
  compaction
    adiabatic shear, 37
    density, 373, 386
    dynamic, 390
    porous medium, of, 373
    powder, of, 3, 9, 369
    static, 390
    strain rate in, 124
    systems of, 370
  forming
    adiabatic shear, 37
    conical shapes, 364
    die
      closed, 4, 6, 351, 360
      design, 357, 360

Explosive—*contd.*
  forming—*contd.*
    die—*contd.*
      open, 3, 6, 351
    dieless, 351
    domes, 360
    ductility in, 86
    economics, 7
    free. *See* dieless
    parameters, 355
    preforms, 360
    spherical vessels, 360
    systems
      air, 350
      water, 350
    tank, 352
  materials
    composition B, 50
    composition C-3, 9
    composition C-4, 9
    gun cotton, 5
    high, 354
    Metabel, 298, 316
    nitro-dynamite, 5
    PETN, 358
    Trimonite, 236, 298, 316
  operations
    autofrettage, 365
    contact, 3, 353
    cutting, 2
    deep drawing, 3
    flanging, 3
    forming, 3, 6, 347, 362, 365
      miscellaneous, 364
      strain in, 123
    hydrostatic extrusion, 343
    punching, 2, 137, 362, 364
    shearing, 2
    stand-off, 3, 124, 353
    tube expansion, 350, 362
  welding
    adiabatic shear, 37
    assemblies, 166, 170, 171, 177, 292
    attenuator in, 230
    butt, 262
    channel, 265
    compound tubing, 164, 296, 326

Explosive—*contd.*
 welding—*contd.*
  criterion of, 200, 202
  cylinders
   duplex, 296
   multilayered, 296, 326
   triplex, 296
  economics, 162
  energy, 201, 205, 229, 299, 314, 331
  facilities, 162
  foil, 265, 292
   reinforced, 265, 268, 331
  geometry
   inclined, 227, 230, 291, 313
   miscellaneous, 257
   parallel, 224, 230, 291, 314
  honeycomb, 12, 279, 290
  limits, 200, 219, 236, 289, 329
  mechanism, 189, 271, 292
  metal
   combinations, 220
   specifications, 169
  multilaminates, 177, 247, 265, 267, 326
  nozzles, 164, 184
  parameters, 164, 195, 200, 234
   base/anvil, 248
   flyer, 220, 229, 236
   measurement of, 251
  patch, 265
  plates, 159, 219
   shell, 164, 167
   tube, 161, 164, 168, 313
  plugging, 321
  range, 164
  scarf, 262
  slabs, 164
  spot, 264
  surface finish, 234, 236, 249
  system, 290
   explosive, 292
   implosive, 292, 327
   implosive/explosive, 299
   plugging, 322
  techniques, 164, 166, 170, 171, 223, 227, 229, 230, 231, 267, 313

Explosive—*contd.*
 welding—*contd.*
  temperature, 170
  transition joints, 12, 164, 338

Flow
 collision region, 210
 hydrodynamic, 386
 instability, 207
 pattern, 213
Formability, 348, 360

Gurney, equation of, 198, 200, 241, 316

Hardening
 explosive, 9, 187
 precipitation, 349
 shock, 94, 98, 104, 105
Heating
 adiabatic, 111, 205
 compaction, 381
 shock, 98
Hugoniot
 curve, 31, 54, 204
 elastic limit, 54, 204
Hugoniot–Rankine theory, 42, 55, 371

Impedance, 42, 77, 293, 329, 331
Interface
 conditions, 192, 205, 223
 line, 223, 245, 293
 pressure, 193, 211, 328
 wave, 205, 223
  amplitude, 236, 247, 267
  formation, 206, 242
  length, 267
 wavy, 166, 205, 293, 300
 wire-mesh reinforced, 331

Jet
 formation, 193, 207
 role, 200, 234

Kennedy, equation of, 241

Ligament distortion, 318, 324
Loving, equation of, 350

Mie and Grüneisen, equation of, 40
Molybdenum
  dislocations, 65
  twinning, 71

Neumann and Richtmayer, equation of, 48
Nickel
  creep
    Inconel 718, 109
    UDIMET 700, 109
  dislocations, 65, 69, 114
  Monel, 160
  nozzles, 184
  shock
    attenuation, 50, 54
    loading, 68, 98, 116
  welding, 220, 242, 275
Niobium
  dislocations, 61
  welding, 220

Persson, equation of, 328
Platinum, welding, 220, 267
Powder
  aluminium, 376
  barium carbonate, 382
  barium titanate, 382
  boron nitride, 383
  CuBr, 383
  diamond, 383
  graphite, 373, 383
  Mar M-200, 103, 387
  nickel, 376
  niobium, 382
  SAE 304L, 387
  silicon, 385
  Synthesis of, 381
  TiC, 376, 382

Powder—*contd.*
  tin, 382
  tungsten, 382
  UDIMET 700, 387
  zinc ferrite, 381
Product
  bulk, 3
  compaction, of, 390
  compound tubing, 164, 291
  duplex cylinder, 291, 296
  honeycomb, 12, 279, 290
  laminates, 247, 265, 267, 268, 291, 296, 326
  nozzles, 164, 184
  plate, 3, 6, 48, 123, 159, 166, 220, 274, 355
  plugging, of, 321
  powder, 3
  rod array, 291, 341
  sheet, 3, 6, 123, 258
  shell, 164, 167, 357
  slab, 164
  transition joint, 276, 277, 338
  triplex cylinder, 291, 296
  tube plate, 161, 164, 168, 313
Properties
  fatigue, 273
  hardness, 94, 98, 104, 105
  materials, of
    bcc, 37, 43, 64, 73, 102, 139, 146
    bct, 37, 73
    cph, 139
    fcc, 37, 65, 73, 139, 143
    hcp, 43, 65, 102, 143
    orthorhombic, 143
  mechanical
    shock, 94
    yield stress
      shock, 94
      welding, in, 272, 335
  powders, of, 385
  strain rate
    high, 139
    very high, 148
  Young's Modulus, 273, 335

Raleigh line, 43, 54, 372

Rarefaction, 43, 46, 54, 67, 90
Recht, equation of, 38
Reflection
  co-efficient of, 295
  stress waves, of, 4, 77, 200, 292
Refraction, 4, 77, 292
Reynold's Number, 204

Shock
  compression, 90
  deformation, 83, 88, 104
  hardening, 94, 98, 104, 105
  loading, 372, 381
  microstructures, 92, 94, 117
  pulse, 105
  thermal. See Thermal shock
  velocity, 40
  wave, 19, 21, 47, 50, 58, 71, 90, 373
Silver, welding, 275
Stand-off distance, 201, 224, 246, 290, 300, 359, 361
State, equation of
  dynamic, 200
  mechanical, 83, 125
  strain rate, at high, 152
Steel
  manganese
    austenitic, 71
    hardening, 8
    welding, 220, 267
  mild
    En2A, 139
    forming, 349, 355
    hardness, 103
    strain rate, 146
    welding, 166, 205, 220, 242, 278, 301
    yield stress, 84, 139, 148
  stainless, 12, 160
    dislocations, 65
    ductility, 86
    microstructure, 304, 98, 111
    nozzles, 184
    welding, 205, 220, 301
Stivers and Wittman, equation of, 246
Strain
  energy, 199

Strain—contd.
  metal forming, in, 123
  plastic, 5
  shear in copper, 89
Strain-rate, high
  compaction, 124, 380
  deformation, 104
  dislocations, effect on, 61
  features, 83
  grain size, effect on, 93
  metal, in
    bcc, 146
    copper, 61
    fcc, 143
    forming, 123
    hcp, 143
    mild steel, 84
    orthorhombic, 143
  precipitates, effect on, 106
  properties, 139, 146
  sensitivity, 84
  shear, 88
  shock
    microstructure, 89, 92, 94, 117
  testing, 126, 130, 137
Stress
  flow, 123, 146
    aluminium, 148
    brass, 148
    copper, 148
    mild steel, 148
    zinc, 148
  multilaminates, of, 329
  thermal, 142
  relief in welding, 178
  residual, 292, 304
    compacts, in, 388
    determination, 306, 388
  wave
    attenuation, 19, 40, 50
    compressive, 21, 47, 292
    computer code, 21, 47, 292
    dynamic, 19
    elastic, 19, 21, 293
      anisotropic, 27
      isotropic, 22, 29
      precursor, 50, 58, 71
    fracture, 5

Stress—*contd.*
  wave—*contd.*
    hydrodynamic, 20, 39
    longitudinal, 21, 293
    peak, 43, 192
    plastic, 20, 30, 32
      shear, 35, 292
        adiabatic, 37
    porous medium, 373
    propagation, 201
    rarefaction, 43, 46, 54, 67, 90
    reflection, 4, 77, 201, 292
    refraction, 4, 77, 292
    shock, 20, 39, 90, 373
    surface, 21, 27, 292
    transmission, 373
Strohecker, equation of, 359
Surface
  finish, 234, 236, 249, 359
  instability, 292

Takizawa, equation of, 199
Tantalum
  welding, 267
  plate, 160
Taylor, equation of, 124
Testing
  compression, in, 123
    Hopkinson bar, 37, 61, 124
    Kolsky bar, 131, 139
  plugs, 324
  shear, in, 123
  strain rate, at high, 6, 124, 126, 130, 137
  techniques, 124, 251
  tension, in, 123
  weld
    destructive, 181
    quality, 301, 331
    ultrasonic, 180, 320
Thermal shock
  compaction, in, 378
  mechanical treatment, 108
  repeated, 109
  stabilisation, 114
  stress, 142
    cycling, 109

Titanium
  α-phase, 142, 143
  cph, 139
  nozzles, 184
  RMI38644, flow stress of, 96
  Ti6Al4V, strain rate of, 146
  twinning, 71
  welding, 159, 166, 220, 242, 267
Transformation
  diffusionless, 61, 72
  displacive, 61, 72

Uranium
  orthorhombic α, 139, 143

Vanadium, welding of, 220
Velocity
  collision, of, 195, 200, 223, 227, 229, 242, 314
  detonation, of, 165, 371
  equation of, 236
  flyer, of, 230, 314
  impact, of, 35, 125, 195, 205, 224, 234, 236, 239, 247
  particle, 6, 42, 105, 199, 293
  shock, 40
  wave
    crystals, in, 27
    longitudinal, 27, 35
    shear, 27, 29, 36
Volume fraction, 331
von Karman and Duvez, equation of, 32, 124
Vortex
  incidence, 293
  shedding, 210

Weldability
  index, 328
  parameters, 164, 195, 201, 220, 229, 236
  window, 203

Zinc
  flow stress, 148
  welding, 220